S0-BLS-307

Lecture Notes in Physics

Edited by H. Araki, Kyoto, J. Ehlers, München, K. Hepp, Zürich
R. Kippenhahn, München, H.A. Weidenmüller, Heidelberg,
J. Wess, Karlsruhe and J. Zittartz, Köln

Managing Editor: W. Beiglböck

272

Homogenization Techniques for Composite Media

Lectures Delivered at the CISM
International Center for Mechanical Sciences
Udine, Italy, July 1–5, 1985

Edited by E. Sanchez-Palencia and A. Zaoui

Springer-Verlag
Berlin Heidelberg New York London Paris Tokyo

53876878
Seplae
Phys

Editors

Enrique Sanchez-Palencia
Université Paris VI
4, place Jussieu, F-75230 Paris

André Zaoui
Université Paris XIII
av. Jean Baptiste Clément, F-93430 Villetaneuse

SD
5/12/87
cy

ISBN 3-540-17616-0 Springer-Verlag Berlin Heidelberg New York
ISBN 0-387-17616-0 Springer-Verlag New York Berlin Heidelberg

This work is subject to copyright. All rights are reserved, whether the whole or part of the material is concerned, specifically the rights of translation, reprinting, re-use of illustrations, recitation, broadcasting, reproduction on microfilms or in other ways, and storage in data banks. Duplication of this publication or parts thereof is only permitted under the provisions of the German Copyright Law of September 9, 1965, in its version of June 24, 1985, and a copyright fee must always be paid. Violations fall under the prosecution act of the German Copyright Law.

© Springer-Verlag Berlin Heidelberg 1987
Printed in Germany

Printing: Druckhaus Beltz, Hemsbach/Bergstr.;
Bookbinding: J. Schäffer GmbH & Co. KG., Grünstadt
2153/3140-543210

QA808
.2
H671
1987
PHYS

PREFACE

This book contains the written version of a course delivered in July 1985 at the International Centre for Mechanical Sciences, Udine, Italy.

One part of the course (lectures by Caillerie, Lévy, Sanchez-Palencia and Suquet) dealt with periodic media (laminated solids, various kinds of porous media, etc.). Special attention was paid to the local stress field in composite materials, generation of cracks and other forms of damage. The two-scale procedure yields, beside the global properties, a good description of the local state of stress which is responsible for the local failure of a composite. Suggestions were given for the numerical solution of the problems.

Another part of the course (lectures by Willis and Zaoui) was concerned with randomly inhomogeneous media (disordered composites, polycrystals, ...). Variational methods for determining the bounds of the overall moduli from practical statistical information were presented, as well as simplified methods for special cases and applications (plastic anisotropy correlated with texture development, plasticity of multiphase metals, etc.).

The course gathered together scientists and research engineers specializing in the fields of suspensions and sedimentation, composite media, plates, polycrystals, plasticity, etc., and gave everyone an opportunity to exchange ideas and to discuss problems.

It is our pleasant duty to express our sincere thanks to the authorities and staff of CISM for the organization of the course.

Paris, November 1986

E.S.P., A.Z.

CONTENTS

PART I

NON HOMOGENEOUS PLATE THEORY AND CONDUCTION IN FIBERED COMPOSITES

by D. Caillerie

PART II

FLUIDS IN POROUS MEDIA AND SUSPENSIONS

by T. Lévy

PART III

BOUNDARY LAYERS AND EDGE EFFECTS IN COMPOSITES

by E. Sanchez-Palencia

PART IV

ELEMENTS OF HOMOGENIZATION FOR INELASTIC SOLID MECHANICS
by P.M. Suquet

PART V

RANDOMLY INHOMOGENEOUS MEDIA
by J.R. Willis

PART VI

APPROXIMATE STATISTICAL MODELLING AND APPLICATIONS
by A. Zaoui

PART I

NON HOMOGENEOUS PLATE THEORY
AND
CONDUCTION IN FIBERED COMPOSITES

Denis Caillerie
Institut de Mécanique de Grenoble (U.A. 6)
Domaine Universitaire B.P. 68
F-38402 Saint Martin d'Hères Cédex, France

CHAPTER 1

HOMOGENIZATION IN ELASTICITY

1 Introduction

The basic notions of homogenization theory have been presented in the lectures of E.Sanchez-Palencia and T.Levy. It is now applied to equations of elasticity, of course in the framework of periodic media. Examples may be found out composite materials, as fiber-reinforced elastic matrix, but other examples may be found, like periodically bored structures.

The method used to find the homogenized equations of such bodies is the double scale asymptotic expansion, which may be worked up in a very similar way as for diffusion equations. The convergence method of L.Tartar is not developped here, it is presented for instance in Sanchez-Palencia.[20] or Duvaut.[11]. The double scale asymptotic expansion does not give only the homogenized equation, it also yields an approximation of the different physical fields, such as displacements and stresses for elasticity.

In the second paragraph, the framework of elasticity is given with the classical and variationnal formulations of the problem and with the existence and uniqueness theorem.The third paragraph is devoted to the homogenization of an elastic body with a periodic structure.

2 Elasticity

The framework of linear three-dimensional elasticity is just outlined, more important developments may be found in classical handbooks about the subject.

Consider an elastic body in its natural state (that is to say without stresses), occupying a bounded domain of R^3, Ω with a smooth boundary $\partial\Omega$.

In this continuous medium the internal forces are represented by a symmetric tensor $\underline{\underline{\sigma}}$ which satisfies the balance equation :

$$\partial_j \sigma_{ij} + f_i = 0 \quad \text{in } \Omega$$

where $\underline{f} = (f_1, f_2, f_3)$ is the body forces density and where ∂_j denotes the derivatives with respect to x_j. We use here and in sequel the convention of repeated indices.

On the boundary $\partial\Omega$, $\underline{\underline{\sigma}}$ satisfies:

$$\sigma_{ij} n_j = F_i$$

where $\underline{F}$ is the surface forces density (given forces or binding forces) and where $\underline{n}$ is the outer normal to $\partial\Omega$.

Under the action of external forces, the elastic body undergoes deformations, the points M (x_1, x_2, x_3) of the body move, their displacement is $u(x_1, x_2, x_3)$. Under the hypothesis of small perturbations, the deformations of the medium are represented by the symmetric tensor $\underline{\underline{e}}(\underline{u})$ derived from $\underline{u}$ by:

$$e_{ij}(\underline{u}) = (1/2)(\partial_j u_i + \partial_i u_j)$$

What characterizes an elastic medium is the strain-stress relation between the tensors $\underline{\underline{\sigma}}$ and $\underline{\underline{e}}(\underline{u})$, in the framework of linear elasticity this constitutive law is linear:

$$\sigma_{ij} = a_{ijkh}(x) e_{kh}(\underline{u})$$

The a_{ijkh} 's are the elastic coefficients of the medium, if they do not depend of x the medium is homogeneous, else it is heterogeneous.

These coefficients satisfy the following relations:

$$a_{ijkh} = a_{jikh} = a_{khij} \qquad \text{(I-1)}$$

$$\exists m>0 \text{ s.t. } \forall \; \underline{\underline{\tau}} \text{ s.t. } (\tau_{ij} = \tau_{ji}) \; : \; a_{ijkh}\tau_{kh}\tau_{ij} \geq m \; \tau_{ij}\tau_{ij} \tag{I-2}$$

The relation (1) comes from the symmetry of $\underline{\underline{\sigma}}$ and $\underline{\underline{e}}(\underline{u})$ and from the thermodynamics, (2) is assumed to ensure the strong ellipticity of the partial differential equations for $\underline{u}$.

An elasticity problem is to find the displacement field $\underline{u}$ and the stress tensor field $\underline{\underline{\sigma}}$ in the body when it is submitted to external body or surface forces and/or when $\underline{u}$ is given on some part of $\partial\Omega$.

For instance, we consider the following problem:

Let $\partial\Omega$ be divided into two disjoint parts $\partial_1\Omega$ and $\partial_2\Omega$, the elastic body is clamped on $\partial_1\Omega$ (i.e. $\underline{u} = 0$ on $\partial_1\Omega$), the surface forces are given on $\partial_2\Omega$, the body forces are $\underline{f}$.

The <u>classical formulation</u> of the problem is then:

Find $\underline{u}$ and $\underline{\underline{\sigma}}$ such that:

$$\partial_j\sigma_{ij} + f_i = 0$$

$$\sigma_{ij} = a_{ijkh}e_{ij}(\underline{u}) \tag{I-3}$$

$$u_i = 0 \quad \text{on } \partial_1\Omega \qquad \sigma_{ij}n_j = 0 \quad \text{on } \partial_2\Omega$$

wherethe elastic coefficients a_{ijkh} satisfy (1) and (2).

This problem may be state under a variationnal form, following a very classical way.

Let $V(\Omega) = \{ \underline{v} \in [H^1(\Omega)]^3 \text{ s.t. } \underline{v}|_{\partial_1\Omega} = 0 \}$ provided with the norm:

$$\|\underline{v}\|_V = [\int_\Omega (v_i v_i + \partial_j v_i \, \partial_j v_i) \, dx]^{\frac{1}{2}} = \|\underline{v}\|_{[H^1]^3}$$

<u>Variationnal formulation</u>

Find $\underline{u} \in V(\Omega)$ s.t.:

$$\forall \; \underline{v} \in V(\Omega) \quad a(\underline{u},\underline{v}) = L(\underline{v}) \tag{I-4}$$

where $a(\underline{u},\underline{v}) = \int_\Omega a_{ijkh}\, e_{kh}(\underline{u})\, e_{ij}(\underline{v})\, dx$ and $L(\underline{v}) = \int_\Omega f_i v_i\, dx$

The existence and uniqueness of $\underline{u}$ is studied with the variationnal formulation.

Proposition I-1. The elasticity problem has one and only one solution.

The proof is just outlined, it is for instance carried out in Duvaut-Lions.[13] or Sanchez-Palencia.[20].

In order to use the Lax-Milgram theorem we need to prove that the bilinear form $a(\underline{u},\underline{v})$ which is bicontinuous on $V(\Omega)$ is also coercive on this space that is to say:

$$\exists\ \beta > 0 \ \text{ s.t. : } \quad \forall\ \underline{v} \in V(\Omega) \qquad a(\underline{v},\underline{v}) \geq \beta \|\underline{v}\|_V$$

This coercivity property is a consequence (see Duvaut-Lions.[13] or Sanchez-Palencia.[20] for the proof) of the Korn's lemma, the proof of which may be found in Duvaut-Lions.[13].

Korn's lemma. Let Ω be a bounded domain of R^3 with a smooth boundary $\partial\Omega$, then there exists a constant (depending on Ω) such that :

$$\forall\ \underline{v} \in [H^1(\Omega)]^3 \quad \int_\Omega [e_{ij}(\underline{v}) e_{ij}(\underline{v}) + v_i v_i]\, dx \geq C\, \|\underline{v}\|_V^2$$

With the coercivity property, the Lax-Milgram theorem may be applied and the variationnal problem has an unique solution.

It is possible to prove the equivalence of the classical and variationnal formulations of the elasticity problem, it is done for instance in Sanchez-Palencia.[20].

3 Homogenization in elasticity

Now we study an elasticity problem set in the domain Ω with, for instance, the boundary conditions of (3). The elastic coefficients are periodic functions of x, the period being homothetic with the ratio ϵ of a given cell $Y =]0,Y_1[\times]0,Y_2[\times]0,Y_3[$, ϵ is a small paramater.

The coefficients $a_{ijkh}(x)$ depend on ϵ, they are defined in the folowing way:

Let $a_{ijkh}(y)$ be bounded functions defined on Y and extended to R^3 by periodicity, they are assumed to satisfy the following relations:

$$a_{ijkh}(y) = a_{jikh}(y) = a_{khij}(y)$$

$$\exists\, m > 0 \quad \text{s.t.} \quad \forall\, \underline{\underline{\tau}} \quad (\tau_{ij} = \tau_{ij}) \quad a_{ijkh}(y)\tau_{ij}\tau_{kh} \geq m\, \tau_{ij}\tau_{ij} \tag{I-5}$$

The coefficients $a^{\epsilon}_{ijkh}(x)$ of the elastic body are then :

$$a^{\epsilon}_{ijkh}(x) = a_{ijkh}(x/\epsilon) \tag{I-6}$$

They are ϵY-periodic.

For instance, for a fiber-reinforced elastic body the cell is:

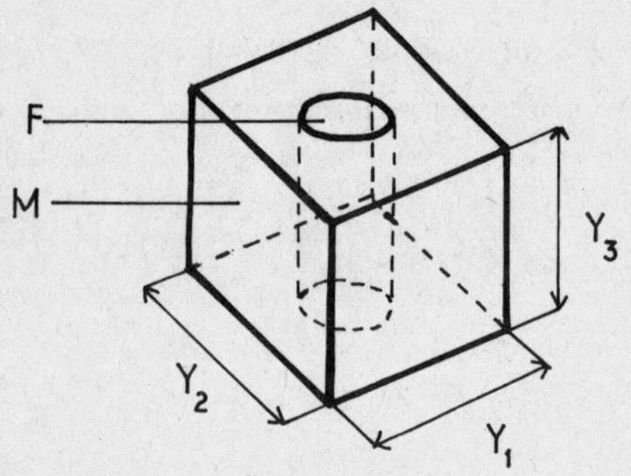

The functions $a_{ijkh}(y)$ are defined by:

$$a_{ijkh}(y) = \begin{cases} a^F_{ijkh} & y \in F \\ a^M_{ijkh} & y \in M \end{cases}$$

The coefficients $a_{ijkh}(x)$ are ϵY-periodic, they are such that:

$$a^{\epsilon}_{ijkh}(x) = \begin{cases} a^{\epsilon}_{ijkh} & x \in \mathfrak{F}^{\epsilon} \\ a^{\epsilon}_{ijkh} & x \in \mathfrak{M}^{\epsilon} \end{cases}$$

The elastic reinforced body is then:

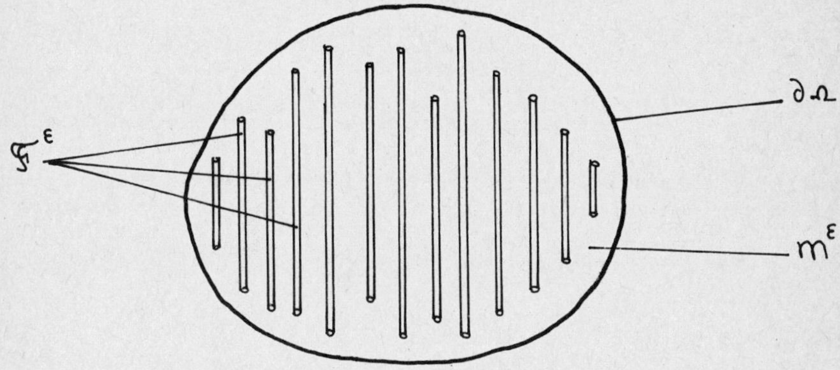

The elastic coefficients being defined by (6), let $\underline{u}$ and $\underline{\underline{\sigma}}$ be the displacements and stresses of the elastic body, these fields satisfy:

$$\partial_j \sigma^{\epsilon}_{ij} + f_i = 0 \qquad \text{(I-7)}$$

$$\sigma^{\epsilon}_{ij} = a_{ijkh}(x/\epsilon)\ e_{kh}(\underline{u}^{\epsilon}) \qquad \text{(I-8)}$$

$$u^{\epsilon}_i = 0 \quad \text{on } \partial_1\Omega \qquad \sigma^{\epsilon}_{ij} n_j = 0 \quad \text{on } \partial_2\Omega \qquad \text{(I-9)}$$

<u>Remark I-1.</u> When the coefficients are piecewise continuous, the previous equations have to be understood in the sense of distributions. For the example of the fiber-reinforced body, this yields continuity equations on the boundary $\partial\mathfrak{F}^{\epsilon} \cap \partial\mathfrak{M}^{\epsilon}$:

$$[\sigma^{\epsilon}_{ij} n_j] = 0$$

$$[u_i^\varepsilon] = 0$$

where [ ] denotes the jump between the fiber and the matrix. ∎

From the proposition I-1, the equations (7), (8), (9) have an unique solution depending on ϵ. The homogenization consists now in finding the limit of this solution as ϵ tends to zero. This study is carried out with the method of double scale expansion developped in the previous lectures of T.Levy and E.Sanchez. Its application to elasticity is straighforward.

We look for expansions of $\underline{u}$ and $\underline{\underline{\sigma}}$ under the form:

$$\left.\begin{aligned} u_i^\epsilon &= u_i^o(x,y) + \epsilon\, u_i^1(x,y) + \epsilon^2\, u_i^2(x,y) + \ldots \\ \sigma_{ij}^\epsilon &= \epsilon^{-1}\sigma_{ij}^{-1}(x,y) + \sigma_{ij}^o(x,y) + \ldots \end{aligned}\right\} \quad y = x/\epsilon$$

where $\underline{u}^n(x,y)$, $\underline{\underline{\sigma}}^n(x,y)$ ... are Y-periodic functions of y.

Putting these expansions in equations (7), (8) and equalling powers of ϵ, we get the following equations for $\underline{u}^o$, $\underline{u}^1$,...

From the balance equation (7):

$$\begin{aligned} &\partial_j^y \sigma_{ij}^{-1} = 0 \\ &\partial_j^y \sigma_{ij}^o = 0 \\ &\partial_j^y \sigma_{ij}^1 + \partial_j^x \sigma_{ij}^o + f_i = 0 \end{aligned} \qquad \text{(I-10)}$$

From the constitutive equation (8):

$$\begin{aligned} &\sigma_{ij}^{-1} = a_{ijkh}(y)\; e_{kh}^y(\underline{u}^o) \\ &\sigma_{ij}^o = a_{ijkh}(y)[\; e_{kh}^y(\underline{u}^1) + e_{kh}^x(\underline{u}^o)] \end{aligned} \qquad \text{(I-11)}$$

where $\partial_j^y = \partial/\partial y_j$ and $e_{ij}^y(\underline{v}) = (1/2)\;(\partial_j^y v_i + \partial_i^y v_j)$.

These equations are partial differential equations in y, x being a parameter.

The suitable Hilbert space for the study of the variationnal formulations of problem for $\underline{u}^o$ and $\underline{u}^1$ is:

$$W(Y) = \{\underline{v} \in [H^1_{loc}(R^3)]^3 ,\underline{v} \text{ Y-periodic}\}$$

provided with the norm:

$$\|\underline{v}\|_W = [\int_Y v_i v_i \, dy + \int_Y \partial_j v_i \partial_j v_i \, dy]^{½}$$

<u>Determination of $\underline{u}^o$</u>. The variationnal formulation of the problem for $\underline{u}^o$ is:

Find $\underline{u}^o \in W(Y)$ such that:

$$\forall \underline{v} \in W(Y) \quad a^Y(\underline{u}^o,\underline{v}) = 0$$

$$\text{where} \quad a^Y(\underline{u}^o,\underline{v}) = \int_Y a_{ijkh}(y) \, e^y_{ij}(\underline{u}^o) \, e^y_{kh}(\underline{v}) \, dy$$

It is obvious that the bilinear form $a(\underline{u},\underline{v})$ is not coercive on W(Y) indeed, for $\underline{v} \neq 0$ such that $e_{ij}(\underline{v}) = 0$, $a(\underline{v},\underline{v})$ is null. Now $e_{ij}(\underline{v})$ is null if $\underline{v}$ is a solid body displacement, but $\underline{v}$ belongs to W(Y) and the periodicity condition yields:

$\underline{v} = \underline{c}$ a constant vector

It may be proved (see Duvaut.[11] or Sanchez-Palencia.[20]) that the bilinear form a $(\underline{u},\underline{v})$ is coercive on the space:

$$\tilde{W}(Y) = \{ \underline{v} \in W(Y) \text{ s.t. } \int_Y \underline{v} \, dy = 0 \}$$

This space is too the space $W(Y)/\mathcal{R}$, where $\mathcal{R}$ is the equivalence relation:

$$\underline{v} \, \mathcal{R} \, \underline{w} = \underline{w} - \underline{v} = \underline{c} \quad \text{a constant vector.}$$

Then, the variationnal problem for $\underline{u}^o$ is well-posed in $\tilde{W}(Y)$ and $\underline{u}^o$ is a constant with respect to y, i.e. a function of x:

$$\underline{u}^o = \underline{u}^o(x) \tag{I-12}$$

<u>Determination of</u> $\underline{u}^1$. From (12) and the linearity of(10) and (11) it is obvious that $\underline{u}^1$ may be looked for under the form:

$$u_i^1(x,y) = \chi_i^{kh}\, e_{kh}^x(\underline{u}^o) + \tilde{u}_i^1(x) \tag{I-13}$$

The partial differential equations and variationnal formulation for $\underline{\chi}^{kh}$ are :

$$\partial_j^y(a_{ijlm}(y)\, e_{lm}^y(\underline{\chi}^{kh})) = -\partial_j^y a_{ijkh}(y)$$

or (I-14)

$$\partial_j^y(a_{ijlm}(y)\, e_{lm}^y(\underline{P}^{kh} + \underline{\chi}^{kh})) = 0$$

where $P_i^{kh} = \delta_{ik} y_h$

And: Find $\underline{\chi}^{kh} \in \tilde{W}(Y)$ such that:

$$\forall\, v \in \tilde{W}\,(Y) \quad a^Y(\underline{\chi}^{kh}, \underline{v}) = -\int_Y a_{ijkh}(y)\, \partial_j^y v_i \, dy$$

(I-15)

$$\text{or} \qquad a^Y(\underline{P}^{kh} + \underline{\chi}^{kh}, \underline{v}) = 0$$

The linear form $L^Y(\underline{v}) = \int_Y a_{ijkh}(y)\, e_{ij}^y(\underline{v})\, dy$ is well defined on $\tilde{W}(Y)$, inded, for $\underline{v} = \underline{c}$, a constant vector, $L^Y(\underline{c})$ vanishes, it is also continuous on $\tilde{W}(Y)$, then from Lax-Milgram theorem, $\underline{\chi}^{kh}$ exists and is unique.

<u>Macoscopic equations</u>. Now we have to determine the equations satisfied by $\underline{u}^o$.

Putting (13) in (11) yields:

$$\sigma_{ij}^o = [a_{ijlm}(y)\, e_{lm}^y(\underline{\chi}^{kh}) + a_{ijkh}(y)]\, e_{kh}^x(u^o)$$

Then integrating on Y, we get:

$$\tilde{\sigma}^{o}_{ij} = a^{H}_{ijkh}\, e^{x}_{kh}(\underline{u}^{o}) \qquad \text{(I-16)}$$

where $$\tilde{\sigma}^{o}_{ij} = (1/|Y|) \int_Y \sigma^{o}_{ij}(x,y)\, dy \qquad \text{(I-17)}$$

and $$a^{H}_{ijkh} = (1/|Y|) \int_Y [a_{ijkh}(y) + a_{ijlm}(y)\, e^{y}_{lm}(\underline{\chi}^{kh})]\, dy \qquad \text{(I-18)}$$

The equation (16) is the strain stress relation for the homogenized elastic body, the macroscopic balance equation is got by integrating (11):

$$\partial^{x}_{j}\tilde{\sigma}^{o}_{ij} + f_i + (1/|Y|) \int_Y \partial^{y}_{j}\sigma^{1}_{ij}\, dy = 0$$

Now, as $\underline{\underline{\sigma}}^{1}$ is periodic, $\sigma^{1}_{ij}n_j$ are opposite on opposite sides of Y and this equation becomes:

$$\partial^{x}_{j}\tilde{\sigma}^{o}_{ij} + f_i = 0 \qquad \text{(I-19)}$$

$(\underline{u}^{o},\underline{\underline{\sigma}}^{o})$ is then solution of an elasticity problem, that is to say that the homogeneized body is linearly elastic,the elastic coefficients being the a^{H}_{ijkh}'s.

The boundary conditions for $\underline{u}^{o}$ and $\underline{\underline{\sigma}}^{o}$ are the same as for $\underline{u}^{\epsilon}$ and $\underline{\underline{\sigma}}^{\epsilon}$, that may be seen from a direct analysis or from a convergence proof (see Duvaut.[11] and Sanchez-Palencia.[20]).

Remark.I-2 The macroscopic stress tensor $\tilde{\underline{\underline{\sigma}}}^{o}$ is defined as the volumic mean of $\underline{\underline{\sigma}}^{o}$,in this example , $\underline{\underline{\sigma}}^{o}$ is such that :

$$\partial^{y}_{k}\sigma^{o}_{ik} = 0 \qquad \text{(I-20)}$$

then $$\int_Y \partial^{y}_{k}\sigma^{o}_{ik}\, y_j\, dy = 0$$

and by integration by parts :

$$\int_Y \sigma^{o}_{ij}\, dy = \int_{\partial Y} \sigma^{o}_{ik}\, n_k\, y_j\, dy$$

Then $\underline{\underline{\sigma}}$ may be defined by the surfacic mean:

$$\tilde{\sigma}^{o}_{ij} = (1/|Y|) \int_Y \sigma^{o}_{ik} n_k y_j \, dy$$

These two means are equal only because (20) holds true. If the body force density $\underline{f}$ is of the order $O(1/\epsilon)$ then (20) does not hold true anymore and so does not the equivalence between the two means. ■

<u>Properties of the homogenized coefficients.</u> From (18) and (15) a^H_{ijkh} may be written :

$$a^H_{ijkh} = (1/|Y|) \int_Y a_{pqrs}(y)\, e^y_{pq}(\underline{P}^{ij} + \underline{\chi}^{ij})\, e^y_{rs}(\underline{P}^{kh} + \underline{\chi}^{kh})\, dy$$

which proves that the following symmetry and coercivity relations hold true :

* $a^H_{ijkh} = a^H_{jikh} = a^H_{khij}$

* $\forall\, \underline{\underline{\tau}}$ s.t. $\tau_{ij} = \tau_{ji} \quad a^H_{ijkh}\, \tau_{kh}\, \tau_{ij} \geq m\, \tau_{ij}\, \tau_{ij}$

where m is the same as in (5).

Even if theelastic material is isotropic, the homogenized one is not. Indeed let :

$$a_{ijkh}(y) = \lambda(y)\, \delta_{ij}\, \delta_{kh} + \mu(y)(\delta_{ik}\, \delta_{jh} + \delta_{ih}\, \delta_{jk})$$

The homogenized coefficients are :

$$a^H_{ijkh} = \tilde{\lambda}\, \delta_{ij}\, \delta_{kh} + \tilde{\mu}\, (\delta_{ik}\, \delta_{jh} + \delta_{ih}\, \delta_{jk}) + [\lambda(y)\, e^y_{ll}(\underline{\chi}^{kh})]^{\sim} \delta_{ij}$$
$$+ [2\, \mu(y)\, e^y_{ij}(\underline{\chi}^{kh})]^{\sim}$$

And in general, the a^H_{ijkh} are not the elastic coefficients of an isotropic material. So that the homogenized material present some symmetries, some invariance hypothesis are to be assumed at a microscopic level. The next results are proved in Léné.[15] and in the peculiar case of applications to plates in Caillerie.[4].

<u>Geometric invariance.</u> Let L be the matrix of an isometric space transformation (a rotation). The net of cells εY is assumed to be invariant under this transformation that is to say that Y', the transformed by L of Y is another cell : Y' = LY.

<u>Mechanical invariance</u> The cells εY are assumed to be mechanically undistinguishable from each other under the transformation L, that is to say that the coefficients $a_{ijkh}(y')$ at the point y' transformed of y, the value of which is :

$$a'_{ijkh}(y') = L_{il}\, L_{jm}\, L_{kr}\, L_{hs}\, a_{lmrs}(L^{-1}y')$$

are equal to the coefficients at the same point before transformation :

The mechanical invariance may be then written :

$$a_{ijkh}(y') = L_{il}\, L_{jm}\, L_{kr}\, L_{hs}\, a_{lmrs}(L^{-1}y')$$

or $$a_{ijkh}(y) = L_{li}\, L_{mj}\, L_{rk}\, L_{sh}\, a_{lmrs}(Ly)$$

for L is an isometry and $L_{ij}\, L_{kj} = \delta_{ik}$.

<u>PropositionI-2.</u> Under the two previous invariance hypothesis, the homogenized coefficients verify :

$$a^H_{ijkh} = L_{il}\, L_{jm}\, L_{kr}\, L_{hs}\, a^H_{lmrs}$$

That is to say that the homogenized material owns the symetry L.

CHAPTER 2

MODELS OF PLATES

1. Introduction

In this chapter we consider a plate as a three dimensional elastic body, a dimension of which, the thickness is very much smaller than the others ones. And by the study of the limit $(t \to 0)$, we derive models of plates as limits of three dimensional elasticity ; according to the behaviour of the elastic coefficients with respect to t , we find the Love Kirchoff's model of plates or a model of thick plates.

If the material is not homogeneous, the stretching and bending of the plate are generally coupled, even in the Love Kirchoff's model. These results generalize those of Ciarlet Destuynder.[8] and Destuynder.[9], they are partly exposed in Caillerie.[4] and .[5].

2. Statement of the problem. Equations and a priori estimates

Let Ω_t be the bounded domain of R^3 occupied by the plate, it is defined in the following way :

$$\Omega_t = \omega \times]-t,t[\qquad t \text{ a real positive number}$$

ω is a bounded domain of R^2, with a smooth boundary $\partial\omega$.

The boundary of Ω is divided in three parts :

$$\Gamma_t^{\pm} = \omega \times \{-t,t\} = \{ x \in \Omega_t \ , \ x_3 = \pm t\}$$

$$\Gamma_t^{o} = \partial\omega \times]-t,t[$$

The current point in Ω_t is denoted by $x = (x_1, x_2, x_3)$.

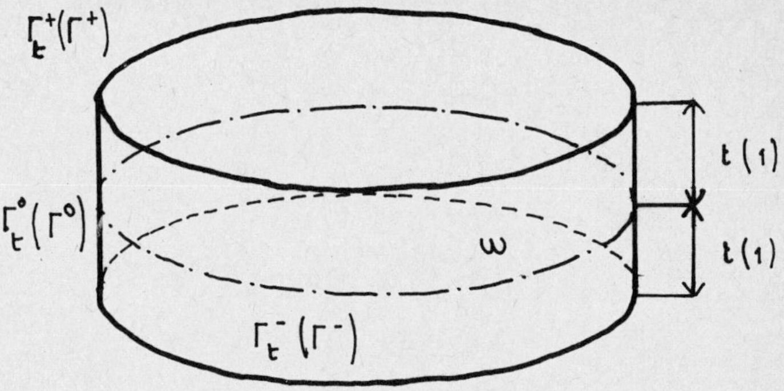

Definition of the elastic moduli. The plate under consideration is made of a linearly elastic anisotropic and heterogeneous material. The elastic moduli are supposed to depend on t and to tend to infinity as t tends to zero, indeed, as the thickness t of the plate is smaller and smaller, the plate has to be more and more rigid in order to support the external given forces. This dependance is not precised now, according to it, different models of plates are found. The plate is heterogeneous, then the elastic moduli depend on x, the dependance on x_3 is defined in the following way (the reason why is going to become obvious in the sequel):

$$\tilde{a}^t_{ijkh}(x_1, x_2, x_3) = a^t_{ijkh}(x_1, x_2, x_3/t)$$

where the functions $a_{ijkh}(z_1, z_2, z_3)$ satisfy symmetry and coercivity relations (I-1) and (I-2).

Three dimensional equations. The elastic plate is submitted to body forces $\underline{f}$, and the faces $\Gamma_t^{\pm}$ to surface forces $\underline{g}^{\pm} = (g_1^{\pm}/t, g_2^{\pm}/t, g_3^{\pm})$. The plate is embedded on its lateral boudary Γ_t^o. $\underline{f}$ is assumed to belong to $[L^2(\Omega_t)]^3$ and $\underline{g}^+$ and $\underline{g}^-$ to $[L^2(\omega)]^3$.

The stress tensor $\underline{\underline{\sigma}}$ and the displacement $\underline{u}$ satisfy the equations of linear anisotropic elasticity :

$$\partial^x_j \tilde{\sigma}^t_{ij} + \tilde{f}_i = 0$$

$$\tilde{\sigma}^t_{ij} = \tilde{a}^t_{ijkh}(x)\ e^x_{kh}(\tilde{\underline{u}}^t) = \tilde{a}^t_{ijkh}(x)\ \partial^x_k \tilde{u}^t_h$$

$$\tilde{\sigma}^t_{\alpha 3}\ n_3 = g^{\pm}_{\alpha}/t \quad \text{and} \quad \tilde{\sigma}^t_{33}\ n_3 = g^{\pm}_3 \quad \text{on} \quad \Gamma^{\pm}_t \ (n_3 = \pm 1 \text{ on } \Gamma^{\pm}_t) \tag{II-1}$$

$$\tilde{\underline{u}}^t = 0 \quad \text{on } \Gamma^o_t$$

From the results of the chapter about elasticity these equations have an unique solution. In these equations as in the following ones, the Greek indices take the values 1 and 2, the Latin ones the values 1, 2 and 3.

<u>Problem on the open set Ω.</u> In order to avoid the dependance on t of Ω_t, this domain is expanded in Ω by the change of variables :

$$z_\alpha = x_\alpha \qquad z_3 = x_3 / t$$

Under this change of variables Γ^o_t (resp. Γ^+_t, Γ^-_t) becomes Γ^o (resp. Γ^+, Γ^-).

Let $\tilde{h}$ be any function defined on Ω_t we set:

$$h(z_1, z_2, z_3) = \tilde{h}(z_1, z_2, tz_3).$$

In the sequel we shall omit the superior index z in ∂^z and $\underline{\underline{e}}^z(\underline{u})$, the whole study being made in Ω.

The equations (1) are written :

$$\partial_\alpha \sigma^t_{i\alpha} + (1/t)\ \partial_3 \sigma^t_{i3} + f_i = 0$$

$$\sigma^t_{ij} = a^t_{ijk\alpha}(z)\ \partial_\alpha u^t_k + (1/t)\ a^t_{ijk3}(z)\ \partial_3 u^t_k \tag{II-2}$$

$$\sigma^t_{\alpha 3}\ n_3 = g^{\pm}_{\alpha} / t \quad \text{and} \quad \sigma^t_{33}\ n_3 = g^{\pm}_3 \quad \text{on} \quad \Gamma^{\pm}$$

$$\underline{u}^t = 0 \quad \text{on } \Gamma^o$$

This problem set on Ω obviously has an only solution ,it is denoted $\mathcal{E}$

A priori estimates. In order to study the limit of $\underline{u}$ and $\underline{\sigma}$ when t tends to zero, we need a priori estimates. First we set up the variational formulation of the problem.

Let $V(\Omega)$ be the space $\{\underline{v} \in [H^1(\Omega)]^3$ s.t. $\underline{v}=0$ on $\Gamma^0\}$, provided with the usual norm on $[H^1(\Omega)]^3$.

Let $\underline{v}$ belonging to $V(\Omega)$, multiplying the first equation of (2) by v_α for i=α and v_3/t for i=3, then integrating on Ω yields :

$$\int_\Omega [\sigma^t_{\alpha\beta}\, e_{\alpha\beta}(\underline{v}) + (2/t)\, \sigma^t_{\alpha 3}\, e_{\alpha 3}(\underline{v}) + (1/t^2)\, \sigma^t_{33}\, e_{33}(\underline{v})]\, dz$$

$$= \int_\Omega (f_\alpha v_\alpha + f_3 v_3/t)\, dz + \int_{\Gamma^\pm} (g^\pm_\alpha v_\alpha/t^2 + g^\pm_3 v_3/t^2)\, d\Gamma \qquad \text{(II-3)}$$

Let $\hat{\underline{u}}^t$ be the vector $(u^t_1/t, u^t_2/t, u^t_3)$ then the strain stress relation of (2) becomes :

$$\sigma^t_{ij} = t\, a^t_{ij\alpha\beta}\, e_{\alpha\beta}(\hat{\underline{u}}^t) + 2\, a_{ij\alpha 3}\, e_{\alpha 3}(\hat{\underline{u}}^t) + (1/t)\, a^t_{ij33}\, e_{33}(\hat{\underline{u}}^t) \qquad \text{(II-4)}$$

Putting (4) in (3) where $\underline{v}$ is taken equal to $\hat{\underline{u}}^t$ and multiplying by t yields :

$$\int_\Omega [t^3\, a^t_{\alpha\beta\gamma\delta}\, e_{\gamma\delta}(\hat{\underline{u}}^t)\, e_{\alpha\beta}(\hat{\underline{u}}^t) + 4\, t^2\, a^t_{\alpha\beta\delta 3}\, e_{\delta 3}(\hat{\underline{u}}^t)\, e_{\alpha\beta}(\hat{\underline{u}}^t)$$

$$+ 2\, t\, a^t_{\alpha\beta 33}\, e_{33}(\hat{\underline{u}}^t)\, e_{\alpha\beta}(\hat{\underline{u}}^t) + 4\, t\, a^t_{\alpha 3\beta 3}\, e_{\beta 3}(\hat{\underline{u}}^t)\, e_{\alpha 3}(\hat{\underline{u}}^t)$$

$$+ 2\, a^t_{\alpha 333}\, e_{33}(\hat{\underline{u}}^t)\, e_{\alpha 3}(\hat{\underline{u}}^t) + (1/t)\, a^t_{3333}\, e_{33}(\hat{\underline{u}}^t)\, e_{33}(\hat{\underline{u}}^t)]\, dz$$

$$= \int_\Omega (t^2\, f_\alpha\, \hat{u}^t_\alpha + t\, f_3\, \hat{u}^t_3)\, dz + \int_{\Gamma^\pm} g^\pm_i\, \hat{u}^t_i\, d\Gamma \qquad \text{(II-5)}$$

For a completely anisotropic material there are 21 different coefficients, then we might consider 21 different behaviours for these 21 coefficients. But we limit ourselves to the study of behaviours that ensure $\hat{\underline{u}}^t$ to be bounded in $V(\Omega)$. As the right hand side of the relation (5) is bounded by $M\, \hat{u}^t$, in order $\hat{u}^t$ to be bounded, the a^t_{ijkh} have to be such as :

$$a^t_{\alpha\beta\gamma\delta} = a_{\alpha\beta\gamma\delta}/t^3 \qquad a^t_{\alpha\beta\delta 3} = a_{\alpha\beta\delta 3}/(t^2\, o_1(t))$$

$$a^t_{\alpha\beta 33} = a_{\alpha\beta 33}/(t\, o_2(t)) \qquad a^t_{\alpha 3\beta 3} = a_{\alpha 3\beta 3}/(t\, o_1^2(t)) \qquad \text{(II-6)}$$

$$a^t_{\alpha 333} = a_{\alpha 333}/(o_1(t)\, o_2(t)) \qquad a^t_{3333} = t\, a_{3333}/o_2^2(t)$$

where the a_{ijkh} do not depend on t and satisfy symmetry and coercivity relations like (I-1) and (I-2).

Most of these coefficients tend to infinity as t tends to zero, this corroborates and precises the mechanical argument stated in the paragraph defining the elastic moduli.

The two functions $o_1(t)$ and $o_2(t)$ are unprecised, they are assumed to be bounded.

The behaviour of the coefficients is different of that chosen in Caillerie.[4] and .[5], it generalizes the study of these two references.

We carry over (6) in (5), we use the coercivity relation satisfied by the a_{ijkh}'s and we get :

$$m \int_\Omega [e_{\alpha\beta}(\hat{\underline{u}}^t)\, e_{\alpha\beta}(\hat{\underline{u}}^t) + (e_{\alpha 3}(\hat{\underline{u}}^t)/o_1(t))(e_{\alpha 3}(\hat{\underline{u}}^t)/o_1(t))$$

$$+ (e_{33}(\hat{\underline{u}}^t)/o_2(t))(e_{33}(\hat{\underline{u}}^t)/o_2(t)]\, dz \leq M \left\|\hat{\underline{u}}^t\right\|_V \qquad \text{(II-7)}$$

As $o_1(t)$ and $o_2(t)$ are bounded we may write :

$$m \int_\Omega e_{ij}(\hat{\underline{u}}^t)\, e_{ij}(\hat{\underline{u}}^t)\, dz \leq M \left\|\hat{\underline{u}}^t\right\|_V$$

As explained in the chapter about elasticity, the Korn's inequality proves that in $V(\Omega)$, the quantity $[\int_\Omega e_{ij}(\underline{v})\, e_{ij}(\underline{v})\, dz]^{1/2}$ is a norm equivalent to $\|\underline{v}\|_V$. Then the previous inequality yields :

$$\left\| \hat{\underline{u}}^t \right\|_V \leq C \tag{II-8}$$

Furthermore, from (7) and (8) we get :

$$\left\| e_{\alpha\beta}(\hat{\underline{u}}^t) \right\|_{L^2} \leq C \qquad \left\| e_{\alpha 3}(\hat{\underline{u}}^t) \right\|_{L^2} \leq C o_1(t) \qquad \left\| e_{33}(\hat{\underline{u}}^t) \right\|_{L^2} \leq C o_2(t) \tag{II-9}$$

As $V(\Omega)$ is an Hilbert space, from a bounded set it is possible to extract a sequence that converges weakly in $V(\Omega)$. And we may state :

Proposition II-1. There exists some $\hat{\underline{u}}^*$ belonging to $V(\Omega)$, such that a subsequence, that converges weakly in $V(\Omega)$ to $\hat{\underline{u}}^*$, may be taken out of $\hat{\underline{u}}^t$.

In order to find the equations verified by $\hat{\underline{u}}^*$, we study the stress tensor $\underline{\underline{\sigma}}$. From (5) and (6) we get :

$$t^2 \sigma^t_{\alpha\beta} = a_{\alpha\beta\gamma\delta} e_{\gamma\delta}(\hat{\underline{u}}^t) + 2\, a_{\alpha\beta\delta 3} e_{\delta 3}(\hat{\underline{u}}^t)/o_1(t) + a_{\alpha\beta 33} e_{33}(\hat{\underline{u}}^t)/o_2(t)$$

$$t\, o_1(t)\, \sigma^t_{\alpha 3} = a_{\alpha 3\gamma\delta} e_{\gamma\delta}(\hat{\underline{u}}^t) + 2\, a_{\alpha 3\delta 3} e_{\delta 3}(\hat{\underline{u}}^t)/o_1(t)$$

$$+ a_{\alpha 333} e_{33}(\hat{\underline{u}}^t)/o_2(t) \tag{II-10}$$

$$o_2(t)\, \sigma^t_{33} = a_{33\beta\delta} e_{\beta\delta}(\hat{\underline{u}}^t) + 2\, a_{33\delta 3} e_{\delta 3}(\hat{\underline{u}}^t)/o_1(t)$$

$$+ a_{3333} e_{33}(\hat{\underline{u}}^t)/o_2(t)$$

We set :

$$\hat{\sigma}^t_{\alpha\beta} = t^2 \sigma^t_{\alpha\beta} \quad ; \quad \hat{\sigma}^t_{\alpha 3} = t\, o_1(t)\, \sigma^t_{\alpha 3} \quad ; \quad \hat{\sigma}^t_{33} = o_2(t)\, \sigma^t_{33} \tag{II-11}$$

And (9) yields : $\left\| \hat{\sigma}^t_{ij} \right\|_{L^2} \leq C$

$L^2(\Omega)$ is an Hilbert space then a weakly converging subsequence may be taken out of $\hat{\underline{\underline{\sigma}}}^t$, let $\hat{\underline{\underline{\sigma}}}^*$ be its limit.

<u>Proposition</u> <u>II-2</u>. $\hat{\sigma}^t_{\alpha 3}/o_1(t)$ and $\hat{\sigma}^t_{33}/o_2(t)$ converge in $H^{-1}(\Omega)$ weak * to q^*_α and q^*_3 ($L^2(\Omega)$ is identified with its dual space).

Furthermore, the weak limit (in $L^2(\Omega)$) $\hat{\underline{\underline{\sigma}}}^*$ and q^*_i satisfy the following equations:

$$\partial_\beta \sigma^*_{\alpha\beta} + \partial_3 q^*_\alpha = 0 \quad \text{in } \Omega \tag{II-12}$$

$$\partial_\alpha q^*_3 + \partial_3 q^*_3 = 0 \quad \text{in } \Omega \tag{II-13}$$

$$q^*_i n_3 = g^\pm_i \quad \text{on } \Gamma^\pm \tag{II-14}$$

These relations are satisfied in the sense of distributions.

<u>Proof</u>. (3) and (11) yields:

$$\int_\Omega [\hat{\sigma}^t_{\alpha\beta} e^z_{\alpha\beta}(\underline{v}) + 2(\hat{\sigma}^t_{\alpha 3}/o_1(t))\, e^z_{\alpha 3}(\underline{v}) + (\hat{\sigma}^t_{33}/o_2(t))\, e^z_{33}(\underline{v})]\, dz$$

$$= \int_\Omega (t^2 f_\alpha v_\alpha + t f_3 v_3)\, dz + \int_{\Gamma^\pm} (g^\pm_\alpha v_\alpha + g^\pm_3 v_3)\, d\Gamma \tag{II-15}$$

Now we take (15) for $v_3 = 0$:

$$\int_\Omega [\hat{\sigma}^t_{\alpha\beta} e^z_{\alpha\beta}(\underline{v}) + (\hat{\sigma}^t_{\alpha 3}/o_1(t))\, \partial^z_3 v_\alpha]\, dz = \int_\Omega t^2 f_\alpha v_\alpha\, dz + \int_{\Gamma^\pm} g^\pm_\alpha v_\alpha\, d\Gamma$$

Making t tend to zero, we see that q^*_α exists and that:

$$\int_\Omega [\hat{\sigma}^*_{\alpha\beta} e^z_{\alpha\beta}(\underline{v}) + q^*_\alpha \partial_3 v_\alpha]\, dz = \int_{\Gamma^\pm} g^\pm_\alpha v_\alpha\, d\Gamma \tag{II-16}$$

The integration by parts prove that (12) and (14) for $\alpha = 0$ holds true.

In the same way, the limit of (15) is:

$$\int_\Omega [q^*_\alpha \, {}^z_\alpha v_3 + q^*_3 \partial^z_3 v_3]\, dz = \int_{\Gamma^\pm} g^\pm_3 v_3\, d\Gamma \tag{II-17}$$

Which yields (13) and (14) for i = 3. ■

Now, following the behaviour of $o_1(t)$ and $o_2(t)$ as t tend to zero, we distinguish different models of plates.

3 Love-Kirchoff plate

We first define the general equations of an embedded Love-Kirchoff plate.

Definition II-1. A displacement field $\underline{w}$ defined in $\Omega = \omega \times]-1,1[$ satisfies the Love-Kirchoff condition if there exists $\underline{w}^!$ defined in $H^2(\omega)$, and which extended to Ω to functions independant of z_3 is such that:

$$w_\alpha = w^!_\alpha - z_3 \partial_\alpha w^! \qquad\qquad w_3 = w^!_3 \tag{II-18}$$

Definition II-2 . Let $A_{\alpha\beta\ \delta}(z_1,z_2)$ be coefficients defined and bounded on Ω such that :

$$A^{\mu\nu}_{\alpha\beta\gamma\delta}(z_1,z_2) = A^{\mu\nu}_{\beta\alpha\gamma\delta}(z_1,z_2) = A^{\nu\mu}_{\gamma\delta\alpha\beta}(z_1,z_2) \tag{II-19}$$

Let $\underline{p} = (p_i)$ be the external force density and $\underline{m} = (m_\alpha)$ the external moment density. The stretching stresses $\underline{\underline{N}}^1$ and bending moments $\underline{\underline{N}}^2$ satisfy the balance equations :

$$\partial_\beta N^1_{\alpha\beta} + p_\alpha = 0 \qquad \partial_{\alpha\beta} N^2_{\alpha\beta} + \partial_\alpha m_\alpha + p_3 = 0 \tag{II-20}$$

The displacements $\underline{w}$ of the points of the plate satisfy the Love Kirchoff condition (18). The constitutive equations of the plate are

$$N^\mu_{\alpha\beta} = A^{\mu 1}_{\alpha\beta\gamma\delta} e_{\gamma\delta}(\underline{w}^!) - A^{\mu 2}_{\alpha\beta\gamma\delta} \partial_{\gamma\delta} w^!_3 \tag{II-21}$$

The plate is embedded, then :

$$w^!_i = \partial_\alpha w^!_3 = 0 \quad \text{on } \partial\omega \quad (i = 1,2,3 \quad \alpha = 1,2) \tag{II-22}$$

We denote $P(A,\underline{p},\underline{m})$ the problem of a Love-Kirchoff plate, the classical formulation of which is constituted by the equations (20), (21) and (22).

__Variational formulation__ of $P(A,\underline{p},\underline{m})$. $\underline{p}$ is assumed to belong to $[L^2(\omega)]^3$ and $\underline{m}$ to $[H^1(\omega)]^2$. Let $U(\omega)$ be the space :

$$U(\omega) = \{ \underline{v} \in H_0^1(\omega) \times H_0^2(\omega)\}$$

provided with the norm :

$$\|v\|_U = [\int (\partial_\alpha v_\beta \partial_\alpha v_\beta + \partial_{\alpha\beta} v_3 \partial_{\alpha\beta} v_3)\, dz_1 dz_2]^{\frac{1}{2}}$$

The __variational formulation__ is then :

Find $\underline{w}^! \in U(\omega)$ such that :

$$\forall\, \underline{v} \in U(\omega) \qquad A(\underline{w}^!, \underline{v}) = P(\underline{v}) \tag{II-23}$$

where : $A(\underline{w}^!, \underline{v}) = \int [A^{11}_{\alpha\beta\gamma\delta} e_{\gamma\delta}(\underline{w}^!)\, e_{\alpha\beta}(\underline{v}) - A^{12}_{\alpha\beta\gamma\delta} \partial_{\gamma\delta} w^!_3\, e_{\alpha\beta}(v)$

$$- A^{21}_{\alpha\beta\gamma\delta} e_{\gamma\delta}(\underline{w}^!) \partial_{\alpha\beta} v_3 + A^{22}_{\alpha\beta\gamma\delta} \partial_{\gamma\delta} w^!_3 \partial_{\alpha\beta} v_3]\, dz_1 dz_2 \tag{II-24}$$

and $$P(v) = \int [p_i v_i - m_\alpha \partial_\alpha v_3]\, dz_1 dz_2 \tag{II-25}$$

The following lemma may be easily proved by using usual methods in elasticity and in theory of plates, see Duvaut-Lions.[13].

__Lemma II-1__. The classical and variational formulations of the problem $P(A,\underline{p},\underline{m})$ are equivalent. And, if the matrix $A = (A_{\alpha\beta\gamma\delta})$ satisfies the following coercivity relation :

$$\exists\, q>0 \text{ s.t. } \forall\, (\tau^\mu_{\alpha\beta})\ (\tau^\mu_{\alpha\beta} = \tau^\mu_{\alpha\beta}) : A^\mu_{\alpha\beta\gamma\delta} \tau^\mu_{\gamma\delta} \tau_{\alpha\beta} \geq q\, \tau^\mu_{\alpha\beta} \tau^\mu_{\alpha\beta} \tag{II-26}$$

then the problem $P(A,\underline{p},\underline{m})$ has one and only one solution.

Notation II-1. In order to simplify the writting (but may be not the understanding) of the formulas let us introduce the following notations:

The partial differential operators $l^{\mu}_{\alpha\beta}$ $(\alpha,\beta,\mu = 1,2)$ denote:

$$l^{1}_{\alpha\beta}(\underline{u}) = e_{\alpha\beta}(\underline{u}) = (1/2)(\partial_{\alpha}u_{\beta} = \partial_{\beta}u_{\alpha}) \qquad l^{2}_{\alpha\beta}(\underline{u}) = -\partial_{\alpha\beta}u_{3}$$

with this notations the constitutive relation (II-21) and the bilinear form (II-24) may be written:

$$N^{\mu}_{\alpha\beta} = A^{\mu\nu}_{\alpha\beta\gamma\delta}l^{\nu}_{\gamma\delta}(\underline{w}^{!}) \tag{II-27}$$

$$A(\underline{w}^{!},\underline{v}) = \int [A^{\mu\nu}_{\alpha\beta\gamma\delta}\, l^{\nu}_{\gamma\delta}(\underline{w}^{!})\ l^{\mu}_{\alpha\beta}(\underline{v})]\, d\omega \tag{II-28}$$

The three dimensional plate under consideration satisfies the Love-Kirchoff condition (18) if :

$$\lim_{t\to 0} o_1(t) = \lim_{t\to 0} o_2(t) = 0 \tag{II-29}$$

Indeed, under these assumptions the following proposition holds true:

Proposition II-3. The weak limit $\hat{\underline{u}}^*$ of $\hat{\underline{u}}^t$ satisfies the Love-Kirchoff condition (18):

$$u_{\alpha} = u^{!}_{\alpha} - z_3\partial_{\alpha}u^{!}_{3} \qquad\qquad u_3 = u^{!}_{3} \tag{II-30}$$

where $u^{!}_{\alpha}$ belongs to $H^1_o(\omega)$ and $u^{!}_{3}$ to $H^2_o(\omega)$.

Then, the main result concerning the derivation of the Love-Kirchoff plate from three dimensional elasticity is stated in the following theorem :

Theorem II-1. Let $o_1(t)$ and $o_2(t)$ tend to zero with t, then the weak limit $\hat{\underline{u}}^*$ of $\hat{\underline{u}}^t$ satisfies the Love-Kirchoff conditions (30) and $\hat{\underline{u}}^*$

is the unique solution of the problem $P(C,\underline{g}^+ + \underline{g}^-,\underline{m})$

where $m_\alpha = g_\alpha^+ - g_\alpha^-$

and where the matrix $(C^{\mu\nu}_{\alpha\beta\gamma\delta})$ is defined by :

$$C^{\mu\nu}_{\alpha\beta\gamma\delta}(z_1,z_2) = \int_{-1}^{1} z_3^{\mu+\nu-2}\; c_{\alpha\beta\gamma\delta}(z)\; dz_3 \tag{II-31}$$

$$c_{\alpha\beta\gamma\delta} = a_{\alpha\beta\gamma\delta} - a_{\alpha\beta i3}\; f_{ij}\; a_{j3\gamma\delta} \tag{II-32}$$

(f_{ij}) is the inverse matrix of the 3X3 matrix (a_{i3j3}).
Furthermore, the stresses $\underline{\underline{\sigma}}^t$ are such that :

$$\lim_{t\to 0} t^2\sigma^t_{\alpha\beta} = c_{\alpha\beta\gamma\delta}[e_{\gamma\delta}(u') - z_3\partial_{\gamma\delta}u_3')] \tag{II-33}$$

<u>Proof</u>. Define $N^{\mu*}_{\alpha\beta}$ by :

$$N^{\mu*}_{\alpha\beta} = \int_{-1}^{1} z_3^{\mu-1}\; \hat{\sigma}^{*}_{\alpha\beta}\; dz_3$$

Then from proposition II-2, $\underline{\underline{N}}^{\mu*}$ satisfy :

$$\partial_\beta N^{1*}_{\alpha\beta} + g_\alpha^+ + g_\alpha^- = 0 \qquad \partial_{\alpha\beta}N^{2*}_{\alpha\beta} + \partial_\alpha(g_\alpha^+ - g_\alpha^-) + g_3^+ + g_3^- = 0$$

These equations are the balance equations (20) for $\underline{\underline{N}}^{\mu*}$.

From proposition II-3, we know that $\hat{\underline{u}}^*$ satisfies the Love-Kirchoff condition. To complete the proof, it remains to settle the constitutive equation.

The relation (10), taking (11) into account , may be inverted in :

$$\begin{aligned} &e_{\alpha\beta}(\hat{\underline{u}}^t) = b_{\alpha\beta ij}\;\hat{\sigma}^t_{ij} \\ &e_{\alpha 3}(\hat{\underline{u}}^t)/o_1(t) = b_{\alpha 3ij}\;\hat{\sigma}^t_{ij} \\ &e_{33}(\hat{\underline{u}}^t)/o_2(t) = b_{33ij}\;\hat{\sigma}^t_{ij} \end{aligned} \tag{II-34}$$

where $b = (b_{ijkh})$ is the inverse of $a = (a_{ijkh})$ i.e. :

$$b_{ijkh}\, a_{khlm} = a_{ijkh}\, b_{khlm} = -\ (1/2)(\delta_{il}\, \delta_{jm} + \delta_{im}\, \delta_{jl})$$

Remark II-1. The existence of the tensor b is a consequence of the coercivity relation satisfied by the tensor a. ■

Proposition II-2 and (29) prove that $\hat{\sigma}^*_{\alpha 3} = \hat{\sigma}^*_{33} = 0$, then the limit (t->0) of the first equation of (34) yields:

$$e_{\alpha\beta}(\hat{\underline{u}}^*) = b_{\alpha\beta\gamma\delta}\, \hat{\sigma}^*_{\gamma\delta}$$

It may be easily proved that this relation is invertible into:

$$\hat{\sigma}^*_{\alpha\beta} = c_{\alpha\beta\gamma\delta}\, e_{\gamma\delta}(\hat{\underline{u}}^*)$$

which, taking (30) into account yields:

$$\hat{\sigma}^*_{\alpha\beta} = c_{\alpha\beta\gamma\delta}\, e_{\gamma\delta}(\underline{u}^!) - z_3\, c_{\alpha\beta\gamma\delta}\, \partial_{\gamma\delta} u^!_3$$

This ends the proof of (32).

Then integrating with respect to z_3 we get:

$$N^{\mu *}_{\alpha\beta} = C^{\mu 1}_{\alpha\beta\gamma\delta}\, e_{\gamma\delta}(\underline{u}^!) - C^{\mu 2}_{\alpha\beta\gamma\delta}\, \partial_{\gamma\delta} u^!_3 = C^{\mu\nu}_{\alpha\beta\gamma\delta} 1^{\nu}_{\gamma\delta}(u^!)$$

which are the constitutve equations of the Love-Kirchoff plate.

It is proved in Caillerie.[4] that the coefficients $C_{\alpha\beta\gamma\delta}$ satisfy the symmetry relations (19)

The existence and uniqueness of $\underline{u}^!$ follow from lemma II-1, indeed, let $\underline{\underline{\theta}} = (\theta_{\alpha\beta})$ be a symmetric 2x2 tensor and define $\underline{\underline{\tau}} = (\tau_{ij})$ by:

$$\tau_{\alpha\beta} = \theta_{\alpha\beta} \qquad \tau_{\delta 3} = -(f_{\delta k}\, a_{k3\alpha\beta}\, \theta_{\alpha\beta})/2 \qquad \tau_{33} = -f_{3k}\, a_{k3\alpha\beta}\theta_{\alpha\beta}$$

The coercivity relation for $a = (a_{ijkh})$ yields:

$$c_{\alpha\beta\gamma\delta}\ \theta_{\alpha\beta}\ \theta_{\gamma\delta} = a_{ijkh}\ \tau_{ij}\ \tau_{kh} \geq m\ \tau_{ij}\ \tau_{ij} \geq m\ \theta_{\alpha\beta}\ \theta_{\alpha\beta}$$

And the coercivity relation for $C = (C_{\alpha\beta\gamma\delta})$ follows from this previous one.

4 Model of "thick" plate.

Let $o_1(t)$ converge to a constant (which is taken equal to 1) and $o_2(t)$ tends to zero when t tends to zero.

Theorem II-2. Suppose $o_1(t)$ and $o_2(t)$ be such that :

$$\lim_{t\to 0} o_1(t) = 1 \qquad \lim_{t\to 0} o_2(t) = 0$$

Then the weak limits $\hat{\underline{u}}^*$ and $\hat{\underline{\underline{\sigma}}}^*$ of $\hat{\underline{u}}^t$ and $\hat{\underline{\underline{\sigma}}}^t$ such that $\hat{u}^*$ is independant on z_3 and $\underline{u}$, $\underline{\underline{\sigma}}$ is the unique solution of the following problem.

Classical formulation :

Find $\underline{u}$, $\underline{\underline{\sigma}}$ such that :

Balance equation :

$$\partial_\beta \hat{\sigma}^*_{\alpha\beta} + \partial_3 \hat{q}^*_\alpha = 0 \qquad \text{in } \Omega$$

$$\partial_\alpha \hat{q}^*_\alpha + \partial_3 \hat{q}^*_3 = 0 \qquad \text{in } \Omega \tag{II-35}$$

$$q^*_\alpha = \sigma^*_{\alpha 3}$$

Constitutive relation :

$$\hat{\sigma}^*_{\alpha j} = d_{\alpha j\gamma\delta}\ e_{\gamma\delta}\ (\hat{\underline{u}}^*) + 2\ d_{\alpha j\gamma 3}\ e_{\gamma 3}(\hat{\underline{u}}^*) \tag{II-36}$$

where $d_{ijkh} = a_{ijkh} - a_{ij33}\ a_{33kh} / a_{3333}$ (II-37)

(only $d_{\alpha i\beta j}$ are not null)

Boundary conditions :

$\hat{\underline{u}}^* = 0$ on Γ^0 $\quad \sigma^*_{\alpha 3} n_3 = g^{\pm}_{\alpha}$ and $q_3\, n_3 = g^{\pm}_3$ on $\Gamma^{\pm}$

Variationnal formulation:

Find $\hat{\underline{u}}^*$ belonging to $T = \{\underline{v} \in [H^1_0(\Omega)]^2 \times H^1_0(\quad) \text{ s.t. } \underline{v} = 0 \text{ on } \Gamma^0\}$ such that:

$$\forall\, \underline{v} \in T \quad D(\hat{\underline{u}}^*, \underline{v}) = \int_{\Gamma^{\pm}} g^{\pm}_{\alpha} v_{\alpha}\, d\Gamma + \int_{\omega} (g^{+}_{3} + g^{-}_{3})\, v_3\, dz \qquad \text{(II-38)}$$

where $D(\hat{\underline{u}}^*, \underline{v}) = \int_{\Omega} d_{ijkh}\, e_{kh}(\hat{u}^*)\, e_{ij}(v)\, dz$

In this definition of $D(\hat{\underline{u}}^*, \underline{v})$, $e_{33}(\hat{u}^*)$, $e_{33}(\underline{v})$ and d_{333j} are null

Proof. As $o_2(t)$ tends to zero from (9), $e_{33}(\hat{\underline{u}}^*) = 0$ and $\hat{u}^*_3$ does not depend on z_3, then $\hat{u}^*_3$ may be identified with a function of $H^1_0(\omega)$.

The proposition II-2 with $\lim o_1(t) = 1$ prove that (35) and the boudary conditions about $\underline{\underline{\sigma}}$ hold true.

The stress strain relation may be proved in a similar way as this one of the previous theorem, but we use another method here.

We have from the third equation of (10) :

$$e_{33}(\hat{\underline{u}}^t)/o_2(t) = [\hat{\sigma}^t_{33} - a_{33\alpha\beta}\, e_{\alpha\beta}(\hat{\underline{u}}^t) - 2\, a_{33\alpha 3}\, e_{\alpha 3}(\hat{\underline{u}}^t)]/a_{3333}$$

Reporting this relation in (10) yields :

$$\hat{\sigma}^t_{\alpha j} = [\, a_{\alpha j\beta\delta} - a_{\alpha j33}\, a_{33\beta\delta} \,/\, a_{3333}\,]\, e_{\beta\delta}(\hat{u}^t)$$

$$+ 2\, [a_{\alpha j\beta 3} - a_{\alpha j33}\, a_{33\beta 3}/\, a_{3333}]\, e_{\beta 3}(\hat{\underline{u}}^t) + a_{\alpha j33}\, \hat{\sigma}^t_{33}/\, a_{3333}$$

And the limit (t->0) yields (36).

The weak formulation (38) is got in an obvious way, the linear form involved is continuous on T, the bilinear form is bicontinuous, then the Lax-Milgram theorem may be applied under the condition that the bilinear form $D(\underline{u},\underline{v})$ is coercive on T.

First we prove the coercivity relation:

$\forall\ \underline{\underline{\tau}} = (\tau_{ij})$ s.t. $(\tau_{ij} = \tau_{ji}$, τ_{33} being now undefined)

$$d_{\alpha\beta\gamma\delta}\tau_{\gamma\delta}\tau_{\alpha\beta} + 2\ d_{\alpha\beta\delta3}\tau_{\delta3}\tau_{\alpha\beta} + 2\ d_{\alpha3\beta\delta}\tau_{\beta\delta}\tau_{\alpha3} + 4\ d_{\alpha3\beta3}\tau_{\beta3}\tau_{\alpha3} \geq$$

$$m\ [\tau_{\alpha\beta}\tau_{\alpha\beta} + 2\ \tau_{\alpha3}\tau_{\alpha3}]$$

For a such tensor $\underline{\underline{\tau}}$ we define:

$$\tau_{33} = -(a_{33\alpha\beta}\tau_{\alpha\beta} + 2\ a_{33\alpha3}\tau_{\alpha3})/\ a_{3333}$$

and we apply the coercivity relation for $a = (a_{ijkh})$:

$$a_{ijkh}\tau_{kh}\tau_{ij} \geq m\ \tau_{ij}\tau_{ij}$$

which yields the enonced coercivity relation for d.

Then for the form $D(\underline{u},\underline{v})$ we have:

$$D(\underline{v},\underline{v}) \geq m \int_\Omega [e_{\alpha\beta}(\underline{v})\ e_{\alpha\beta}(\underline{v}) + 2\ e_{\alpha3}(\underline{v})\ e_{\alpha3}(\underline{v})]\ dz$$

$\underline{v}$ belongs to T then v_3 does not depend on z_3 and $e_{33}(\underline{v}) = 0$ then:

$$D(\underline{v},\underline{v}) \geq m \int_\Omega e_{ij}(\underline{v})\ e_{ij}(\underline{v})\ dz$$

And from the study of elasticity of chapter I we get:

$$D(\underline{v},\underline{v}) \geq m\ \|\underline{v}\|^2_{H^1(\Omega)} = m\ [\sum_{\alpha=1}^{2} \|v_\alpha\|^2_{H^1(\Omega)} + \|v_3\|^2_{H^1(\omega)}]$$

As v_3 is independant on z_3:

$$D(\underline{v},\underline{v}) \geq m\ [\sum_{\alpha=1}^{2} \|v_\alpha\|^2_{H^1(\Omega)} + \|v_3\|^2_{H^1(\omega)}]$$

$D(\underline{u},\underline{v})$ is then coercive on T and the problem has one and only one solution.

Remark II-2. This model of thick plate is near that of a so called "natural" theory of plates, but here, the problem remain set on Ω, indeed $\hat{u}^*_\alpha$ depends on z_3, and it does not seem that $e_{\alpha 3}(\hat{\underline{u}}^*)$ might be independant on z_3 as in the over-mentionned theory.

We may assume that $o_1(t)$ and $o_2(t)$ tend to one when t tends to zero, $\hat{\underline{u}}^t$ is yet bounded and it is possible to find the equations satisfied by the weak limit $\hat{\underline{u}}^*$, but under this hypothesis, even $\hat{\underline{u}}^*$ depends on z_3 and the whole problem for $\hat{\underline{u}}^*$ is three-dimensionnal, it is hardly possible to speak of a plate.

CHAPTER 3

PERIODIC PLATES

1 Introduction

In the two previous lectures, two models of plates are derived from three dimensional elasticity. Now we study plates having a periodic structure, the periodicity being in the two directions of the mean plane, as ever in homogenization theory the period is assumed to be of the order of a small parameter ϵ. Then we deal with a perturbation problem with two small parameters, t the thickness of the plate and ϵ the size of the periods.

First, it is proved that the two limits $(t \to 0)$ and $(\epsilon \to 0)$ do not commute ($o_1(t)$ tending to zero or being equal to one). That is to say that the two successive limits ($t \to 0$ then ϵ->)) and ($\epsilon \to 0$ then $t \to 0$) yields both equation of plates but with different coefficients. Then we study intermediate state when the two parameters are small altogether; for Love-Kirchoff plates we suppose that $o_1(t) = \epsilon$, $o_2(t) = \epsilon^2$, for the model of "thick" plates we suppose that $o_2(t) = \epsilon$. These studies are carried out with the help of asymptotic expansions; a proof concerning the Love-Kirchoff plates may be found in Caillerie.[4].

2 Definition of the periodic coefficients

We go back to the problem of the previous chapter, but we suppose that the coefficients a_{ijkh}'s of the plate are periodic in the variables z_1 and z_2, the period being of the order of ϵ. That is to say that they are defined in the following manner:

Let Y be the rectangle $]0,Y_1[\times]0,Y_2[$ and $\mathcal{Y}$ be $Y \times]-1,1[$.

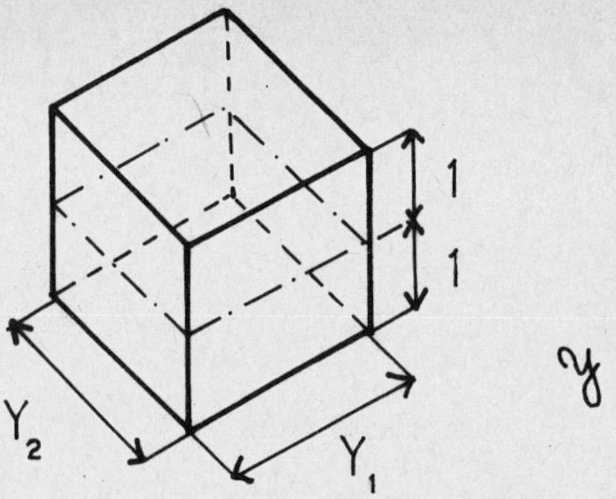

The coefficients $a_{ijkh}(y_1,y_2,y_3)$ are defined on $\mathcal{Y}$, and extended to $R^2 \times]-1,1[$ by periodicity. They are assumed to satisfy the classical symmetry and coercivity relations of elasticity (I-1) and (I-2).

We define $a^{\epsilon}_{ijkh}(z)$ by:

$$a^{\epsilon}_{ijkh}(z) = a_{ijkh}(z_1/\epsilon, z_2/\epsilon, z_3)$$

And the coefficients $a^{\epsilon t}_{ijkh}(z)$ are defined from the a^{ϵ}_{ijkh}'s by relations analogous to (II-6).

3 Study of the successive limits

We first study the limit $\epsilon \to 0$, that is to say that we homogenize the three dimensionnal plate.

This homogenization is a little bit different from that of the chapter one, for the elasticity problem is three dimensionnal and the periodicity is only in two directions. Yet, the method is still valid and the following proposition holds true.

<u>Proposition III-1.</u> The homogenized coefficients of the three dimensionnal body are defined in the following way:

Let $\tilde{W}(Y)$ be the sace:

$$\tilde{W}(Y) = \{ \underline{\Phi} \in [H^1(Y)]^3 \text{ , } \underline{\Phi} \text{ periodic, } \int_Y \underline{\Phi} \, dy_1 dy_2 = 0 \}$$

Let $\underline{\theta}^{kh}$ be the vector of $\tilde{W}(Y)$ such that:

$$\forall \underline{\Phi} \in \tilde{W}(Y) \quad \int_Y [(a_{\alpha\beta\tau\delta}\, e^y_{\tau\delta}(\underline{\theta}^{kh}) + a_{\alpha\beta\delta 3}\, \partial^y_\delta \theta^{kh}_3 + a_{\alpha\beta kh})\, e^y_{\alpha\beta}(\underline{\Phi})$$

$$+ (a_{\alpha 3\beta\delta}\, e^y_{\beta\delta}(\underline{\theta}^{kh}) + a_{\alpha 3\delta 3}\, \partial^y_\delta \theta^{kh}_3 + a_{\alpha 3kh})\, \partial^y_\alpha \Phi_3]\, dy_1 dy_2 = 0 \qquad \text{(III-1)}$$

The coefficients a^H_{ijkh} are defined by:

$$a^H_{ijkh} = \frac{1}{Y}\int_Y [a_{ijkh} + a_{ij\mu\delta}\, e^y_{\mu\delta}(\underline{\theta}^{kh}) + a_{ij\mu 3}\, \partial^y_\mu \theta^{kh}_3]\, dy_1\, dy_2 \qquad \text{(III-2)}$$

And the coefficients a^{Ht}_{ijkh} of the three dimensional plate are given from a^H_{ijkh} by a relation analogous to (II-6).

Proof. Assuming some conditions on the functions $a_{ijkh}(y)$, it is possible to apply to elasticity the study of p.77 and following of Bensoussan & al.[1] .

Let the coefficients $a^t_{ijkh}(y)$ be defined on Y from $a_{ijkh}(y)$ by a relation analogous to (II-6).

The coefficients a^{Ht}_{ijkh} are then defined by :

$$a^{Ht}_{ijkh} = \frac{1}{|Y|}\int_Y [a^t_{ijkh} + a^t_{ij\mu\delta}\, e^y_{\mu\delta}(\underline{\theta}^{kht}) + a^t_{ij\mu 3}\, \partial^y_\mu \theta^{kht}_3]\, dy_1 dy_2 \qquad \text{(III-3)}$$

where $\underline{\theta}^{kht}$ belongs to $\tilde{W}(Y)$ and is such that :

$$\forall \underline{\Phi} \in \tilde{W}(Y) \quad \int_Y [(a^t_{\alpha\beta\tau\delta}\, e^y_{\tau\delta}(\underline{\theta}^{kht}) + a^t_{\alpha\beta\delta 3}\, \partial^y_\delta \theta^{kht}_3 + a^t_{\alpha\beta kh})\, e^y_{\alpha\beta}(\underline{\Phi})$$

$$+ (a^t_{\alpha 3\beta\delta}\, e^y_{\beta\delta}(\underline{\theta}^{kht}) + a^t_{\alpha 3\delta 3}\, \partial^y_\delta \theta^{kht}_3 + a^t_{\alpha 3kh})\, \partial^y_\alpha \Phi_3]\, dy_1 dy_2 = 0$$

From the linearity of this variational formulation, it is easy to see that :

$$\theta^{\alpha\beta t}_\delta = \theta^{\alpha\beta}_\delta \qquad \theta^{\alpha\beta t}_3 = (o_1(t)/t)\,\theta^{\alpha\beta}_3$$

$$\theta^{\alpha 3t}_\delta = (t/o_1(t))\,\theta^{\alpha 3}_\delta \qquad \theta^{\alpha 3t}_3 = \theta^{\alpha 3}_3$$

$$\theta_{\delta}^{33t} = t^2/o_2(t)\ \theta_{\delta}^{33} \qquad \theta_3^{33t} = t\ o_1(t)/o_2(t)\ \theta_3^{33}$$

Then reporting these expressions of $\underline{\theta}^{kht}$ as also these of $a_{ijkh}^{t}(y)$ in (3) we find the relation between a_{ijkh}^{Ht} and a_{ijkh}^{H}.

Remark III-1. This homogenization is a two dimensional one, z_3 is a parameter and the homogenized coefficients a_{ijkh}^{H} depend on z_3. ∎

The limit t->0 may be now easily carried out, indeed the behaviour with respect to t of the coefficients a_{ijkh}^{Ht} is that of (II-6), then we may state :

Proposition III-2. The coefficients of the Love-Kirchoff plate ($o_1(t)$->0), resp. the thick plate ($o_1(t)=1$), are related to a_{ijkh}^{H} by relations analogous to (II-31, 32) and resp. (II-38).

We study now the second successive limits (t->0 then ϵ->0). The coefficients $a_{ijkh}^{\epsilon}(z)$ being periodic with respect to z_1, z_2, it is obvious that the coefficients $C_{\alpha\beta\mu\delta}(z_1,z_2)$ and $d_{\alpha k\beta h}(z_1,z_2,z_3)$ of the Love-Kirchoff plate and the thick plate defined respectively by (II-31), (II-32) and (II-38) are periodic too, then the homogenization method may be applied.

Homogenization of a Love Kirchoff plate. We use the notation II-1.

Let $\tilde{L}(Y)$ be the space :

$$\tilde{L}(Y) = [\underline{\Phi} \in [H^1(Y)]^2 \times H^2(Y) \text{ s.t. } \underline{\Phi} \text{ periodic and } \int_Y \underline{\Phi}\, dy = 0]$$

provided with the usual norm.

Let $\underline{\zeta}^{\mu\pi}$ be the unique vector of $\tilde{L}(Y)$ such that :

$$\forall\, \underline{\Phi} \in \tilde{L}(Y) \qquad \int_Y C_{\alpha\beta\delta\delta}^{\tau\theta}(y)\, 1_{\alpha\beta}^{\theta}(\underline{\zeta}^{\mu\nu\pi} - \underline{P}^{\mu\nu\pi})\, 1_{\delta\delta}^{\tau}(\underline{\Phi})\, dy_1 dy_2 = 0$$

where $P_{\alpha}^{\mu\nu\pi} = \delta_{1\pi} y_{\nu} \delta_{\alpha\mu}$ $\qquad P_3^{\mu\nu\pi} = \delta_{2\pi} y_{\mu} y_{\nu}/2$

Then the homogenized coefficients of the Love Kirchoff plate are :

$$C_{\alpha\beta\delta\delta}^{\mu\nu H} = \frac{1}{|Y|} \int_Y C_{\pi\tau\delta\delta}^{\theta\nu}\, 1_{\pi\tau}^{\theta}\, (\underline{P}^{\alpha\beta\mu} - \underline{\zeta}^{\alpha\beta\mu})\, dy_1 dy_2 \qquad \text{(III-4)}$$

where the $C^{\mu\nu}_{\alpha\beta\gamma\delta}(y)$ are defined from $a_{ijkh}(y)$ by relations analogous to (II-27).

It may be noticed that :

$$1^{\theta}_{\pi\tau}(\underline{P}^{\alpha\beta\nu}) = \delta_{\theta\nu}/2\ (\delta_{\pi\alpha}\ \delta_{\tau\beta} + \delta_{\pi\beta}\ \delta_{\tau\alpha})$$

The coefficients $C^{\mu\nu H}_{\alpha\beta\gamma\delta}$ satisfy the symetry relations (II-19) and coercivity relation (II-26). These points may be proved using the method developped in I for the study of the properties of homogenized coefficients. This generalizes the results of coefficients of Duvaut-Metellus.[13] and Duvaut.[11], which were obtained for uncoupled plates.

<u>Homogenization of a "thick" plate.</u> This homogenization may be carried out by an asymptotic double scale expansion, only the results are given here.

Let $\underline{\Psi}^{\mu k}$ be the unique vector of $\tilde{W}(Y)$ such that :

$$\forall\ \underline{\delta} \in \tilde{W}(Y) \quad \int_Y [(d_{\alpha\beta\tau\delta}\ e^y_{\tau\delta}(\underline{\Psi}^{\mu k}) + d_{\alpha\beta\delta 3}\ \partial^y_\delta \Psi^{\mu k}_3 + d_{\alpha\beta\mu k})e^y_{\alpha\beta}(\underline{\delta})$$

$$+ (d_{\alpha 3\beta\delta}\ e^y_{\beta\delta}(\underline{\Psi}^{\mu k}) + d_{\alpha 3\delta 3}\ \partial^y_\delta \Psi^{\mu k}_3 + d_{\alpha 3\mu k})\ \partial^y_\alpha \delta_3]\ dy_1\ dy_2 = 0$$

The homogenized coefficients of the thick plate are :

$$d^H_{\alpha i\beta j} = \frac{1}{|Y|}\int_Y [d_{\alpha i\beta j} + d_{\alpha i\mu\delta}\ e^y_{\mu\delta}(\underline{\Psi}^{\beta j}) + d_{\alpha i\mu 3}\ \partial^y_\mu \Psi^{\beta j}_3]\ dy_1\ dy_2 \qquad \text{(III-5)}$$

where the $d_{\alpha i\beta j}(y)$ are defined from the $a_{ijkh}(y)$ by relations analogous to (II-32).

The expressions of the homogenized coefficients are quite different following the considered limit ($\epsilon \to 0$ then $t \to 0$) or ($t \to 0$ then $\epsilon \to 0$), either for Love Kirchoff or thick plates. And they are actually different, this is proved for the equations of stationnary heat conduction, see Caillerie.[3]. These studies of successive limits prove that the two limits ($t \to 0$) and ($\epsilon \to 0$) do not commute and show the importance of the relative orders of t and ϵ for the determination of the homogenized coefficients. Now we study limits for which ϵ and t are small altogether.

4. Simultaneous limits

We consider the equations (II-2), with the $a^{t\epsilon}_{ijkh}$'s defined in paragraph 2, $o_1(t)$ and $o_2(t)$ are assumed to tend to zero with t. This problem has an unique solution $\underline{u}^{\epsilon t}$, let $\hat{\underline{u}}^{\epsilon t}$ be $(u^{\epsilon t}_1/t,\ u^{\epsilon t}_2/t\ , u^{\epsilon t}_3)$, the analysis about a priori estimates may be carried out as in II-8, the constant being independant on t and ϵ, then $\hat{\underline{u}}^{\epsilon t}$ (at least a subsequence) converges in $[H^1(\Omega)]^3$ weakly to a certain limit when t and ϵ tend to zero. We determine this limit by the mean of double scale asymptotic expansion, for Love Kirchoff plates and "thick" plates.

Love-Kirchoff plates. We assume that $o_1(t)$ and $o_2(t)$ are such that:

$$o_1(t) = \epsilon \qquad o_2(t) = \epsilon^2$$

The balance equations, boundary conditions and constitutive law are:

$$\partial_\beta \hat{\sigma}^{\epsilon t}_{i\beta} + \partial_3 \hat{\sigma}^{\epsilon t}_{i3} / \epsilon + f^{\epsilon t}_i = 0 \quad \text{with} \quad f^{\epsilon t}_\alpha = t^2 f_\alpha,\ f^{\epsilon t}_3 = \epsilon t f_3 \tag{III-7}$$

$$\hat{\sigma}^{\epsilon t}_{\alpha 3} n_3 = \epsilon\, g^{\pm}_\alpha \qquad \hat{\sigma}^{\epsilon t}_{33} n_3 = \epsilon^2 g^{\pm}_3 \qquad \text{on } \Gamma^{\pm}$$

$$\hat{\sigma}^{\epsilon t}_{ij} = a_{ij\alpha\beta} e_{\alpha\beta}(\hat{\underline{u}}^{\epsilon t}) + 2a_{ij\alpha 3} e_{\alpha 3}(\hat{\underline{u}}^{\epsilon t})/\epsilon + a_{ij33} e_{33}(\hat{\underline{u}}^{\epsilon t})/\epsilon^2 \tag{III-8}$$

Then we look for expansion of $\hat{\underline{u}}^{\epsilon t}$ and $\hat{\underline{\underline{\sigma}}}^{\epsilon t}$ under the form:

$$\hat{u}^{\epsilon t}_\alpha = u^o_\alpha(z) - z_3 \partial^z_\alpha u^o_3 + \epsilon\, u^1_\alpha(z, y_1, y_2) + \dots$$

$$\hat{u}^{\epsilon t}_3 = u^o_3(z_1, z_2) + \epsilon\, u^1_3(z_1, z_2) + \dots$$

$$\hat{\underline{\underline{\sigma}}}^{\epsilon t} = \underline{\underline{\sigma}}^o(z, y_1, y_2) + \epsilon\, \underline{\underline{\sigma}}^1(z, y_1, y_2) + \dots.$$

where $y_\alpha = z_\alpha/\epsilon$; the periodicity being only in the two directions z_1, z_2, the double scale is taken only in these directions. We may consider that $z_3 = y_3$ and , sometimes, z_3 is considered as z_3 or as y_3 and then $y = (y_1, y_2, y_3)$ is a point of $\mathcal{Y}$.

The $\underline{\underline{\sigma}}^n$ and $\underline{u}^n$ are obviously Y periodic functions of y_1, y_2.

In order to simplify the reckoning, we anticipate the results by taking $\underline{u}^o$ independant on y_1, y_2 and u_3^1 independant on (y_1, y_2, y_3), to the same end, the a priori estimates (II-9) holding true for $u^{\kappa\epsilon t}$, we take $e^z_{\alpha 3}(u) = e^z_{33}(u) = 0$. This reckoning is done completely in Caillerie.[4].

Then carrying the expansion (6) in (7) and (8) (∂^z_α being replaced by $\partial^z_\alpha + \epsilon^{-1}\partial^y_\alpha$) we get :

$$\partial^y_j \sigma^o_{ij} = 0$$

$$\partial^y_j \sigma^1_{ij} + \partial^z_\beta \sigma^o_{i\beta} = 0 \qquad \sigma^1_{\alpha 3} n_3 = g^\pm_\alpha \qquad \sigma^1_{33} n_3 = 0 \quad \text{on } \Gamma^\pm \tag{III-9}$$

$$\partial^y_j \sigma^2_{3j} + \partial^z_\beta \sigma^1_{3\beta} = 0 \qquad \sigma^2_{33} n_3 = g^\pm_3 \qquad \text{on } \Gamma^\pm$$

and
$$\sigma^o_{ij} = a_{ij\alpha\beta} e^y_{\alpha\beta}(\underline{u}^1) + a_{ij\alpha\beta} e^z_{\alpha\beta}(\underline{u}^o) - y_3\, a_{ij\alpha\beta} \partial^z_{\alpha\beta} u^o_3$$

$$+ a_{ijk3} \partial^y_k u^2_3 + a_{ij\alpha 3} \partial^z_\alpha u^1_3 + a_{ij\alpha 3} \partial^y_3 u^1_\alpha \tag{III-10}$$

Let us define :

$$N^1_{\alpha\beta} = \frac{1}{|Y|} \int_{\mathcal{Y}} \sigma^o_{\alpha\beta}\, dy \qquad N^2_{\alpha\beta} = \frac{1}{|Y|} \int_{\mathcal{Y}} y_3 \sigma^o_{\alpha\beta}\, dy \qquad \text{and } Q_\alpha = \frac{1}{|Y|} \int_{\mathcal{Y}} \sigma^1_{\alpha 3}\, dy$$

$\underline{\underline{\sigma}}^o$ and $\underline{\underline{\sigma}}^1$ are periodic then, the integration on $\mathcal{Y}$ of the second and third equations yields :

$$\partial^z_\beta N^1_{\alpha\beta} + g^+_\alpha + g^-_\alpha = 0$$

$$\partial^z_\beta N^2_{\alpha\beta} - Q_\alpha + g^+_\alpha - g^-_\alpha = 0 \tag{III-11}$$

$$\partial^z_\alpha Q_\alpha + g^+_3 + g^-_3 = 0$$

these balance equations of the plate are identical with (II-20).

Equations (III-10) and the first of (III-9) show that $\underline{u}^1$ and $\underline{u}^2$ may be written :

$$u^1_\alpha = \chi^{\beta\delta 1}_\alpha \; e^z_{\beta\delta}(u^o) + \chi^{\beta\delta 2}_\alpha \; \partial^z_{\beta\delta} u^o_3 \; + \chi^{3\delta 1}_\alpha \; \partial^z_\delta u^1_3$$

$$u^2_3 = \chi^{\beta\delta 1}_3 \; e^z_{\beta\delta}(u^o) + \chi^{\beta\delta 2}_3 \; \partial^z_{\beta\delta} u^o_3 + \chi^{3\delta 1}_\alpha \; \partial^3_\delta u^1_3$$

where $\underline{\chi}^{l\delta\mu}(y_1,y_2,y_3)$ is Y-periodic and satisfies :

$$\partial^y_j \left[a_{ijkh} \; e^y_{kh}(\underline{\chi}^{l\delta\mu}) + (-y_3)^{\mu-1} \; a_{ijl\delta}\right] = 0$$

$$\left[a_{i3kh} \; e^y_{kh}(\underline{\chi}^{l\delta\mu}) + (-y_3)^{\mu-1} \; a_{i3l\delta}\right] n_3 = 0 \quad \text{for } y_3 = \pm 1 \qquad \text{(III-12)}$$

The variational formulation for $\underline{\chi}^{l\delta\mu}$ is :

Find $\underline{\chi}^{l\delta\mu}$ belonging to $\mathcal{W}(\mathcal{Y}) = \{ \underline{\underline{\phi}} \in [H^1(\mathcal{Y})]^3$], Y-periodic in y_1, y_2 and s.t. $\int_{\mathcal{Y}} \underline{\underline{\phi}} dy = 0 \}$ such that:

$$\forall \underline{\underline{\phi}} \in \mathcal{W}(\mathcal{Y}): \quad \int_{\mathcal{Y}} \left(a_{ijkh} \; e^y_{kh}(\underline{\chi}^{l\delta\mu}) + y_3^{\mu-1} \; a_{ijl\delta} \right) e^y_{ij}(\underline{\phi}) \; dy = 0$$

It may be easily proved that $\chi^{3\delta 1}_h$ is equal to $-y_3\delta_{h\delta}$, and then, using the notations II-1, $\underline{\underline{\sigma}}^o$ may be written :

$$\sigma^o_{ij} = \left[a_{ijkh} e^y_{kh}(\underline{\chi}^{\alpha\beta\mu}) + (-y_3)^{\mu-1} \; a_{ij\alpha\beta}\right] \; l^{\mu z}_{\alpha\beta}(\underline{u}^o)$$

Then, the integration on $\mathcal{Y}$ yields:

$$N^\mu_{\alpha\beta} = S^{\mu\nu}_{\alpha\beta\tau\delta} \; l^{\nu z}_{\tau\delta}(\underline{u}^o)$$

$$\text{where } S^{\mu\nu}_{\alpha\beta\;\delta} = \frac{(-1)^{\mu+\nu}}{Y} \int_{\mathcal{Y}} y_3^{\mu-1} \left[a_{\alpha\beta kh} \; e^y_{kh}(\underline{\chi}^{\gamma\delta\nu}) + (-y_3)^{\nu-1} \; a_{\alpha\beta\gamma\delta}\right] dy \qquad \text{(III-13)}$$

It may be proved (this done in Caillerie.[4]) that the tensor S satisfies symmetry and coercivity relations similar to those of A (II-19) and (II-26). Then the following proposition holds true:

<u>Proposition III-3</u>. u^o is the unique solution of the Love-Kirchoff plate problem $P(S,\underline{g}^+ + \underline{g}^-, m)$ ($m_\alpha = g^+_\alpha - g^-_\alpha$)

<u>Remark III-2.</u> The vectors $\underline{\chi}^{\alpha}_{1\ell\mu}$ are defined by three dimensionnal partial differential equations and the matrix of coefficients of the plate S are means on $\mathcal{Y}$ which is a three dimensionnal cell, though the periodicity is only in the directions y_1 and y_2 and though the two homogenizations of the previously studied limits are two dimensional.

It should be possible to consider different relative behaviours between $o_1(t)$, $o_2(t)$ and ϵ, it is likable that some of them should yield the same results as the successive limits ($\epsilon \to 0$ then $t \to 0$) and ($t \to 0$ then $\epsilon \to 0$), but it seems that the case $o_1(t) = o_2(t) = \epsilon$ should give another type of homogenization, a three dimensionnal one for $\hat{u}_1$ and $\hat{u}_2$, a two dimensional one for $\hat{u}_3$, the both being coupld. Some points are not completely solved for this problem so it will not be developped here.

■

<u>Thick plates.</u> Now we consider the same problem as for the Love-Kirchoff plates but $o_1(t)$ and $o_2(t)$ are such that:

$$o_1(t) = 1 \qquad o_2(t) = \epsilon$$

The balance equations, boundary conditions and constitutive law are:

$$\partial_\beta \hat{\sigma}^{\epsilon t}_{\alpha\beta} + \partial_3 \hat{\sigma}^{\epsilon t}_{\alpha 3} + t^2 f_\alpha = 0 \qquad \partial_\alpha \hat{\sigma}^{\epsilon t}_{3\alpha} + \partial_3 \hat{\sigma}^{\epsilon t}_{33} /\epsilon + t\, f_3 = 0$$

$$\hat{\sigma}^{\epsilon t}_{\alpha 3} n_3 = g^\pm_\alpha \qquad \hat{\sigma}^{\epsilon t}_{33} n_3 = \epsilon\, g^\pm_3 \qquad \text{on } \Gamma^\pm$$

$$\hat{\sigma}^{\epsilon t}_{ij} = a_{ij\alpha\beta} e^z_{\alpha\beta}(\underline{\hat{u}}^{\epsilon t}) + 2\, a_{ij\alpha 3} e^z_{\alpha 3}(\underline{\hat{u}}^{\epsilon t}) + a_{ij33} e^z_{33}(\underline{\hat{u}}^{\epsilon t})/\epsilon$$

The expansion for $\underline{\hat{u}}^{\epsilon t}$ and $\underline{\underline{\hat{\sigma}}}^{\epsilon t}$ are looked under the form:

$$\hat{u}^{\epsilon t}_\alpha = u^o_\alpha(z) + \epsilon\, u^1_\alpha(z, y_1, y_2) + \ldots$$

$$\hat{u}_3^{\epsilon t} = u_3^o(z_1,z_2) + \epsilon\, u_3^1(z,y_1,y_2) + \dots$$

$$\hat{\underline{\underline{\sigma}}}^{\epsilon t} = \underline{\underline{\sigma}}^o(z,y_1,y_2) + \epsilon\, \underline{\underline{\sigma}}^1(z,y_1,y_2) + \dots.$$

where $y_\alpha = z_\alpha/\epsilon$ and the functions $\underline{u}^n$ and $\underline{\underline{\sigma}}^n$ being Y-periodic in y_1,y_2.

As previously for the study of Love-Kirchoff plates, we assume that $\underline{u}^o$ does not depend on y_1,y_2 and that $\underline{\underline{\sigma}}^{-1}$ is null, we may not assume that $e^z_{\alpha 3}(\underline{u}^o)$ is null because $o_1(t) = 1$, but $\underline{u}^o$ is necessarily independant on z_3.

Now we carry over the expansions in the balance and strain stress equations:

$$\partial^y_\beta \sigma^o_{\alpha\beta} = 0 \qquad \partial^y_j \sigma^o_{3j} = 0 \tag{III-14}$$

$$\partial^y_\beta \sigma^1_{\alpha\beta} + \partial^z_j \sigma^o_{\alpha j} = 0 \qquad \partial^y_j \sigma^1_{3j} + \partial^z_\beta \sigma^o_{3\beta} = 0 \tag{III-15}$$

$$\sigma^o_{\alpha 3} n_3 = g^\pm_\alpha \quad \sigma^1_{33} n_3 = g^\pm_3 \quad \text{on } \Gamma^\pm \quad \sigma^o_{33} n_3 = 0 \quad \text{for } y_3 = \pm 1 \tag{III-16}$$

and $\sigma^o_{ij} = a_{ij\alpha\beta} e^y_{\alpha\beta}(\underline{u}^1) + (a_{ijk3}\partial^y_k u^1_3\,)$

$$+\, a_{ij\alpha\beta} e^z_{\alpha\beta}(\underline{u}^o) + 2\, a_{ij\alpha 3} e^z_{\alpha 3}(\underline{u}^o) \tag{III-17}$$

We define $\tilde{\sigma}^n_{ij}$ as the mean over Y of σ^n_{ij}:

$$\tilde{\sigma}^n_{ij} = \frac{1}{|Y|} \int_Y \sigma^n_{ij}\, dy_1 dy_2$$

then the integration of (15) over Y and (16) yields (taking into account the Y-periodicity of $\underline{\underline{\sigma}}^n$)

$$\partial^z_j \tilde{\sigma}^o_{\alpha j} = 0 \qquad \partial^z_\alpha \tilde{\sigma}^o_{3\alpha} + \partial^z_3 \tilde{\sigma}^1_{33} = 0$$

$$\tilde{\sigma}^o_{\alpha 3} n_3 = g^\pm_\alpha \qquad \tilde{\sigma}^1_{33} n_3 = g^\pm_3 \quad \text{on } \Gamma^\pm \tag{III-18}$$

These equations are the balance and boundary equations of the "thick" plate (see II-36).

From (14) and (17) it is easy to deduce that $\underline{u}^1$ is equal to:

$$\underline{u}^1 = \underline{\xi}^{\alpha\beta} \, e^z_{\alpha\beta}(\underline{u}^o) + \underline{\xi}^{\alpha 3} \, (2 \, e^z_{\alpha 3}(u^o) \qquad \text{(III-19)}$$

where $\underline{\xi}^{\mu k}$ is a Y-periodic vector defined on $\mathcal{Y}$ and verifying:

$$\partial^y_\beta(a_{\alpha\beta\tau\delta} \, e^y_{\tau\delta}(\underline{\underline{\xi}}^{\mu k}) + a_{\alpha\beta j3} \, \partial^y_j \xi^{\mu k}_3 + a_{\alpha\beta\mu k}) = 0$$

$$^y_i(a_{3i\alpha\beta} \, e^y_{\alpha\beta}(\underline{\underline{\xi}}^{\mu k}) + a_{3ij3} \, \partial^y_j \xi^{\mu k}_3 + a_{3i\mu k}) = 0 \qquad \text{(III-20)}$$

$$(a_{33\tau\delta} \, e^y_{\tau\delta}(\underline{\underline{\xi}}^{\mu k}) + a_{33j3} \, \partial^y_j \xi^{\mu k}_3 + a_{33\mu k})n_3 = 0 \qquad y_3 = \pm 1$$

<u>Remark III-3.</u> These equations are two dimensionnal for $\xi^{\mu h}_1$ and $\xi^{\mu h}_2$ z_3 being a parameter for these functions but they are three dimensionnal for $\xi^{\mu h}_3$ and the problems are coupled.

If we suppose that the $a_{\alpha\beta j3}$'s are null, then the eqations are no more coupled:

$$\partial^y_\beta(a_{\alpha\beta\tau\delta} \, e^y_{\tau\delta}(\underline{\xi}^{\mu k}) + a_{\alpha\beta\mu k}) = 0$$

$$\partial^y_i(a_{3ij3} \, \partial^y_j \xi^{\mu k}_3 + a_{3i\mu k}) = 0$$

$$a_{33j3} \, \partial^y_j \xi^{\mu k}_3 + a_{33\mu k})n_3 = 0 \qquad y_3 = \pm 1$$

And it may be proved with usual methods that $\underline{\xi}^{\mu k}$ exists (up to an addditive constant). It is quite obvious that $\xi^{\mu 3}_\alpha = 0$ and $\xi^{\mu\tau}_3 = 0$. The equations are two dimensionnal for $\xi^{\mu\tau}_\alpha$ and three dimensionnal for $\xi^{\mu 3}_3$ that is to say that the homogenization for u_α is of the kind of that of the limit ($\epsilon \to 0$ then $t \to 0$) and for u_3 of the that of the Love-Kirchoff plate when $o_1(t) = \epsilon$, $o_2(t) = \epsilon^2$. ■

When the equation remain coupled, the existence of $\underline{\xi}^{\mu k}$ is <u>assumed</u>, it had not been proved and the following assertions remain <u>formal</u>.

The report of (19) in the expression (17) yields:

$$\sigma^{o}_{ij} = [a_{ij\mu\delta}\, e^{y}_{\mu\delta}(\underline{\zeta}^{\alpha\beta}) + a_{ijk3}\, \partial^{y}_{k}\zeta^{\alpha\beta}_{3} + a_{ij\alpha\beta}\,]\, e^{z}_{\alpha\beta}(u^{o})$$

$$+ [a_{ij\mu\delta}\, e^{y}_{\mu\delta}(\underline{\zeta}^{\alpha 3}) + a_{ijk3}\, \partial^{y}_{k}\zeta^{\alpha 3}_{3} + a_{ij\alpha 3}\,]\; 2\, e^{z}_{\alpha 3}(u^{o})$$

The mean over Y of this equations gives:

$$\tilde{\sigma}^{o}_{ij} = R_{ij\alpha\beta}\, e^{z}_{\alpha\beta}(u^{o}) + 2\, R_{ij\alpha 3}\, e^{z}_{\alpha 3}(u^{o})$$

where $$R_{ij\alpha k} = \frac{1}{|Y|}\int_{Y} [a_{ij\alpha k} + a_{ij\beta\delta}\, e^{y}_{\beta\delta}(\underline{\zeta}^{\alpha k}) + a_{ijh3}\, \partial_{h}\zeta^{\alpha k}_{3}]\, dy_{1}\, dy_{2}$$

These strain stress relations are analogous to (II-37) , they are the strain stress relation of a "thick" plate.

Remark III-4. In order to prove symmetry and coercivity relations for $R_{ij\alpha k}$, it would be necessary to study the equations III-20 and this has not been done. ∎

CHAPTER 4

MACROSCOPIC HEAT CONDUCTION IN A FIBERED BODY

Case of highly conducting fibers at dilute concentration

1 Introduction

We consider a problem of stationnary heat conduction in a domain Ω of R^3, the boundary of which, $\partial\Omega$, is smooth. This domain is split up into two subdomains, the first one $\mathfrak{F}^{e\epsilon}$ is the union of parallel "fibers" periodically distributed, the period being of the order of a small parameter ϵ, the radius of these fibers is of the order of another small parameter e; the second subdomain $\mathfrak{M}^{e\epsilon}$ is the interior of $\Omega \setminus \mathfrak{F}^{e\epsilon}$ and called the "matrix". The conduction coefficient of the matrix is independant on e and ϵ, it is of the order of $(\epsilon/e)^2$ in $\mathfrak{F}^{e\epsilon}$, so the total conductivity of the fibers is equivalent to that of the matrix.

The problem is a problem of perturbation with two small parameters, it is studied with an asymptotic expansion.

Previously, the study of the limits (e ->0 then ϵ ->0) and (ϵ ->0 then e ->0) for the similar problem of elasticity (see Caillerie.[2] and .[3]) proved the importance of the relative orders of e and ϵ, indeed the two limits ϵ ->0 and e ->0 do not commute. The aim of this work is to study the problem of perturbation when the two parameters are small altogether, and to classify the different "limit problem" according to the relative orders of the parameters.

The used method is based on a double scale asymptotic expansion. We accurately study the unknown function in the neighbourhood of a fiber by a change of variables in the two directions perpendicular to that of the fibers, and we build a double scale asymptotic expansion on the unknown function. The first terms of the expansion being determined, by matching the development with this one got for a next fiber, we determine the function (or functions) describing the macroscopic conduction. At this point, essential distinctions have to be made between different relative orders of e and ϵ. Then the "flux method" used in homogenization theory yields the conduction equation satisfied by that or those functions.

We find three different "limit problem" following the order of magnitude of $\epsilon^2 |\log e|$ with respect to one.

If ϵ and e are such that $\epsilon^2 |\log e| \gg 1$ or $\epsilon^2 |\log e| \ll 1$, the limit problem is a classical heat conduction problem involving only one temperature. For the first case, the conductivity is the same in all the directions and is that of the matrix, the importance of the fibers vanishes completely, for the second case, the conductivity is increased in the direction of the fibers, it remains that of the matrix in the two perpendicular ones. These two results fit with those ones got by the study of the limits ($e \to 0$ then $\epsilon \to 0$) and ($\epsilon \to 0$ then $e \to 0$).

The more interesting case is when $\epsilon^2 |\log e| = \mu$ (μ being a positive number). The limit conduction problem involves two unknown functions, one being the temperature far from the fibers, the other is the temperature in the fibers. The derivatives of the last function with respect to the two directions perpendicular to the fibers do not occur in the limit problem.

The convergence proof is not considered in this paper for it had not been carried out completely. That mathematical problem is very near from that one approached by Cioranescu and Murat.[7]. From another point of view, the subject of the present work reminds the studies of Phan-Thien.[17] and of Russel.[18] on elasticity problems.

Notations. We use the convention of repeated indices. The Latin letters i,j,k,... denote the indices 1,2 and 3, the Greek ones $\alpha,\beta,\ldots$ the indices 1 and 2.

The derivatives with respect to x_i and x_α are denoted ∂_i and ∂_α, the last one being sometimes denoted ∂_α^x in order to distinguish it from the derivative with respect to z which is denoted ∂_α^z.

∂_{ij} and ∂^2_{ij} denotes $\partial^2/\partial x_i \partial x_j$

$\partial^{zx}_{\alpha\beta}$ denotes $\partial^2/\partial z_\alpha \partial x_\beta$

Δ^z is the two dimensionnal Laplacian operator $\partial^{zz}_{\alpha\alpha}$

Δ^x is the three dimensional Laplacian operator ∂_{ii} ∎

2 Statement of the problem

We consider a domain Ω of R^3 with a smooth boundary $\partial\Omega$. The domain is composed of two parts, one is the union $\mathfrak{F}^{e\epsilon}$ of cylinders parallel to the direction Ox_3, they are periodically distributed in the directions Ox_1 and Ox_2, the period being homothetic in the ratio ϵ of a given period $Y =]0,Y_1[\times]0,Y_2[$, ϵ is a small parameter. We denote $|Y|$ the surface of Y.

The section S^e of each fiber is a disk of radius e (e is the second small parameter, it is supposed to be such that $e \ll \epsilon$).

The second part of Ω is the interior $\mathfrak{M}^{e\epsilon}$ of $\Omega \backslash \mathfrak{F}^{e\epsilon}$ and called the "matrix".

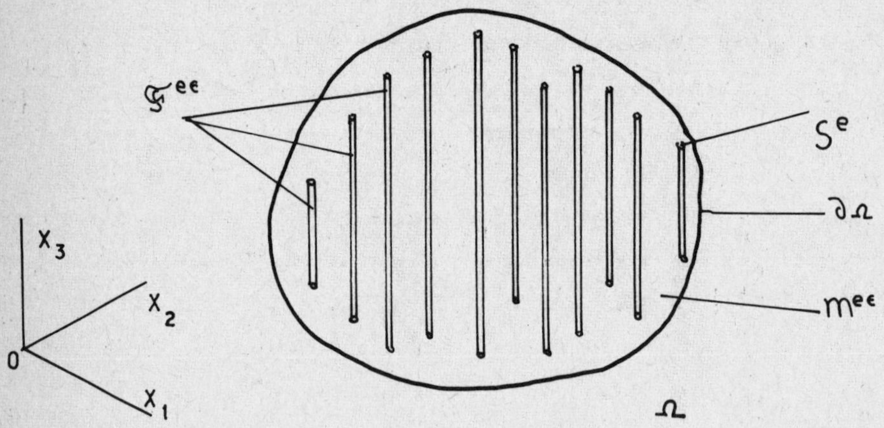

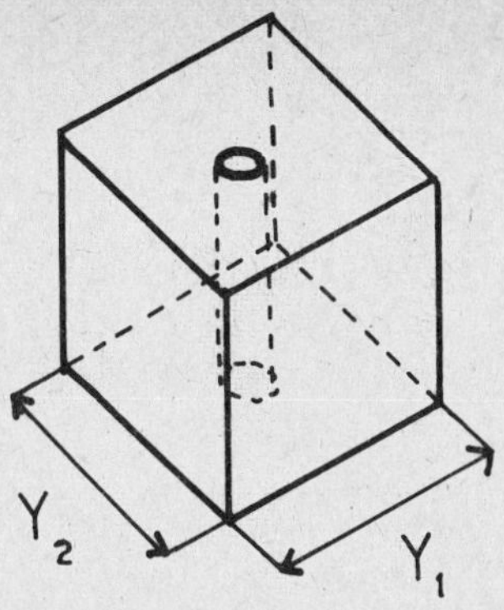

In this domain Ω, we consider the following problem:

Find $u^{e\epsilon}$ such that:

$$q_i^{e\epsilon} = (\epsilon/e)^2\ k\ \partial_i u^{e\epsilon} \qquad \text{in} \quad \mathcal{F}^{e\epsilon} \tag{IV-1}$$

$$q_i^{e\epsilon} = K\ \partial_i u^{e\epsilon} \qquad \text{in} \quad \mathcal{M}^{e\epsilon} \tag{IV-2}$$

$$\partial_i q_i^{e\epsilon} + f = 0 \qquad \text{in } \Omega \tag{IV-3}$$

$$q_i^{e\epsilon} n_i \ \text{ and } \ u^{e\epsilon} \ \text{ continuous on } \partial\mathcal{F}^{e\epsilon} \cap \partial\mathcal{M}^{e\epsilon} \tag{IV-4}$$

(n_i , i = 1,2,3, are the components of the normal to $\partial\mathcal{F}^{e\epsilon}$)

$$u^{e\epsilon} = 0 \qquad \text{on } \partial\Omega \tag{IV-5}$$

In these formulas, k and K are two constants.

This is a stationnary conduction problem, the conductivity being equal to K in the matrix $\mathcal{M}^{e\epsilon}$ and $k(\epsilon/e)^2$ in the fibers $\mathcal{F}^{e\epsilon}$, which are then very conducting; f is the volumic heat source.

It is easy to prove that this problem has one unique solution $u^{e\epsilon}$, and we aim to find the possible limits of this solution when e and ϵ are small.

3 Results

When e and ϵ are small, the conduction of the fibered body is governed by equations that depend on the order of $\epsilon^2 \log e$ with respect to one.

These equations are:

$\epsilon^2 |\log e| \ll 1$

$$Q_{ij} \partial_{ij} U + f = 0 \qquad \text{(IV-6)}$$

where U is the "limit" of $u^{e\epsilon}$

$$\text{and} \quad Q_{ij} = \begin{bmatrix} K & 0 & 0 \\ 0 & K & 0 \\ 0 & 0 & K+k\pi/|Y| \end{bmatrix}$$

$\epsilon^2 |\log e| \gg 1$

$$K\partial_{ii} U + f = 0 \qquad \text{(IV-7)}$$

$\epsilon^2 |\log e| = \mu$ (μ is a real number)

The conduction problem is now described by two temperatures U and u:

$$K\partial_{ii} U + (k\pi/|Y|)\partial_{33} u + f = 0$$

$$\partial_{33} u - (2K/\mu k)(u - U) = 0 \qquad \text{(IV-8)}$$

4 Asymptotic expansion

In order to study the function $u^{e\epsilon}$ in the neighbourhood of a fiber, we expand this region by the following change of variables:

$$z_\alpha = x_\alpha / e \qquad \alpha = 1,2$$

Thus, in the variables z_1, z_2, the fiber has a circular section S of radius 1 and it is embedded in a matrix M which is unbounded, for e is very small.

We look for a double-scale expansion of $u^{e\epsilon}$, the terms of which are functions of $x = (x_1,x_2,x_3)$ and $z = (z_1,z_2)$. The equations (1) to (3) have to be modified by replacing the operator ∂_α by $\partial^x_\alpha + (1/e)\partial^z_\alpha$, which yields:

$$\partial^x_i q^{e\epsilon}_i + (1/e)\partial^z_\alpha q^{e\epsilon}_\alpha + f = 0 \qquad \text{in } S \cup M \qquad \text{(IV-9)}$$

$$q^{e\epsilon}_\alpha = (\epsilon/e)^2\, k\,\partial^x_\alpha u^{e\epsilon} + (\epsilon^2/e^3)\, k\,\partial^z_\alpha u^{e\epsilon} \qquad \text{in } S \qquad \text{(IV-10)}$$

$$q^{e\epsilon}_3 = (\epsilon/e)^2\, k\,\partial^x_3 u^{e\epsilon} \qquad \text{in } S \qquad \text{(IV-11)}$$

$$q^{e\epsilon}_\alpha = K\,\partial^x_\alpha u^{e\epsilon} + (K/e)\partial^z_\alpha u^{e\epsilon} \qquad \text{in } M \qquad \text{(IV-12)}$$

$$q^{e\epsilon}_3 = K\,\partial^x_3 u^{e\epsilon} \qquad \text{in } M \qquad \text{(IV-13)}$$

$$q^{e\epsilon}_\alpha n_\alpha \text{ and } u^{e\epsilon} \text{ continuous on } \partial S \qquad \text{(IV-14)}$$

(n_α, $\alpha = 1,2$, are the components of the normal to ∂S)

In these formulas, the Greek indices take the values 1 and 2.

We see from (10), (12) and (14) that, if we look for an expansion of $u^{e\epsilon}$ in S in the form:

$$u^{e\epsilon} = g(e,\epsilon)\,[u_o(x,z) + e\,u_1(x,z) + \ldots] \qquad z \in S \qquad \text{(IV-15)}$$

we have to look for an expansion of $u^{e\epsilon}$ in M in the form:

$$u^{e\epsilon} = g(e,\epsilon)\,[\,(U^o_o(x,z) + e\,U^1_o(x,z) + \ldots)$$

$$+ \epsilon^2\,(U^o_1(x,z) + e\,U^1_1(x,z) + \ldots) + \ldots] \qquad z \in M \qquad \text{(IV-16)}$$

In order to distinguish the expansions of $u^{e\epsilon}$ in S and in M, we use the capital letter U for the terms in M and the small letter u for the terms in S.

Remark IV-1

- The forms (15) and (16) for the expansions of u are of course "suggested" by the results of the reckoning of the terms, which is done only for the first ones, then it is not obvious that the expansion may be proceeded under this form.

- $g(e,\epsilon)$ is a gauge function which have to be fitted.

- From (16) we may see that we have to distinguish between $\epsilon \ll e$, $\epsilon = e$ and $\epsilon \gg e$, the calculus shows that the third case is the more interesting one, its results hold those of the two first ones, then we limit ourselves to the study of the case $\epsilon^2 \gg e$. ∎

For $\epsilon \gg e$, the expansion (16) may be reorganized in:

$$u^{e\epsilon} = g(e,\epsilon) \left[(U_o^o(x,z) + \epsilon^2 U_1^o(x,z) + e\, U_o^1(x,z) + \dots \right] \qquad \text{(IV-17)}$$

Putting the expansions (15) and (17) in the equations (9) ... (14) we get:

* For $z \in S$

$$\Delta^z u_o = 0 \qquad \text{(IV-18)}$$

$$\Delta^z u_1 + 2\,\partial^{xz}_{\alpha\alpha} u_o = 0 \qquad \text{(IV-19)}$$

$$\Delta^z u_2 + 2\,\partial^{xz}_{\alpha\alpha} u_1 + u_o = 0 \qquad \text{(IV-20)}$$

* For $z \in M$

$$\Delta^z U_o^o = 0 \qquad \text{(IV-21)}$$

$$\Delta^z U_1^o = 0 \qquad \text{(IV-22)}$$

* On ∂S

$$U_o^o = u_o \quad ; \quad U_1^o = 0 \qquad \text{(IV-23)}$$

$$\partial^z_\alpha u_o n_\alpha = 0 \; ; \quad (\partial^z_\alpha u_1 + \partial^x_\alpha u_o \,)\, n_\alpha = 0 \qquad \text{(IV-24)}$$

$$\partial^z_\alpha U^o_o n_\alpha = 0 \; ; \quad K\, \partial^z_\alpha U^o_1 n_\alpha = k(\partial^z_\alpha u_1 + \partial^x_\alpha u_o \,)\, n_\alpha \qquad \text{(IV-25)}$$

All these equations have to be considered as partial differential equations in $z = (z_1, z_2)$, x being a parameter.

Δ^z denotes the two dimensionnal Laplacian operator $\partial^{zz}_{\alpha\alpha}$ and Δ^x the three dimensionnal one ∂^{xx}_{ii}.

<u>Determination if u_o and u_1</u>

u_o is solution of (18) and (24), then it is obvious that u_o does not depend on z:

$$u_o = u_o(x) \qquad \text{(IV-26)}$$

Then from (19) and (24), we see that u_1 satisfied:

$$\Delta^z u_1 = 0 \quad \text{and} \quad \partial^z_\alpha u_1 \, n_\alpha = -\partial^x_\alpha u_o \, n_\alpha$$

As u_o does not depend on z, the solution of these equations is obviously:

$$u_1 = -\, \partial^x_\alpha u_o \, z_\alpha + \tilde{u}_1(x) \qquad \text{(IV-27)}$$

<u>Determination if U^o_o and U^o_1</u>

The equations satisfied by U^o_o and U^o_1 are set on the unbounded domain M.

$U^o_o = u_o(x)$ is obviously solution of (21), (23) and (25), from the Holmgren unicity theorem, it is the unique one. Then:

$$U^o_o = u_o(x) \quad \text{in M} \qquad \text{(IV-28)}$$

U^o_1 has to satisfied the equations (22), (23) and (25). u_2 is not determined, the second equation of (25) may be considered as a Neumann condition for u_2, then, from (20) and (25), we derive the compatibility condition for u_2:

$$\int_S (2\,\overset{zx}{\partial}_{\alpha\alpha} u_1 + \overset{x}{\Delta} u_o)\,dz + (K/k)\int_{\partial S} \overset{z}{\partial}_\alpha U_1^o\, n_\alpha\, d\Gamma - \int_{\partial S} \overset{x}{\partial}_\alpha u_1\, n_\alpha\, d\Gamma = 0$$

which yields:

$$(K/k)\int_{\partial S} \overset{z}{\partial}_\alpha U_1^o\, n_\alpha\, d\Gamma = -\int_S (\overset{zx}{\partial}_{\alpha\alpha} u_1 + \overset{x}{\Delta} u_o)\,dz$$

now $u_1 = -\partial_\alpha u_o\, z_\alpha + \tilde{u}_1(x)$ then:

$$(K/k)\int_{\partial S} \overset{z}{\partial}_\alpha U_1^o\, n_\alpha\, d\Gamma = -\int_S \partial_{33} u_o\, dz$$

As u_o does not depend on z, the compatibility condition for the existence of u_2 is:

$$(K/k)\int_{\partial S} \overset{z}{\partial}_\alpha U_1^o\, n_\alpha\, d\Gamma = -\pi\,\partial_{33} u_o$$

Therefore, U_1^o has to satisfy the following equations:

$$\begin{aligned}
&\overset{z}{\Delta} U_1^o = 0 \qquad \text{in } M \\
&U_1^o = 0 \qquad \text{on } S \qquad\qquad \text{(IV-29)} \\
&\int_{\partial S} \overset{z}{\partial}_\alpha U_1^o\, n_\alpha\, d\Gamma = -(k\pi/K)\,\partial_{33} u_o
\end{aligned}$$

It is proved in the sequel that (29) does not determine U_1^o uniquely but we may state the following proposition:

Proposition IV-1. The solution of (29) that is the least singular for large $z = (z_1, z_2)$ is:

$$U_1^o = -(k/2K)\,\partial_{33} u_o \log|z|$$

■

<u>Proof</u> U_1^o is an harmonic function, its general form is then:

$$U_1^o = a\log(r) + b + \sum_{\substack{-\infty \\ n\neq 0}}^{+\infty} r^n\,(c_n\cos n\theta + d_n \sin n\theta)$$

where a, b, c_n, d_n are constants and r, θ are the polar coordinates of the current point in M.

If we commute derivation, integration and sommation we get:

$$\int_{\partial S} \partial^z_\alpha U_1^o\, n_\alpha\, d\Gamma = a\int_0^{2\pi} d\theta + \sum_{\substack{-\infty \\ n\neq 0}}^{+\infty} [\,c_n \int_0^{2\pi} \cos n\theta\, d\theta + d_n \int_0^{2\pi} \sin n\theta\, d\theta] = 2\pi a$$

Then:

$$a = -(k/2K)\,\partial_{33}u_o$$

The Dirichlet condition on ∂S gives:

$$0 = b + \sum_{n=1}^{+} [(c_n + c_{-n})\cos n\theta + (d_n - d_{-n})\sin n\theta\,]$$

Then the general form of U_1^o is:

$$U_1^o = -(k/2K)\,\partial_{33}u\,\log(r) + \sum_{n=1}^{+} [c_n(r^n - r^{-n})\cos n\theta$$

$$+\, d_n(r^n + r^{-n})\sin n\theta\,]$$

If one of c_n or d_n is not null, then U_1^o is $O(r^n)$ for large r and the most regular solution of (29) is such that $c_n = d_n = 0$ for every n.

<u>Result of the expansion.</u>

In the neighbourhood of a fiber we may write (for $\epsilon^2 \gg e$):

for $z \in S$:

$$u^{e\epsilon} = g(e,\epsilon)\,[u_o(x) + e\,(-\partial_\alpha u_o\, z_\alpha + \tilde{u}_1(x)) + \ldots] \qquad \text{(IV-30)}$$

for $z \in M$:

$$u^{e\epsilon} = g(e,\epsilon)\,[u_o(x) + \epsilon^2(-(k/2K)\partial_{33}u_o \log z) + \ldots] \qquad \text{(IV-31)}$$

5 Matching of the expansions

The development (31) holds true for z belonging to M which is an expanded neighbourhood of a fiber, in order to be able to construct an expansion of $u^{e\epsilon}$ in the whole domain Ω, we have to match two expansions of $u^{e\epsilon}$ valid for two neighbouring fibers. The matching has to be carried out at a distance of the order of ϵ of the two fibers, then for $|z| \approx \epsilon/e$, therefore we require that the greater order of the terms of the expansion (31) should be one for $z \approx \epsilon/e$. The most significant term is called U(x), it represents the temperature far from the fibers; u(x) denotes the temperature in the fibers, it is then the term of order one in the expansion (30).

For $|z| \approx \epsilon/e$ in (31) we get:

$$u^{e\epsilon} = U(x) + \ldots$$

$$U(x) = g(e,\epsilon)\,[u_o(x) - \epsilon^2 |\log e|\,(k/2K)\;_{33}u_o\,] \qquad \text{(IV-32)}$$

<u>Remark IV-2.</u> The matching justifies the choose of U_1^o of the previous paragraph, indeed, if for U_1^o we kept terms such as $c_n r^n \cos(n\theta)$ then the function U(x) should depend on θ and the matching should not be possible. ∎

The function U(x) has to be of the order of one; then three case are to be considered, following the order of $\epsilon^2 |\log e|$:

$\epsilon^2 |\log e| \ll 1$

We take then $g(e,\epsilon) = 1$ and we get from (32) and (30):

$$U(x) = u_o(x) \qquad \text{and} \qquad u(x) = u_o(x) \qquad \text{(IV-33)}$$

In this case, the temperature u in the fibers is equal to the temperature far from them. ∎

$\epsilon^2 |\log e| \gg 1$

We choose $g(e,\epsilon) = 1/(\epsilon^2 |\log e|$ and we find:

$$U(x) = -(k/2K)\partial_{33}u_o \qquad u(x) = 0$$

This seems to show that when $\epsilon^2 |\log e| \gg 1$ the temperature in the fibers $u(x)$ tends to zero with e and ϵ, this is obviously untrue. For $f = 0$ and suitable boundary conditions on Ω we may have $u^{e\epsilon}$ constant and different from zero all over Ω and whatever e and ϵ may be.

Then, the expansions (30) and (31) have to be modified for $\epsilon^2 |\log e| \gg 1$.

It may be proved that the expansions:

$$u^{e\epsilon} = C + [1/(\epsilon^2 |\log e|)] [u_o(x) + \ldots] \qquad \text{in } S \qquad \text{(IV-34)}$$

$$u^{e\epsilon} = C + [1/(\epsilon^2 |\log e|)] [u_o(x) + \epsilon^2(-(k/2K)\partial_{33}u_o \log|z|) + \ldots] \qquad \text{in } M \qquad \text{(IV-35)}$$

are correct, C being a real constant number.

Then the two temperatures U and u are:

$$U(x) = C - (k/2K)\partial_{33}u_o \qquad u(x) = C \qquad \text{(IV-36)}$$

$\epsilon^2 |\log e| = \mu$ (μ a real positive number)

We take $g(e,\epsilon) = 1$ and we get:

$$U(x) = u_o(x) - (\mu k/2K)\partial_{33}u_o \qquad u(x) = u_o(x) \qquad \text{(IV-37)}$$

The, when $\epsilon^2 |\log e|$ is equal to μ, the temperatures U and u are related by the equation:

$$U - u = -(\mu k/2K)\partial_{33}u_o \qquad \text{(IV-38)}$$

■

Now the asymptotic expansions are matched and the relations between the temperatures U and u are known. We note that these relations are quite different following the order of $\epsilon^2 |\log \epsilon|$ with respect to one.

The equations governing U and u are determind in the following paragraph.

6 Flux method

This method was developped for homogenization theory, for more details, see Sanchez-Palencia.[19] and .[20]. The idea of the method is the following:

Let D_ϵ be any subdomain of Ω composed of a some entire cells c_p^ϵ (see figure and notations), as ϵ is small, such domain D_ϵ may approach any "smooth" subdomain D of Ω.

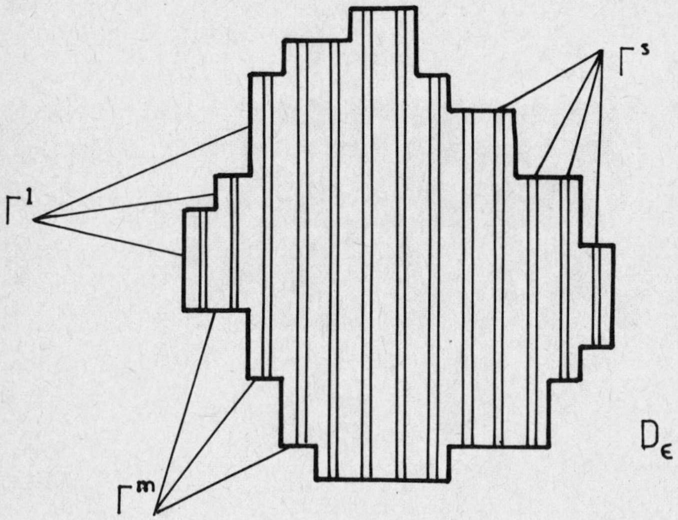

We integrate the equation (3) in D_ϵ:

$$\int_{D_\epsilon} (\partial_i q_i^{e\epsilon} + f)\, dx = 0$$

Integrating by parts we get:

$$\int_{\partial D_\epsilon} q_i^{e\epsilon} n_i \, d\Gamma + \int_{D_\epsilon} f \, dx = 0 \qquad \text{(IV-39)}$$

where $q_i^{e\epsilon}$ is given by (1) and (2);

We have to distinguish between different parts of ∂D_ϵ. Let Γ^l be the part of ∂D_ϵ with a normal $n = (n_1, n_2, 0)$ perpendicular to the direction Ox_3 of the fibers.

Let Γ^s and Γ^m be the parts of ∂D_ϵ with a normal $n = (0,0,\pm 1)$ parallel to the fibers and corresponding respectively to the fibers and to the matrix; and let $\Gamma = \Gamma^s \cup \Gamma^m$.

With these notations, (39) may be written:

$$\int_{\Gamma^l} K \, \partial_\alpha u^{e\epsilon} n_\alpha \, d\Gamma + \int_{\Gamma^s} (\epsilon^2/e^2) k \, \partial_3 u^{e\epsilon} n_3 \, d\Gamma$$
$$+ \int_{\Gamma^m} K \, \partial_3 u^{e\epsilon} n_3 \, d\Gamma + \int_{D_\epsilon} f \, dx = 0 \qquad \text{(IV-40)}$$

Now we use the expansions of $u^{e\epsilon}$ in this equation.

All the points of Γ^l are far from the fibers then the first term of $u^{e\epsilon}$ on Γ^l is equal to $U(x)$.

The points of Γ^s are points of the fibers then $u^{e\epsilon}$ is on Γ^s of the order of $u(x)$ and the following proposition holds true (it is justified in the sequel).

<u>Proposition IV-1.</u> Up to a term $o(1)$ (very much smaller than one), we may write:

$$\int_{\Gamma^s} (\epsilon/e)^2 \, \partial_3 u^{e\epsilon} n_3 \, d\Gamma = \int_{\Gamma} (\pi/|Y|) \partial_3 u \, n_3 \, d\Gamma + o(1)$$

■

Some points of Γ^m are near a fiber, some other are far from them, but the following proposition holds true.

<u>Proposition IV-2.</u> Up to a term $o(1)$, we have:

$$\int_{\Gamma^m} \partial_3 u^{e\epsilon} n_3 \, d\Gamma = \int_{\Gamma} \partial_3 u \, n_3 \, d\Gamma + o(1)$$

■

With these two propositions (40) yields:

$$\int_{\Gamma^l} K \partial_\alpha U \, n_\alpha \, d\Gamma + \int_{\Gamma^s} [(k\pi/ Y) \partial_3 u + K \partial_3 U] \, n_3 \, d\Gamma + \int_{D_\epsilon} f \, dx = o(1)$$

Now integrating by parts we get:

$$\int_{D_\epsilon} (K\partial_{ii} U + (k\pi/|Y|)\partial_{33} u + f) \, dx = o(1)$$

Now we use the following lemmma proved in Sanchez-Palencia.[19] and [20].

<u>Lemma IV-1.</u> If h(x) is a regular function such that:

$$\int_{D_\epsilon} h(x) \, dx = 0$$

for any subdomain D_ϵ composed of cells c_p^ϵ, then h(x) is of order $O(\epsilon)$.

Then (41) yields:

$$K \partial_{ii} U + (k\pi/|Y|)\partial_{33} u + f = 0 \qquad \text{in } \Omega \qquad \text{(IV-42)}$$

This equation takes different forms following the order of $\epsilon^2 |\log e|$ with respect to one.

$\epsilon^2 |\log e| \ll 1$

From (33) and (34), U and u are equal, then (42) becomes:

$$Q_{ij} \, {}_{ij}U + f = 0$$

where $Q_{ij} = \begin{bmatrix} K & 0 & 0 \\ 0 & K & 0 \\ 0 & 0 & K+k\pi/|Y| \end{bmatrix}$

This equation governing the temperature U in Ω is a classical stationnary heat conduction equation, the conductivity in the direction of the fibers is raised up by $k\pi/|Y|$. This result is the same as this one got in the limit (ϵ->0 then e->0). ∎

$\epsilon^2 |\log e| \gg 1$

u is a constant in (35), then (42) becomes:

$$K\partial_{ii}U + f = 0$$

It is the same result as in the limit (e->0 then ϵ->0). The fibers do not conduct heat and the conductivity of the fibered body is that of the matrix.

$\epsilon^2 |\log e| = \mu$

U and u are related by (38). Then, the two functions U and u are solutions of the system (8):

$$K\partial_{ii}U + (k\pi/|Y|)\partial_{33}u + f = 0$$

$$\partial_{33}u - (2K/\mu k)(u - U) = 0$$

This is a non standard conduction problem with two temperature U and u, U being the temperature far from the fibers, u the temperature in the fibers. This problem is not studied here, it is partly done in Caillerie and Dinari.[6] and Dinari.[10].

Justification of proposition IV-1. We want to justify that:

$$(\epsilon^2/e^2)\int_{\Gamma^s} \partial_3 u^{e\epsilon} n_3 \, d\Gamma = (\pi/|Y|)\int_{\Gamma} \partial_3 u \, n_3 \, d\Gamma + o(1)$$

Γ is the part of ∂D_ϵ where the normal is parallel to the fibers, if we number the fibers of D_ϵ with p variyng from 1 to N, we may write $\Gamma = \bigcup_{p=1}^{N} \Gamma_p$, where Γ_p is the part of Γ corresponding to the cell c_p^ϵ. We define Γ_p^s and Γ_p^m, the parts of Γ_p corresponding to the fibers and to the matrix in the same way. Then:

$$\int_{\Gamma^s} (\epsilon^2/e^2)\, \partial_3 u^{e\epsilon} n_3 \, d\Gamma = \sum_{p=1}^{N} \int_{\Gamma^s_p} (\epsilon^2/e^2)\, \partial_3 u^{e\epsilon} n_3 \, d\Gamma$$

now, in the fiber p, $u^{e\epsilon} = u(x) + o(1)$ then:

$$(\epsilon^2/e^2) \int_{\Gamma^s} \partial_3 u \, n_3 d\Gamma = \sum_{p=1}^{N} (\epsilon^2/e^2) \int_{\Gamma^s_p} \partial_3 u \, n_3 d\Gamma + \sum_{p=1}^{N} (\epsilon^2/e^2) \int_{\Gamma^s_p} o(1) \, d\Gamma$$

u(x) is quite constant in Γ^s_p (it is a function of x independant on z), then we may write:

$$\sum_{p=1}^{N} (\epsilon^2/e^2) \int_{\Gamma^s_p} \partial_3 u \; n_3 d\Gamma \approx \sum_{p=1}^{N} (\pi/ Y) \int_{\Gamma_p} \partial_3 u \; n_3 d\Gamma = (\pi/ Y) \int_{\Gamma} \partial_3 u \, n_3 d\Gamma$$

in a similar way:

$$\sum_{p=1}^{N} (\epsilon^2/e^2) \int_{\Gamma^s_p} o(1) \, d\Gamma = o(1)$$

Then:

$$(\epsilon^2/e^2) \int_{\Gamma^s} \partial_3 u^{e\epsilon} n_3 \, d\Gamma = (\pi/ Y) \int_{\Gamma} \partial_3 u \, n_3 \, d\Gamma + o(1)$$

The proposition IV-1 is then justified. ∎

<u>Justification of proposition IV-2.</u> We want to justified that:

$$\int_{\Gamma^m} \partial_3 u^{e\epsilon} n_3 \, d\Gamma = \int_{\Gamma} \partial_3 U \, n_3 \, d\Gamma + o(1)$$

We may write:

$$\int_{\Gamma^m} \partial_3 u^{e\epsilon} n_3 \, d\Gamma = \sum_{p=1}^{N} \int_{\Gamma^m_p} \partial_3 u^{e\epsilon} n_3 \, d\Gamma \qquad \text{(IV-43)}$$

on each of Γ^m_p we may use the expansion of $u^{e\epsilon}$, but, we have to express this expansion in terms of U(x) for the different cases.

$\epsilon^2 |\log e| \ll 1$

From (31) and (33) we have:

$$u^{e\epsilon} = U + o(1) \qquad \text{in each } \Gamma_p^m$$

indeed, the following term of the expansion is very much smaller than U for z is atmost equal to ϵ/e and $\epsilon^2\log|z|$ is very much smaller than one.

Then (43) becomes:

$$\int_{\Gamma^m} \partial_3 u^{e\epsilon} n_3 \, d\Gamma = \sum_{p=1}^{N} \int_{\Gamma_p^m} \partial_3 U \, n_3 \, d\Gamma + o(1)$$

And, as e is very small, the measure of Γ^s is small and:

$$\int_{\Gamma^m} \partial_3 u^{e\epsilon} n_3 \, d\Gamma = \int_{\Gamma} \partial_3 U \, n_3 \, d\Gamma + o(1)$$

The proposition is justified for $\epsilon^2 |\log e| \ll 1$.

$\epsilon^2 |\log e| \gg 1$ and $\epsilon^2 |\log e| = \mu$

For these two case we may write:

$$\partial_3 u^{e\epsilon} = \partial_3 U + A\, \partial_3^3 u \;\; (\log|x-x_p|)/\log e + \dots \qquad \text{in each } \Gamma_p^m \qquad \text{(IV-44)}$$

where x_p is the center of the section Γ_p^m and with:

$A = k/2K$ if $\epsilon^2 |\log e| > 1$

$A = \mu k/2K$ if $\epsilon^2 |\log e| = \mu$

Indeed, if $\epsilon^2 |\log e| \gg 1$, from (34) and (35) we may write (in Γ_p^m)

$$u^{e\epsilon} = U(x) + (k/2K)\partial_{33}u_o \, [1 + \log(|x - x_p|/e) / \log e] + o(1)$$

$$u^{e\epsilon} = U(x) + (k/2K)\partial_{33}u_o \, (\log|x-x_p|)/\log e + o(1)$$

The term $u_o / (\epsilon^2 |\log e|)$ is always very much smaller than one, whereas $(\log |x-x_p|)/\log e$ may be of the order of one (for $|x-x_p|$ of the order of e).

If $\epsilon^2 |\log e| = \mu$ from (37) and (31) we get (in Γ_p^m):

$$u^{e\epsilon} = U(x) + (\mu k/2K)\partial_{33}u_o - (k/2K)\partial_{33}u_o\epsilon^2\log(|x-x_p|/e) + o(1)$$

$$u^{e\epsilon} = U(x) + (\mu k/2K)\partial_{33}u_o [1 + (1/\log e)\log(|x-x_p|/e)] + o(1)$$

$$u^{e\epsilon} = U(x) + (\mu k/2K)\partial_{33}u_o (\log |x-x_p|/\log e) + o(1)$$

The relation (44) is then settled for $\epsilon^2 |\log e| \gg 1$ or $\epsilon^2 |\log e| = \mu$.

Then from (43) and (44) we have:

$$\int_{\Gamma^m} \partial_3 u^{e\epsilon} n_3 \, d\Gamma = \int_{\Gamma^m} \partial_3 U n_3 \, d\Gamma + A \sum_{p=1}^{N} \int_{\Gamma_p^m} \partial_3^3 u_o n_3 (\log |x-x_p|)/\log e \, d\Gamma + o($$

In order to end the justification of the proposition, we just have to prove that:

$$T = \sum_{p=1}^{N} \int_{\Gamma_p^m} \partial_3^3 u_o n_3 (\log |x-x_p|)/\log e \, d\Gamma \quad \text{is} \quad o(1)$$

u_o is a function of x, then quite constant over each Γ_p^m then:

$$T = \sum_{p=1}^{N} \partial_3^3 u_o n_3 \int_{\Gamma_p^m} (\log |x-x_p|)/\log e \, d\Gamma$$

It is obvious that all these integrals are equal, and it may be easily proved that they tend to zero with e and ϵ, that is to say that they are equal to o(1), then:

$$T = \sum_{p=1}^{N} \partial_3 u_o \, o(1)/\log e$$

The measure of Γ_p^m is $\epsilon^2 |Y|$ then:

$$T = o(1)/(\epsilon^2 \log e) \int_{\Gamma} \partial_3^3 u_3 n_3 \, d\Gamma$$

As $\epsilon^2 \log e$ is bounded below this term is o(1), then (45) yields:

$$\int_{\Gamma^m} \partial_3 u^{e\epsilon} n_3 \, d\Gamma = \int_{\Gamma^m} \partial_3 U n_3 \, d\Gamma + o(1)$$

The justification is then ended as in the case $\epsilon^2 |\log e| \ll 1$.

REFERENCES

1. BENSOUSSAN A., LIONS J.L. and PAPANICOLAOU G., Asymptotic structures for periodic media, North-Holland, Amsterdam, 1978.

2. CAILLERIE D., Homogénéisation d'un corps élastique renforcé par des fibres minces de grande rigidité et réparties périodiquement. C.R.A.S. Paris, série II, t.292, 1981, p 477-480.

3. CAILLERIE D., Equations aux dérivées partielles dans des domaines cylindriques applatis, R.A.I.R.O. Analyse numérique, v 15, n° 4, 1981, p 295-270.

4. CAILLERIE D., Thin Elastic and Periodic Plates, Math. Meth. in the Appl. Sci, 6, 1984, p 159-191.

5. CAILLERIE D., Etude de quelques problemes de perturbation en théorie de l'élasticité et de la conduction thermique, Thèse, Université P. et M. Curie Paris, 1982.

6. CAILLERIE D. and DINARI B.,A perturbation problem with two small parameters in the framework of the heat conduction of a fiber reinforced body.,Banach center semester on P.D.E., Banach center publications, Varsovie, (to appear).

7. CIORANESCU D. and MURAT F., Un terme étrange venu d'ailleurs. Non Lin. Part. Diff. Equ. and their Appl., v II and III, Collège de France Seminars, Research notes in Mathematics, Pitman, London, 1982.

8. CIARLET P.G. and DESTUYNDER P. A justification of the two-dimensionnal linear plate model, J. de Mécanique, 18, 1979, p 315-344.

9. DESTUYNDER P. Sur une justification des modèles de plaques et de coques par les méthodes asymptotiques, Thèse, Université P. et M. Curie Paris, 1980.

10. DINARI B., Etude de la conduction stationnaire dans un solide comportant une distribution de fibres fines de grande conductivité. Thèse de 3ème cycle, Université P. et M. Curie, Paris, 1984.

11. DUVAUT G., Analyse fonctionnelle et mécanique des milieux continus. Applications à l'étude des matériaux composites élastiques à structure périodique. Homogénéisation, in _Theorical and Applied Mechanics_, ed. Koiter W.T., North-Holland, Amsterdam, 1978, p 119-131.

12. DUVAUT G. and LENE F., Résultat d'isotropie pour des milieux homogénéisés, C.R.A.S., Paris, série II, t.293, 1981, p 477-480.

13. DUVAUT G. and LIONS J.L., _Les Inéquations en Mécanique et en Physique_, Dunod, Paris, 1972.

14. DUVAUT G. and METELLUS A.M., Homogénéisation d'une plaque mince en flexion de structure périodique, C.R.A.S., Paris, série A, t.283, 1976.

15. LENE F., Contribution à l'étude des matériaux composites et de leur endommagement, Thèse, Université P. et M. Curie, Paris, 1984.

16. LEVY T., Fluid flow past an array of fixed particles, I.J.E.S., v 21, n°1, 1983, p 11-23.

17. PHAN-THIEN N., On the properties of composite materials: slender nearly perfect conductors at dilute concentration., I.J.E.S., v 18, n°7, 1980, p 1325-1331.

18. RUSSEL W.B., On the effective moduli of composite materials: effect of fiber length and geometry at dilute concentration, Z.A.M.P., v 27, 1973, p 581-600.

19. SANCHEZ-PALENCIA E., Comportement local et macroscopique d'un type de milieux physiques hétérogènes., I.J.E.S., v 12, 1974, p 331-351.

20. SANCHEZ-PALENCIA E., _Non Homogeneous Media and Vibration Theory_ Lecture Notes in Physics, Berlin, Springer, 1980.

PART II

FLUIDS IN POROUS MEDIA AND SUSPENSIONS

Thérèse Lévy
Université de Rouen
F-76130 Mont Saint Aignan

and

Laboratoire de Mécanique Théorique (UA 229)
F-75230 Paris Cédex 05, France

CHAPTER 1

1. - INTRODUCTION TO HOMOGENIZATION THEORY

I - GENERALITIES

Composite materials or suspensions are media with microstructure on a scale much smaller than the macroscopic scale of interest. This macroscopic scale length, denoted by L in the following, may be the dimension of a specimen of the medium or a typical wavelength. The characteristic length of the medium configuration is denoted by ℓ ; the small ratio $\varepsilon = \ell/L$ will play a key role in our analysis. In the study of physical or mechanical processes in media with microstructure, known and unknown quantities are depending on ε and an asymptotic analysis is used to determine the unknown field quantities. Furthermore, to make precise the fact that the medium varies rapidly on the small scale ℓ and may also vary slowly on the larger scale L, we assume that every property of the medium is of the form $f(x,y)$ where $y = x/\varepsilon$. Here $x = (x_1,x_2,x_3)$ is the position vector of a point in Cartesian coordinates and $y = (y_1,y_2,y_3)$ is the vector of stretched coordinates. We shall look for each unknown field quantity $u^\varepsilon(x)$ in the form of a double-scale asymptotic expansion :

$$u^\varepsilon(x) = u^o(x,y) + \varepsilon\, u^1(x,y) + \varepsilon^2\, u^2(x,y) + \ldots$$

The two-scales process introduced in the partial differential equations of the problem produces equations in x and y variables. Generally speaking, equations in y are solvable if the microstructure is, in some sense, periodic ; and terms $u^i(x,y)$ in the postulated asymptotic expansions are periodic in the y variable with the same period as that of the structure. Moreover this leads to a rigorous deductive procedure for obtaining the macroscopic equations (in x) for the global behavior of the medium. Thus, homogenization process gives the passage from a microscopic description to a macroscopic description of the studied problem ; periodicity appears as an hypothesis which is very convenient in order to obtain results in a precise form, and which may be used as a simplified model for more general situations. In most problems, a mathematical proof of the convergence of solutions to the homogenized solution as $\varepsilon \to 0$ is available.

II - PRECISE FORMULATION OF PERIODICITY

First, let us consider a medium with a fine periodic structure, the small parameter ε appears in a natural way : it is the ratio of a period dimension ℓ to a characteristic length L of the macroscopic phenomena. The period may be considered homothetic with the ratio ε of a basic period Y in the stretched coordinates $y = x/\varepsilon$; the periodicity of the medium is εY. When we use a function $f(x,y)$ Y-periodic with respect to the y variable, we may consider the y variable in the basic period (by translation of an integer number of periods in the y space variable) and the function f of the y variable takes equal values on the opposite faces of Y. Such a function $f(x,y)$ with $y = x/\varepsilon$ will be called "*locally periodic*".

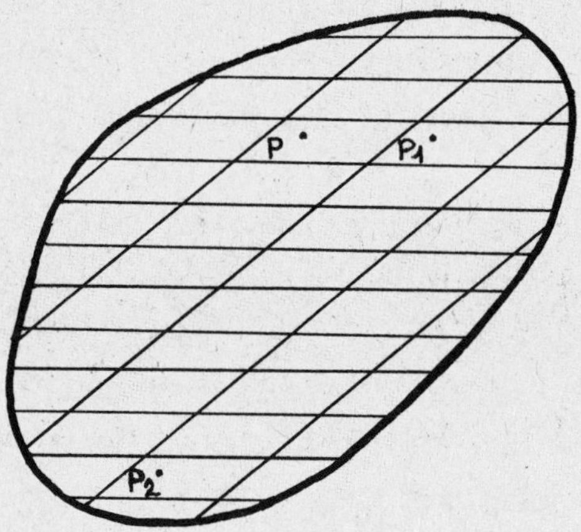

In fact in points such as P and P_1, homologous by periodicity and in neighbouring periods, the values of the function are almost the same ; but in points such as P and P_2, homologous by periodicity but far in the x variable, the values of the function are very different. We assume that any material property of the medium is locally periodic, then it may vary slowly on the larger scale L. We can do the same for the basic period Y and consider it not as a constant but as a smooth function of the x variable. Such a medium will be called "*locally periodic*". For problems which work out with given geometric pattern the hypothesis of local periodicity just extends the class of concerned media, but if we consider problems with large deformations the media may no longer be supposed rigorously periodic because of the structure deformation.

In the following, we emphasis the main properties of the homogenization process in the study of a model problem.

III - ELLIPTIC EQUATION IN DIVERGENCE FORM

1. - Formulation of the problem - We consider a conservation law of the form

$$(1) \qquad -\frac{\partial}{\partial x_i}\left(a^{\varepsilon}_{ij}\frac{\partial u^{\varepsilon}}{\partial x_j}\right) = f^{\varepsilon}$$

with

$$(2) \qquad a^{\varepsilon}_{ij} = a^{\varepsilon}_{ji}$$

$$(3) \qquad a^{\varepsilon}_{ik}\xi_i \ \xi_k \geqq \gamma\ \xi_j\ \xi_j \qquad\qquad \gamma > 0$$

(we adopt here and in what follows the summation convention).

This equation is for instance the equation of steady diffusion, u^{ε} is the temperature and f^{ε} the given source term, the tensor a^{ε}_{ij} is the thermal conductivity. The small parameter ε characterises the fine periodicity of the medium. We assume that the material has a locally periodic structure, the basic period Y(x), (in the following we denote it by Y for simplicity), is a parallelepiped cell and the a^{ε}_{ij} coefficients are of the form : $a_{ij}(x,y)$ with $y = x/\varepsilon$, Y-periodic in y. The source term f^{ε} is of the same form.

First we only consider a formal expansion, when $\varepsilon \ll 1$, of u^ε out of a neighbourhood of the medium, consequently we may consider an infinite tridimensional medium. Related to equation (1), it is possible to consider transmission problems with coefficients a_{ij} which are not smooth but piecewise constant, or piecewise smooth, in the period. For example the period εY is divided in two regions εY_1 and εY_2 separated by the smooth surface $\varepsilon\Gamma$ where the a_{ij} are discontinuous ; equation (1) must be verified in εY_1 and εY_2, and for the discontinuities accross $\varepsilon\Gamma$, one must have

$$[\![u^\varepsilon]\!] = 0 \quad , \quad [\![a^\varepsilon_{ij} \frac{\partial u^\varepsilon}{x_j} n_i]\!] = 0$$

Then equation (1) must be interpretated in the sense of distributions. So we do in the following.

As we have seen before, we look for an asymptotic expansion of u^ε, when $\varepsilon \ll 1$, in the form

$$(4) \qquad u^\varepsilon(x) = u^o(x,y) + \varepsilon\, u^1(x,y) + \ldots \quad , \quad y = x/\varepsilon$$

where functions u^i are Y-periodic with respect to y. The idea of the method is to insert (4) in equation (1) and to identify powers of ε. In order to present these computations in a simple form, it is useful to consider first x and y as independent variables and to replace next y by x/ε. Applied to a function $\phi(x,x/\varepsilon)$, the operator $\partial/\partial x_j$ becomes $\partial/\partial x_j + (1/\varepsilon)(\partial/\partial y_j)$. With this in mind, we deduce successive differential problems with respect to the y variable, x being a parameter.

2. Asymptotic expansion of the solution - From (1) at $O(\varepsilon^{-2})$, we obtain

$$(5) \qquad \frac{\partial}{\partial y_i} \left(a_{ij} \frac{\partial u^o}{\partial y_j}\right) = 0 \quad ;$$

this is a differential problem for the Y periodic function $u^o(y)$. As we shall see later on, the only periodic solution of (5) is u^o = constant where x is a parameter, i.e.

$$(6) \quad u^o(x,y) = u^o(x).$$

From (1) at $O(\varepsilon^{-1})$, taking (6) into account :

$$(7) \quad \frac{\partial}{\partial y_i}\left(a_{ij}\left(\frac{\partial u^1}{\partial y_j} + \frac{\partial u^o}{\partial x_j}\right)\right) = 0 \quad ,$$

which is an equation to obtain $u^1(y)$ if u^o is supposed to be known ; (7) is then an elliptic equation in Y (with Y-periodicity instead of boundary conditions). We introduce W_y, the appropriate space of Y periodic functions :

$$W_y = \{\phi, \ \phi \in H^1(Y), \ \phi \text{ Y-periodic}\}.$$

To obtain a variational formulation of (7) we multiply it by a test function $\phi \in W_y$, and integrate over Y taking into account the Y-periodicity of u^1 and ϕ. The variational formulation of (7) is :

$$(8) \begin{cases} \text{Find } u^1 \in W_y \text{ such that} \\ \int_Y a_{ij} \frac{\partial u^1}{\partial y_j} \frac{\partial \phi}{\partial y_i} dy = -\frac{\partial u^o}{\partial x_j} \int_Y a_{ij} \frac{\partial \phi}{\partial y_i} dy \quad \forall \phi \in W_y \ . \end{cases}$$

Conversely, if u^1 satisfies (8) it can easily be proved that u^1 is solution of (7) in the sense of distributions. In order to avoid difficulties associated to the fact that u^1 is only determinated up to an additive constant, we introduce the space $\tilde{W}_y$ of the functions of W_y with zero mean value, ($\tilde{W}_y$ is also the space W_y/R):

$$\tilde{W}_y = \{\phi \in W_y \ , \ \tilde{\phi} = \frac{1}{|Y|} \int_Y \phi(y)\, dy = 0\}$$

(where $|Y|$ = measure of Y) with the scalar product

$$(u,v)_{\tilde{W}_y} = \int_Y \frac{\partial u}{\partial y_i} \frac{\partial v}{\partial y_i} dy \ .$$

The left hand term in (8) is a continuous linear form on $\tilde{W}_y$ which is coercive as a result of the positivity condition (3). Then, by Lax Milgram lemma (8) admits a unique solution in $\tilde{W}_y$ (i.e. u^1 defined up to an additive constant which may depend on the parameter x) if the right hand term is a continuous linear form on $\tilde{W}_y$; this condition is satisfied iff the right hand term vanishes when ϕ is a constant, which is clearly true, (let us remark that the same arguments were also valid for the solution of (5)). Moreover, let w_k be the unique solution of :

$$(9) \begin{cases} \text{Find } w_k \in \tilde{W}_y \text{ such that} \\ \int_Y a_{ij} \frac{\partial w_k}{\partial y_j} \frac{\partial \phi}{\partial y_i} dy = - \int_Y a_{ik} \frac{\partial \phi}{\partial y_i} dy \qquad \forall \phi \in \tilde{W}_y \ , \end{cases}$$

then, by virtue of the linearity of problem (8), we have

$$(10) \quad u^1(x,y) = \frac{\partial u^o}{\partial x_k} w_k + \tilde{u}^1(x)$$

$\tilde{u}^1(x)$ is an arbitrary function of x and represents the mean value of $u^1(x,y)$ because $\tilde{w}_k = 0$.

3. - **Macroscopic equation** - Equation (1) can be written in term of the heat flux $\vec{q}^\varepsilon$:

$$- \operatorname{div} \vec{q}^\varepsilon = f^\varepsilon \quad , \quad q_i^\varepsilon = a_{ij}^\varepsilon \frac{\partial u^\varepsilon}{\partial x_j} \ .$$

The asymptotic expansion of $\vec{q}^\varepsilon$ is, taking (6) into account

$$\vec{q}^\varepsilon(x) = \vec{q}^o(x,y) + \varepsilon \vec{q}^1(x,y) + \dots$$

(11) $q_i^o(x,y) = a_{ij}\left(\frac{\partial u^o}{\partial x_j} + \frac{\partial u^1}{\partial y_j}\right)$, ...

We now apply the conservation law (1) in a macroscopic domain D, made of whole periods, but otherwise arbitrary

$$- \int_D \mathrm{div}\, \vec{q}^{\varepsilon}\, d\,v = \int_D f^{\varepsilon}\, d\,v \quad \text{i.e.} \quad - \int_{\partial D} \vec{q}^{\varepsilon} . n\, d\sigma = \int_D f^{\varepsilon}\, d\,v .$$

At the first order when $\varepsilon \ll 1$:

$$\int_D f^{\varepsilon}\, d\,v = \int_D \tilde{f}\, d\,x \quad , \quad \int_{\partial D} \vec{q}^{\varepsilon} . \vec{n}\, d\,\sigma = \int_{\partial D} \vec{q}^{o} . \vec{n}\, d\sigma \quad .$$

The boundary ∂D is made of faces of the neighbourhood periods of D. Let us suppose (this will be proved later on) that the integral of $\vec{q}^o . \vec{n}$ on a face of the basic period Y is the same as that of its mean value $\tilde{\vec{q}}^o . \vec{n}$, then at the first order when $\varepsilon \ll 1$

$$\int_{\partial D} \vec{q}^{o} . \vec{n}\, d\,\sigma = \int_{\partial D} \tilde{\vec{q}}^{o} . \vec{n}\, d\,\sigma_x$$

Thus the conservation law for the function $\tilde{\vec{q}}^o$ of the mascroscopic variable x :

$$- \int_{\partial D} \tilde{\vec{q}}^{o} . \vec{n}\, d\,\sigma_x = \int_D \tilde{f}\, dx \quad .$$

This implies the macroscopic equation :

(12) $- \mathrm{div}\, \tilde{\vec{q}}^o = \tilde{f}$.

The macroscopic function $\tilde{\vec{q}}^o(x)$ is related to the other macroscopic function $u^o(x)$, in virtue of (11) and (10), by the constitutive relation :

(13) $\tilde{q}_i^o = a_{ik}^h \frac{\partial u^o}{\partial x_k}$

$$a_{ik}^{h} = \frac{1}{|Y|} \int_Y \left(a_{ik} + a_{ij} \frac{\partial w_k}{\partial y_j}\right) dy \quad .$$

The homogenized coefficients a_{ik}^{h} are completly determined by the microstructure.

We now prove the assumed property on the integrals of $q_k^{o} n_k$ and $\tilde{q}_k^{o} n_k$. Denoting by Σ_i a face of the basic period Y with outward normal unit vector $\vec{n}^{i}$, we shall establish :

$$\int_{\Sigma_i} q_k^{o} n_k^{i} \, d\sigma_y = \int_{\Sigma_i} \tilde{q}_k^{o} n_k^{i} \, d\sigma_y \quad .$$

Let us remark that

$$\tilde{q}_k^{o} = \frac{1}{|Y|} \int_Y q_k^{o} \, dy = \frac{1}{|Y|} \int_Y \frac{\partial}{\partial y_j} (q_j^{o} y_k) \, dy - \frac{1}{|Y|} \int_Y \frac{\partial q_j^{o}}{\partial y_j} y_k \, dy$$

$$= \frac{1}{|Y|} \int_{\partial Y} q_j^{o} y_k n_j \, d\sigma_y \qquad \text{taking (7) into account.}$$

And for example

$$\int_{\Sigma_1} \tilde{q}_k^{o} n_k^{1} \, d\sigma_y = \frac{|\Sigma_1| \, n_k^{1}}{|Y|} \int_{\partial Y} q_j^{o} n_j y_k \, d\sigma_y =$$

$$\frac{|\Sigma_1|}{|Y|} \Big(\int_{\Sigma_1} q_j^{o} n_j^{1} \vec{n}^{1} . (\vec{y} - \vec{y}') \, d\sigma_y + \int_{\Sigma_2} q_j^{o} n_j^{2} \vec{n}^{1} . (\vec{y} - \vec{y}') \, d\sigma_y +$$

$$+ \int_{\Sigma_3} q_j^{o} n_j^{3} \vec{n}^{1} . (\vec{y} - \vec{y}') \, d\sigma_y \Big)$$

taking into account the Y-periodicity and denoting by y and y' the homologous points of opposite faces. Denoting now by $\vec{\alpha}_1$, $\vec{\alpha}_2$, $\vec{\alpha}_3$ the 3 oriented sides of Y ($\vec{\alpha}_1$, $\vec{\alpha}_2$ in Σ_3,...), for $\vec{y} \in \Sigma_3$ $\vec{y} - \vec{y}' = \vec{\alpha}_3$ is normal to $\vec{n}^{1}$ and the last integral vanishes. So do the second one and

$$\int_{\Sigma_1} \tilde{q}_k^{o} n_k^{1} \, d\sigma_y = \frac{|\Sigma_1| \, \vec{n}^{1} . \vec{\alpha}_1}{|Y|} \int_{\Sigma_1} q_j^{o} n_j^{1} \, d\sigma_y = \int_{\Sigma_1} q_j^{o} n_j^{1} \, d\sigma_y \quad ,$$

which demonstrates the proposition. This property is general for a vector or a tensor which is divergence free in the y variable.

Finally, by (12) and (13), $u^o(x)$ is solution of the equation

$$- \frac{\partial}{\partial x_i} (a^h_{ij} \frac{\partial u^o}{\partial x_j}) = \tilde{f} .$$

4. - **Properties of the homogenized coefficients** - We now search for symmetry and positivity relations for the a^h_{ij}. We can write the homogenized coefficients in the form

$$a^h_{ik} = \frac{1}{|Y|} \int_Y a_{ij} \frac{\partial}{\partial y_j} (w_k + y_k) dy = \frac{1}{|Y|} \int_Y a_{mj} \frac{\partial}{\partial y_j} (w_k + y_k) \frac{\partial y_i}{\partial y_m} dy .$$

Moreover, by (9)

$$\int_Y a_{mj} \frac{\partial}{\partial y_j} (w_k + y_k) \frac{\partial \phi}{\partial y_m} dy = 0 \qquad \forall \phi \in \tilde{W}_y ,$$

taking $\phi = w_i$ and adding this null quantity to the last expression of a^h_{ik} we obtain :

$$a^h_{ik} = \frac{1}{|Y|} \int_Y a_{mj} \frac{\partial}{\partial y_j} (w_k + y_k) \frac{\partial}{\partial y_m} (w_i + y_i) dy .$$

Then the symmetry of a^h_{ik} follows from (2).

We now prove that the coefficients a^h_{ik} satisfy an ellipticity condition analogous to (3) :

$$a^h_{ik} \xi_i \xi_k \geqq \delta \xi_j \xi_j \quad , \quad \delta > 0 \qquad \forall \vec{\xi} \in R^3 .$$

It suffices to establish that the left hand side is positive for any $\vec{\xi} \neq 0$

$$a^h_{ik}\,\xi_i\,\xi_k = \frac{1}{|Y|}\int_Y a_{mj}\frac{\partial}{\partial y_j}\big(\xi_k(w_k+y_k)\big)\frac{\partial}{\partial y_m}\big(\xi_i(w_i+y_i)\big)\,dy$$

$$= \frac{1}{|Y|}\int_Y a_{mj}\frac{\partial W}{\partial y_j}\frac{\partial W}{\partial y_m}\,dy \qquad \text{with } W = \xi_j(w_j + y_j).$$

By (3) :

$$a^h_{ik}\,\xi_i\,\xi_k \geqq \frac{1}{|Y|}\int_Y \gamma\,\Sigma\Big(\frac{\partial W}{\partial y_j}\Big)^2\,dy$$

so that $a^h_{ik}\,\xi_i\,\xi_k \geqq 0$ and $a^h_{ik}\,\xi_i\,\xi_k = 0$ iff $\partial W/\partial y_j = 0\ \forall\, j$, i.e. $W = C$; but then $\xi_j\,y_j = C - \xi_j\,w_j$ is Y-periodic i.e. $\xi_j = 0\ \forall\, j$, so that $a^h_{ik}\,\xi_i\,\xi_k$ is positive definite.
Homogenized coefficients a^h_{ik} are determined by (13), their form is interesting

$$a^h_{ik} = \tilde{a}_{ik} + \Big(a_{ij}\frac{\partial w_k}{\partial y_j}\Big)^{\sim} ,$$

the second term appears as a corrector to the mean value. With a locally periodic structure of the medium ($Y = Y(x)$, $a_{ij} = a_{ij}(x,y)$), the homogenized medium is not homogeneous : the a^h_{ik} depend on x. If the microstructure is strictly periodic (Y = Constant, $a_{ij} = a_{ij}(y)$) the a^h_{ik} are constant.

The homogenized medium is anisotropic even if at the microstructure level the behavior is isotropic. Indeed if $a_{ij} = a\,\delta_{ij}$, for $i \neq k$ one has

$$a^h_{ik} = \frac{1}{|Y|}\int_Y a\,\frac{\partial w_k}{\partial y_i}\,dy$$

which is not necessarily equal to zero.

BIBLIOGRAPHY

. **Bensoussan, A., Lions, J.L., Papanicolaou, G.** - *Asymptotic Analysis for periodic Structures*. North-Holland, Amsterdam, 1978.

. **Sanchez-Palencia, E.** - *Non-homogeneous Media and Vibration Theory*. Springer-Verlag, Berlin, 1980.

CHAPTER 2

2. - FLUIDS IN POROUS MEDIA
DARCY'S LAW

I - FORMULATION OF THE PROBLEM

Let us consider a solid permeated by pores containing a viscous incompressible fluid. We assume that the pore configuration has a scale length ℓ which is small compared to a typical macroscopic scale length L which may be the dimension of a specimen of the porous solid. We suppose the medium locally periodic. The period is a parallelepiped cell homothetic with the small ratio $\varepsilon = \ell/L$ of a basic period Y (in the stretched coordinates $y = x/\varepsilon$) in which the fluid domain Y_F and the solid one Y_S have a smooth boundary Γ. The medium configuration is such that the fluid part is of one piece, so all the εY_F parts are connected.

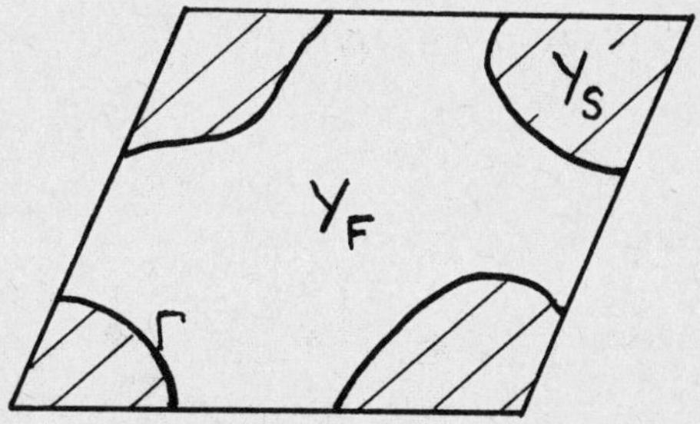

Fluid flow in porous media is often very slow and inertia effects may be neglected. This is the reason why we use the Stokes rather than the Navier-Stokes equations. The velocity and the pressure in the fluid are depending on ε and satisfy :

(1) $\text{div}\ \vec{V}^{\varepsilon} = 0$

(2) $0 = -\ \text{grad}\ P^{\varepsilon} + \mu\ \Delta\ \vec{V}^{\varepsilon} + \vec{f}$

with the boundary condition

(3) $\vec{V}^{\varepsilon} = 0$ on the solid.

We consider that the only small parameter in the problem is ε. The viscosity coefficient μ and the applied density of forces $\vec{f}(x)$ are fixed independent of ε. First we only consider a formal expansion of $\vec{V}^{\varepsilon}$ and P^{ε}, when $\varepsilon << 1$, out of a neighbourhood of the porous medium, consequently we may consider an infinite tridimensional medium. We apply homogenization method to this problem with the standard macroscopic variable x and the microscopic one $y = x/\varepsilon$ just as in the first lecture. The medium under consideration now is a mixture of a rigid fixed solid and of a viscous incompressible fluid. We look for an asymptotic expansion of each field quantity (V_i^{ε} and P^{ε} in the fluid), when $\varepsilon << 1$, in the form :

(4) $F^{\varepsilon}(x) = \varepsilon^{\alpha}\left(F^{o}(x,y) + \varepsilon\ F^{1}(x,y) + \ldots\right)$

where the functions F^{i} are Y-periodic in y.

II - ASYMPTOTIC EXPANSIONS

We insert asymptotic expansions (4) in relations (1)-(3) and identify powers of ε. We recall that, applied to a function $\phi(x,x/\varepsilon)$, the operator $\partial/\partial x_j$ becomes $\partial/\partial x_j + (1/\varepsilon)(\partial/\partial y_j)$, and the expansion of Δ is

$$\Delta = \frac{1}{\varepsilon^2}\ \Delta_y + \frac{1}{\varepsilon}\ \ldots ,$$

in the following the sub index y or x specifies the partial derivative, so Δ_y denotes the Laplacian with respect to the variables y_i (x_i being parameters).

The order with respect to ε of $\vec{V}^\varepsilon$ and P^ε is such that the problem to be solved, at the first power of ε, has a physical meaning and is not reduced to the zero solution. The form of the expansions of $\partial/\partial x_i$ and Δ shows that the first significative term of (4) for $\vec{V}^\varepsilon$ must be the ε^2 term and for P^ε the term $\varepsilon^o P^o(x)$. Thus we write :

$$\left.\begin{aligned} \vec{V}^\varepsilon(x) &= \varepsilon^2\,\vec{V}^o(x,y) + \varepsilon^3\,\vec{V}^1(x,y) + \dots \\ P^\varepsilon(x) &= P^o(x) + \varepsilon\,P^1(x,y) + \dots \end{aligned}\right\}\quad y = x/\varepsilon,$$

where the functions $\vec{V}^i$ and P^i are Y-periodic in y.

From (1) at $O(\varepsilon)$, (2) at $O(\varepsilon^o)$, (3) at $O(\varepsilon^2)$, we deduce the differential problem in the y variable, x being a parameter, for the Y-periodic functions $\vec{V}^o$ and P^1

$$\left.\begin{aligned} &(5)\quad \operatorname{div}_y \vec{V}^o = 0 \\ &(6)\quad 0 = -\operatorname{grad}_y P^1 + \mu\,\Delta_y \vec{V}^o + \vec{f} - \operatorname{grad}_x P^o \end{aligned}\right\}\quad \text{in } Y_F$$

$$(7)\quad \vec{V}^o = 0 \quad \text{on } \Gamma.$$

This is the local problem where $\vec{f} - \operatorname{grad}_x P^o$ plays the role of a given force. For the study of this problem, we introduce an appropriate space of Y-periodic functions :

$$V_y = \{\vec{w}\ ,\ \vec{w} \in \left(H^1(Y_F)\right)^3\ ,\ \vec{w}\ \text{Y-periodic},\ \vec{w} = 0\ \text{on } \Gamma,\ \operatorname{div}_y \vec{w} = 0\}$$

which is a Hilbert space with the scalar product

$$(\vec{u},\vec{v})_{V_y} = \int_{Y_F} \frac{\partial u_j}{\partial y_k} \frac{\partial v_j}{\partial y_k} dy .$$

To obtain a variational formulation of equations (5)-(7), we multiply (6) by a test function $\vec{w} \in V_y$; we integrate over Y_F, taking into account the Y-periodicity of P^1, $\vec{V}^o$, $\vec{w}$, and the properties $\vec{w} = 0$ on Γ, $\mathrm{div}_y\, \vec{w} = 0$. This leads to the virtual power equation :

$$(8) \quad \mu(\vec{V}^o,\vec{w})_{V_y} = (f_i - \frac{\partial P^o}{\partial x_i}) \int_{Y_F} w_i \, dy \qquad \forall\ \vec{w} \in V_y .$$

Conversely, if $\vec{V}^o \in V_y$ and satisfies (8), by integrating by parts, it verifies

$$\int_{Y_F} (\mu \, \Delta_y \vec{V}^o + \vec{f} - \mathrm{grad}_x P^o).\vec{w} \, dy = 0 \qquad \forall\ \vec{w} \in V_y ;$$

thus we know (see Temam[1] in references) that there exists P^1 of class L^2 satisfying (6) in the distributional sense. And it can be proved that the function P^1 just found is Y-periodic (using the hypothesis that Y_F is connected). Consequently, the local problem (5)-(7) is equivalent to the following variational problem :

Find $\vec{V}^o \in V_y$ satisfying (8).

The existence and uniqueness of the solution $\vec{V}^o$ is immediate consequence of the Lax Milgram lemma, because the right hand side of (8) is linear and bounded functional on V_y. And using the linearity property, we can write this solution $\vec{V}^o$ in the form :

$$(9) \quad \vec{V}^o = (f_i - \frac{\partial P_o}{\partial x_i}) \vec{v}^i$$

where $\vec{v}^i$ is the only solution of the problem :

$$(10)\quad \begin{cases} \text{Find } \vec{v}^i \in V_y \text{ such that} \\ \mu(\vec{v}^i,\vec{w})_{V_y} = \int_{Y_F} w_i \, dy \qquad \forall \, \vec{w} \in V_y \, , \end{cases}$$

$\vec{v}^i$ is, in the weak sense, the solution of the local problem

$$\left.\begin{aligned} &\operatorname{div}_y \vec{v}^i = 0 \\ &0 = - \operatorname{grad}_y q^i + \mu \, \Delta_y \vec{v}^i + \vec{e}_i \end{aligned}\right\} \text{ in } Y_F$$

$$\vec{v}^i = 0 \quad \text{on } \Gamma,$$

$\vec{v}^i$ and q^i Y-periodic, where $\vec{e}_i$ is the unit vector in the direction of the y_i axis.

III - MACROSCOPIC LAWS

Let us note that $\vec{V}^o$, $\vec{V}^1$, $\vec{v}^i$ as functions of y are defined on Y_F, it is natural to extend them to Y with zero value on Y_S since they are zero on Γ as a consequence of (3). Then we can define their mean value as usual

$$\tilde{\cdot} = \frac{1}{|Y|} \int_Y \cdot \, dy$$

Applied to (9) the mean value operator leads to :

$$(11) \quad \tilde{V}^o_j = K_{ji} \left(f_i - \frac{\partial P^o}{\partial x_i}\right)$$

with $K_{ji} = \tilde{v}^i_j$.

$\mathbb{K} = (K_{ij})$ is a tensor which only depends on the viscosity coefficient as μ^{-1} (according to (10)) and on the geometry of the period Y. If the porous medium is strictly periodic (Y = constant), $\mathbb{K}$ is a constant tensor ; if the medium is locally periodic ($Y = Y(x)$), $\mathbb{K}$ depends on x. $\mathbb{K}$ is called the permeability tensor and the macroscopic law (11) is the Darcy's law.

The tensor $\mathbb{K}$ is symmetric and positive definite. For, if in (10) we take $\vec{w} = \vec{v}^j$ and in the equation analogous to (10) for $\vec{v}^j$ we take $\vec{w} = \vec{v}^i$, we obtain, by the symmetry of the scalar product

$$K_{ij} = |Y|^{-1} \mu(\vec{v}^i, \vec{v}^j)_{V_y} = K_{ji} .$$

On the other hand, the tensor $\mathbb{K}$ is positive because we have :

$$K_{ij} \xi_i \xi_j = \mu |Y|^{-1} \xi_i \xi_j (\vec{v}^i, \vec{v}^j)_{V_y} = \mu |Y|^{-1} (\xi_i \vec{v}^i, \xi_j \vec{v}^j)_{V_y}$$

$$= \mu |Y|^{-1} \|\xi_i \vec{v}^i\|^2_{V_y} \geq 0 ,$$

and it can also be proved that $K_{ij} \xi_i \xi_j = 0$ iff $\xi_k = 0 \ \forall k$. It suffices to show that $\xi_i \vec{v}^i = 0$ only if $\xi_i = 0$; in order to verify that we multiply (10) by ξ_i and we take a test function $\vec{w}$ such that $\tilde{w}_i = \xi_i$ (this is possible for Y_F is connected), we have

$$\mu(\xi_i \vec{v}^i, \vec{w})_{V_y} = \xi_i \int_{Y_F} w_i \, dy = |Y| (\xi_i \xi_i)$$

and $\xi_i \vec{v}^i = 0 \implies \xi_i = 0$. So that $K_{ij} \xi_i \xi_j$ is positive definite.

We now obtain another macroscopic equation by the method of conservation law. We apply the fluid mass conservation law in a macroscopic domain D consisting of a great discrete number of periods ; D_F denoting the part of D filled with the fluid, we have

$$\int_{D_F} \operatorname{div} \vec{V}^{\varepsilon} \, dv = \int_{\partial D_F} \vec{V}^{\varepsilon} . \vec{n} \, d\sigma = 0 .$$

Thus at the first order when $\varepsilon \to 0$, we have

$$\int_{\partial D} \vec{V}^{o} . \vec{n} \, d\sigma = 0$$

since $\vec{V}^{\varepsilon}$ is zero on the solid boundaries of ∂D_F and can be extended to the whole medium with value zero in the solid. By (5), $\vec{V}^{o}(x,y)$ is a divergence free vector in the y variable, then as we have seen already in the first lecture, the integral of $\vec{V}^{o}.\vec{n}$ on a face of the basic period Y is the same as that of its mean value $\tilde{\vec{V}}^{o}.\vec{n}$ and we can write

$$\int_{\partial D} \vec{V}^{o}.\vec{n}\, d\sigma = \int_{\partial D} \tilde{\vec{V}}^{o}.\vec{n}\, d\,\sigma_x .$$

Therefore the fluid mass conservation law implies, at the first order when $\varepsilon \to 0$

$$\int_{\partial D} \tilde{\vec{V}}^{o}.n\, d\,\sigma_x = 0 \qquad \text{i.e.}$$

(12) $\operatorname{div} \tilde{\vec{V}}^{o} = 0.$

Let us remark that the equality of the volumic mean value of $\vec{V}^{o}$ in Y with its surfacic mean value on ∂Y emphasizes the physical interpretation of $\tilde{\vec{V}}^{o}$ as a filtration velocity.

The macroscopic field quantities are $\tilde{\vec{V}}^{o}(x)$ and $P^{o}(x)$, they are related by the macroscopic equations (11) and (12). Finally $P^{o}(x)$ is the solution of the elliptic equation

$$(13) \quad \frac{\partial}{\partial x_i}\left(K_{ij}\frac{\partial P^{o}}{\partial x_j}\right) = \frac{\partial}{\partial x_i}\left(K_{ij}\, f_j\right) .$$

When P^{o} is obtained, $\tilde{\vec{V}}^{o}$ is given by the Darcy's law and the microscopic velocity $\vec{V}^{o}(x,y)$ by (9).

The homogenization method has furnished a deductive procedure for obtaining the macroscopic equations of the limit phenomenon as the ratio ε of the microstructure to the macrostructure tends to zero. It is to be noticed that for this problem as for most problems in mechanics the micro and macro processes are of very different nature, indeed Darcy's law is the "*homogenized*" form of the Stokes equation.

Finally let us notice that physical velocities are small as ε^2, the macroscopic one is $\varepsilon^2 \vec{V}^o(\vec{x})$ and the microscopic one $\varepsilon^2 \vec{V}^o(x,y)$, whereas the pressure gradient is $O(1)$.

IV - BOUNDARY CONDITIONS

In order to obtain $P^o(x)$ we must adjoin boundary conditions to equation (13). These conditions depend on the nature of the porous medium boundary S, it may be a solid boundary, or the boundary between two different porous media, or the boundary between a fluid in the porous medium and the fluid in a free domain. We can obtain the boundary conditions by expressing the fluid mass conservation and the fluid momentum conservation in a control volume $\mathcal{D}$ flattened on S. For example thickness of $\mathcal{D}$ may be $O(\sqrt{\ell L})$, which is great with respect to ℓ and small with respect to L, while the other dimensions of $\mathcal{D}$ are $O(L)$. We investigate the contribution of each term of the conservation laws in the limit process $\varepsilon \to 0$, and write theses laws at the first order in ε. Because the large face of $\partial\mathcal{D}$ inside the porous medium is very much distant from S than $O(\ell)$, we can suppose that the behavior is locally periodic in the porous medium and express the flux of mass and momentum through the part of $\partial\mathcal{D}$ in the porous medium as for periodic solutions. It appears that volumic integrals and surfacic integrals on the faces of $\partial\mathcal{D}$ which are not parallel to S are small compared to the integrals on the two faces of $\partial\mathcal{D}$ which are parallel to S, when $\varepsilon \to 0$. Boundary conditions or transmission conditions are then obtained.

It is also possible to study in detail the fluid flow in the vicinity of S, we see that the boundary or transmission conditions are the necessary and sufficient conditions for the existence of the boundary layer which matches the locally periodic flow in the porous solid with the medium on the other side of S.

When S is an impervious boundary, the appropriate boundary condition is[2] :

$$\vec{V}^o.\vec{n} = 0 \qquad \text{on } S.$$

When S is surface where the $K_{ij}(x)$ are discontinuous, the transmission conditions between the two porous media are :

$$[\![P^o]\!] = 0 \quad , \quad [\![\vec{V}^o.\vec{n}]\!] = 0 \qquad \text{on } S.$$

When S is the boundary between a fluid in the porous medium and the adjacent free flow, two classes of problems arise[3,4]. First, the porous body is contained in an outer domain filled by the fluid, then :

$$\vec{V}_F = 0 \quad , \quad P_F = P^o(x) \qquad \text{on } S,$$

$\vec{V}_F$ and P_F denoting the velocity and the pressure in the free flow, for this flow S is an impervious surface. There is an other class of problems where the free flow is in a cavity of the porous solid, then the free motion is a consequence of the motion in the porous body. Then

$$P_F = cste = P^o(x) \qquad \text{on } S,$$

$\vec{V}_F$ is $O(\varepsilon^2)$, its tangential components on S are imposed (by the geometrical properties of the problem) and its normal component is $\varepsilon^2 \vec{V}^o.\vec{n}$.

REFERENCES

1. **Temam, R.** - *Navier-Stokes equations.* North-Holland,Amsterdam, 1977, chap. 1.

2. **Sanchez-Palencia, E.** - *Non homogeneous media and vibration theory.* Lect. Notes in Physics 127, Springer Verlag, Berlin, 1980, Chap. 7.

3. **Ené, H.I., Sanchez-Palencia, E.** - Equations et phénomènes de surface pour l'écoulement dans un modèle de milieux poreux, *Journal de Méca.*, 14, 73, 1975.

4. **Lévy, T., Sanchez-Palencia, E.** - On boundary conditions for fluid flow in porous media, *Int. J. Engng. Sc.*, 13, 923, 1975.

CHAPTER 3

3. - ACOUSTICS IN ELASTIC POROUS MEDIA

I - FORMULATION OF THE PROBLEM

Acoustics deal with small perturbations. The motions are assumed to be small enough to be governed by linearized equations.

We consider an infinite porous medium made of an elastic matrix filled with a compressible viscous fluid. An equilibrium state of this medium will be taken as the reference configuration. We assume that the geometry in this configuration and the components behavior are locally periodic. Dimensions of the periods are small compared to the wavelengths arising in the study, the small ratio of these two length scales is denoted by ε. Then the field quantities in a motion of the medium are depending on ε. When thermal effects are not taken into account, for time harmonic perturbations of the equilibrium state with angular frequency ω, the equations are :

$$(1) \quad i\,\omega\,\rho_F^\varepsilon + \rho_o \operatorname{div} \vec{V}^\varepsilon = 0$$

$$(2) \quad i\,\omega\,\rho_o\, V_i^\varepsilon = \frac{\partial\, \sigma_{ij}^{F\varepsilon}}{\partial\, x_j}$$

$$(3) \quad \sigma_{ij}^{F\varepsilon} = -\,P^\varepsilon\,\delta_{ij} + \overline{\lambda}\,\operatorname{div} \vec{V}^\varepsilon\,\delta_{ij} + 2\,\overline{\mu}\,D_{ij}(\vec{V}^\varepsilon)$$

$$\text{with } D_{ij}(\vec{V}) = \frac{1}{2}\left(\frac{\partial\, V_i}{\partial\, x_j} + \frac{\partial\, V_j}{\partial\, x_i}\right)$$

$$(4) \quad P^\varepsilon = a_o^2\,\rho_F^\varepsilon$$

in the fluid ; there are 4 constants : ρ_o the density in the reference state, $\overline{\lambda}$ and $\overline{\mu}$ the viscosity coefficients and a_o the sound velocity. In the solid the equations are :

$$(5) \quad -\rho_S^\varepsilon \ \omega^2 u_i^\varepsilon = \frac{\partial \sigma_{ij}^{S\varepsilon}}{\partial x_j}$$

$$(6) \quad \sigma_{ij}^{S\varepsilon} = a_{ijkh}^\varepsilon \ e_{kh}(\vec{u}^\varepsilon) \quad \text{with} \quad e_{kh}(\vec{u}) = \frac{1}{2}\left(\frac{\partial u_k}{\partial x_h} + \frac{\partial u_h}{\partial x_k}\right)$$

ρ_s^ε is the solid density and a_{ijkh}^ε are the coefficient of elasticity associated with the reference state, they are locally periodic ; a_{ijkh}^ε are piecewise smooth functions and they satisfy the usual conditions of symmetry and positivity :

$$a_{ijkh}^\varepsilon = a_{jikh}^\varepsilon = a_{ijhk}^\varepsilon = a_{khij}^\varepsilon$$

$$\exists \ c_o > 0 \ \text{such that} \ a_{ijkh}^\varepsilon \ e_{ij} \ e_{kh} \geqq c_o \ e_{ij} \ e_{ij} \quad \forall \ e_{ij} = e_{ji}.$$

The interface conditions are :

$$(7) \quad \vec{V}^\varepsilon = i \ \omega \ \vec{u}^\varepsilon$$

$$(8) \quad \sigma_{ij}^{F\varepsilon} \ n_j = \sigma_{ij}^{S\varepsilon} \ n_j \ .$$

Before applying homogenization techniques to this problem, we notice that results are highly dependent on the presence of other small parameters apart from the ε one, so we must specify the order with respect to ε of dimensionless parameters. We assume that the dimensionless viscosity $\overline{\mu}/\omega \rho_o \ell^2$ appropriate to the small scale ℓ is of the order unity (ε^0), therefore the dimensionless viscosity appropriate to the large scale L, $\overline{\mu}/\omega \rho_o L^2$ is small of order ε^2 ; we write in the following $\overline{\mu} = \varepsilon^2 \mu$. The same assumption is taken for $\overline{\lambda}$, $\overline{\lambda} = \varepsilon^2 \lambda$. Furthermore we assume that characteristic quantities in the reference state are of the same order, with respect to ε, in the fluid and in the solid, there are the densities (ρ_o and ρ_s) and the small perturbation velocities ; this implies that the bulk modulus of fluid and solid are comparable. The theory will display the influence of the solid-fluid interaction in the microstructure when the physical conditions are those of acoustics for each of components.

According to the general features of homogenization method, we introduce the microscopic variable $y = x/\varepsilon$ and we can consider the y variable defined on a basic period Y, homothetic to an actual period of the medium with the ratio ε^{-1}. Therefore ρ_s^ε and a_{ijkh}^ε, according to the assumption of local periodicity for the components of the medium are functions $\rho_s(x,y)$, $a_{ijkh}(x,y)$, of the macroscopic variable x and of the microscopic one y, Y-periodic with respect to y. We examine the physical situation where both phases (the elastic solid and the saturating viscous fluid) are connected . So denoting by Y_S the solid part of Y, by Y_F the fluid part of Y (with smooth boundary Γ), the unions of all the εY_S and of all the εY_F parts are connected. We seek solution of equations (1)-(8), when $\varepsilon \ll 1$, in the form :

$$F^\varepsilon(x) = F^o(x,y) + \varepsilon F^1(x,y) + \varepsilon^2 \ldots$$

where functions F^i are Y-periodic in y, for the velocity $\vec{V}^\varepsilon(x)$ in the fluid, the pressure $P^\varepsilon(x)$ in the fluid and the displacement $\vec{u}^\varepsilon(x)$ in the solid. Clearly the displacement in the fluid is $\vec{V}^\varepsilon/i\omega$ and the velocity in the solid is $i\,\omega\vec{u}^\varepsilon$.

II - ASYMPTOTIC EXPANSION OF THE SOLUTION

We insert in equations (1)-(8) the asymptotic expansions of $\vec{V}^\varepsilon$, P^ε, $\vec{u}^\varepsilon$ and identify powers of ε.

1. - Determination of $\vec{u}^o$ in the solid - From (5) at $O(\varepsilon^{-2})$ and (8) at $O(\varepsilon^{-1})$:

$$\frac{\partial}{\partial y_j}\left(a_{ijkh}\, e_{khy}\,(\vec{u}^o)\right) = 0 \quad \text{in } Y_S$$

$$a_{ijkh}\, e_{khy}\,(\vec{u}^o)\, n_j = 0 \qquad \text{on } \Gamma .$$

It is a differential problem for the Y-periodic function $\vec{u}^o(y)$ in Y_S. As we shall see later on (§II.3) this problem admits a unique solution determined up to an additive constant vector which may depend on the parameter x. Then

$$(9) \quad \vec{u}^o(x,y) = \vec{u}^o(x)$$

2. - Determination of P^o in the fluid - From (2) at $O(\varepsilon^{-1})$:

$$(10) \quad \frac{\partial P^o}{\partial y_i} = 0 \quad \text{in } Y_F .$$

Thus :

$$(11) \quad P^o(x,y) = P^o(x).$$

3. - Determination of $\vec{u}^1$ in the solid - From (5) at $O(\varepsilon^{-1})$ and (8) at $O(\varepsilon^{o})$:

$$(12) \quad \frac{\partial}{\partial y_j}\left\{a_{ijkh}\left(e_{khy}\,(\vec{u}^1) + e_{khx}\,(\vec{u}^o)\right)\right\} = 0 \quad \text{in } Y_S$$

$$(13) \quad a_{ijkh}\left(e_{khy}(\vec{u}^1) + e_{khx}(\vec{u}^o)\right) n_j = -P^o\, n_i \quad \text{on } \Gamma.$$

This is an elastostatic problem in Y_S, where the unknown is $\vec{u}^1(y)$ and $e_{khx}(\vec{u}^o)$ and P^o are considered as given constants. We introduce the Hilbert space E :

$$E = \{\vec{v}, \vec{v} \in \left(H^1(Y_S)\right)^3, \ \vec{v} \ \text{Y-periodic}, \ \int_{Y_S} \vec{v}\, dy = 0\}$$

with the scalar product

$$(\vec{u},\vec{v})_E = \int_{Y_S} e_{ijy}(\vec{u})\, e_{ijy}(\vec{v})\, dy$$

in order to avoid difficulties associated to the fact that $\vec{u}^1$ is determined up to a vector with $e_{khy} = 0 \ \forall\, k,h$ that is to say a constant vector taking into account the Y-periodicity of $\vec{u}^1$ and the hypothesis that Y_S is connected. This recalls the model problem of the first lecture, but we are dealing now with an elliptic system. The variational formulation of (12)-(13) is :

$$\begin{cases} \text{Find } \vec{u}^1 \in E \text{ satisfying} \\ \int_{Y_S} a_{ijkh}\, e_{khy}(\vec{u}^1)\, e_{ijy}(\vec{v})\, dy = -\, e_{khx}(\vec{u}^o) \int_{Y_S} a_{ijkh}\, e_{ijy}(\vec{v})\, dy \\ \qquad\qquad - P^o \int_{Y_S} \mathrm{div}_y\, \vec{v}\, dy \quad \forall\, \vec{v} \in E. \end{cases}$$

This recalls the local problem of homogenization in elasticity but we have a supplementary given constant $P^o(x)$ and functions are not defined on Y but only on Y_S. The right hand side term vanishes when $\vec{v}$ is a constant vector, then the existence and uniqueness of solution follow from the Lax-Milgram lemma. Displacement $\vec{u}^1$ is determined up to an additive constant vector and depends on $e_{khx}(\vec{u}^o)$ and P^o in a linear way :

$$(14) \quad \vec{u}^1(x,y) = e_{\ell mx}(\vec{u}^o)\, \vec{w}^{\ell m} + P^o\, \vec{w}^o + \vec{c}^1(x),$$

where $\vec{w}^{\ell m}$ is the unique solution of :

$$(15)\begin{cases}\text{Find } \vec{w}^{\ell m} \in E \text{ such that} \\ \int_{Y_S} a_{ijkh}\, e_{khy}(\vec{w}^{\ell m})\, e_{ijy}(\vec{v})dy = -\int_{Y_S} a_{ij\ell m}\, e_{ijy}(\vec{v})dy \quad \forall \vec{v} \in E\end{cases}$$

and $\vec{w}^o$ the unique solution of :

$$(16)\begin{cases}\text{Find } \vec{w}^{o} \in E \text{ such that} \\ \int_{Y_S} a_{ijkh}\, e_{khy}(\vec{w}^{o})\, e_{ijy}(\vec{v})\, dy = -\int_{Y_S} \mathrm{div}_y\, \vec{v}\, dy \qquad \forall \vec{v} \in E.\end{cases}$$

4. - <u>Determination of P^1 and $\vec{V}^o$ in the fluid</u> - From (1) at $O(\varepsilon^{-1})$, (2) at $O(\varepsilon^{o})$ and (7) at $O(\varepsilon^{o})$:

$$\left.\begin{aligned}&\mathrm{div}_y\, \vec{V}^o = 0 \\ &i\,\omega\,\rho_o\, \vec{V}^o = -\,\mathrm{grad}_y\, P^1 + \mu\,\Delta_y\, \vec{V}^o - \mathrm{grad}_x\, P^o\end{aligned}\right\}\text{in } Y_F$$

$$\vec{V}^o - i\,\omega\,\vec{u}^o = 0 \quad \text{on } \Gamma.$$

Taking (9) into account, we study the relative motion of the fluid with $\vec{W}^o(x,y) = \vec{V}^o(x,y) - i\,\omega\,\vec{u}^o(x)$, we have to solve :

$$\left.\begin{aligned}&(17)\quad \mathrm{div}_y\, \vec{W}^o = 0 \\ &(18)\quad i\,\omega\,\rho_o\, \vec{W}^o = -\,\mathrm{grad}_y\, P^1 + \mu\,\Delta_y\, \vec{W}^o - \mathrm{grad}_x\, P^o + \rho_o\,\omega^2\,\vec{u}^o\end{aligned}\right\}\text{in } Y_F$$

$$(19)\quad \vec{W}^o = 0 \quad \text{on } \Gamma.$$

This recalls the local problem leading to the Darcy's law for Stokes flow in porous media, but there are new complex inertia terms in (18). In order to obtain a variational formulation of this problem, we define the Hilbert space F :

$$F = \{\vec{W},\ \vec{W} \in (H^1(Y_F))^3,\ \vec{W}\ Y\text{-periodic},\ \vec{W} = 0 \text{ on } \Gamma,\ \mathrm{div}_y\, \vec{W} = 0\}$$

with the scalar product

$$(\vec{W},\vec{V})_F = \int_{Y_F} \left(\frac{\partial W_k}{\partial y_j}\frac{\partial \overline{V}_k}{\partial y_j} + W_k \overline{V}_k\right) dy$$

(where — is for the complex conjugate).

The variational formulation of (17)-(19) is easily obtained by multiplying (18) by a test function $\vec{V} \in F$, it is :

$$\begin{cases} \text{Find } \vec{W}^o \in F \quad \text{such that} \\ \int_{Y_F} i\,\omega\rho_o\, W_k^o\, \overline{V}_k dy + \mu \int_{Y_F} \frac{\partial W_k^o}{\partial y_j}\frac{\partial \overline{V}_k}{\partial y_j} dy = (\rho_o \omega^2 u_k^o - \frac{\partial P^o}{\partial x_k})\int_{Y_F} \overline{V}_k dy \;\forall\, \vec{V} \in F. \end{cases}$$

The existence and uniqueness of $\vec{W}^o$ are proved using the Lax-Milgram lemma, and taking into account the linearity property we can write

$$(20) \quad \vec{W}^o = (\rho_o\, \omega^2 u_i^o - \frac{\partial P^o}{\partial x_i})\, \vec{v}^i$$

where $\vec{v}^i$ is the unique solution of :

$$\begin{cases} \text{Find } \vec{v}^i \in F \text{ such that} \\ \int_{Y_F} i\omega\rho_o\, v_k^i\, \overline{V}_k dy + \mu\int_{Y_F} \frac{\partial v_k^i}{\partial y_j}\frac{\partial \overline{V}_k}{\partial y_j} dy = \int_{Y_F} \overline{V}_i\, dy \qquad \forall\, \overline{V} \in F . \end{cases}$$

III - MACROSCOPIC LAWS

1. - **Unstationary Darcy's law** - Let us define the mean velocity $\tilde{\vec{V}}^o$

$$\tilde{\vec{V}}^o = \frac{1}{|Y|}\int_{Y_F} \vec{V}^o(x,y)\, dy ,$$

then we obtain, taking the mean value of (20) on Y_F

(21) $\vec{V}^o - i\,\omega\,\pi\,\vec{u}^o = -\mathbf{K}(\omega)\left(\text{grad}\ P^o - \rho_o\,\omega^2\,\vec{u}^o\right)$

where $\pi = |Y_F|/|Y|$ is the porosity of the medium, (we omit the sub index x when there are only derivatives with respect to the macroscopic variable). (21) is an unstationary Darcy's law for the relative motion of the fluid, the permeability tensor $\mathbf{K}$ is defined by :

(22) $K_{ij} = \tilde{v}_i^{\,j} = \dfrac{1}{|Y|}\displaystyle\int_{Y_F} v_i^{\,j}(x,y)\,dy$.

Hence $\mathbf{K}$ is a complex symmetric tensor depending on ω.

2. - **Averaged state law** - From (4) written at the first order, ε^o, we obtain directly according to (11), that the first term of the asymptotic expansion of ρ_F^ε is $\rho^o(x)$ related to $P^o(x)$ by

(23) $P^o(x) = a_o^2\,\rho^o(x)$.

3. - **Fluid mass conservation law** - We apply the fluid mass conservation law in the fluid part D_F of a macroscopic domain D consisting of a great number of periods. We obtain :

$$i\,\omega\int_{D_F}\rho_F^\varepsilon\,dv + \rho_o\int_{D_F}\text{div}\,\vec{V}^\varepsilon\,dv = 0\ .$$

At the first order when $\varepsilon \to 0$

$$i\,\omega\int_{D_F}\rho_F^\varepsilon\,dv = i\,\omega\int_{D_F}\rho^o\,dv = i\,\omega\int_D \pi\,\rho^o\,dx = i\,\omega\int_D \frac{\pi}{a_o^2}\,P^o\,dx$$

according to (23).

On the other hand, since $\vec{V}^\varepsilon$ on the solid boundaries is not equal to 0 but equal to $i\,\omega\,\vec{u}^\varepsilon$, we write the obvious identity

$$\int_{D_F} \operatorname{div} \vec{V}^{\varepsilon}\, dv = \int_{\partial D_F} (\vec{V}^{\varepsilon} - i\omega\, \vec{u}'^{\varepsilon}).\vec{n}\, d\sigma + i\omega \int_{D_F} \operatorname{div} \vec{u}'^{\varepsilon}\, dv$$

where $\vec{u}'^{\varepsilon}$ is obtained by extending, with local periodicity, the displacement $\vec{u}^{\varepsilon}$, defined in the solid, to the whole medium in the form

$$\vec{u}'^{\varepsilon} = \vec{u}^{o}(x) + \varepsilon\, \vec{u}'^{1}(x,y) + \varepsilon^2\, \vec{u}'^{2}(x,y) + \dots$$

$$\vec{u}'^{1}(x,y) = \vec{u}^{1}(x,y) \quad \text{in the solid,}$$

$$\vec{u}'^{1}(x,y) = \vec{V}^{1}(x,y)/i\omega \quad \text{in the fluid, ...}$$

Then the term $\vec{V}^{\varepsilon} - i\,\omega\,\vec{u}'^{\varepsilon}$ emphasizes at the first order when $\varepsilon \to 0$, the relative velocity of the fluid with respect to the solid. Let us study the first term

$$\int_{\partial D_F} (\vec{V}^{\varepsilon} - i\,\omega\, \vec{u}'^{\varepsilon}).\vec{n}\, d\sigma = \int_{\partial D} (\vec{V}^{\varepsilon} - i\,\omega\, \vec{u}'^{\varepsilon}).\vec{n}\, d\sigma$$

since, by (7) $\vec{V}^{\varepsilon} - i\,\omega\, \vec{u}'^{\varepsilon} = 0$ on the solid boundaries of ∂D_F and can be extended to the whole medium with value zero in the solid. At the first order when $\varepsilon \to 0$, this term is

$$\int_{\partial D} (\vec{V}^{o} - i\,\omega\, \vec{u}^{o}).\vec{n}\, d\sigma = \int_{\partial D} \vec{W}^{o}.\vec{n}\, d\sigma = \int_{\partial D} \tilde{\vec{W}}^{o}.\vec{n}\, d\sigma_x$$

the last equality holds because $\vec{W}^{o}(x,y)$ is a divergence free vector in the y variable (according to (17) in Y_F and to $\vec{W}^{o} = 0$ in Y_S). Furthermore :

$$\tilde{\vec{W}}^{o} = \frac{1}{|Y|} \int_Y \vec{W}^{o}(x,y)\, dy = \tilde{\vec{V}}^{o} - i\,\omega\,\pi\, \vec{u}^{o},$$

thus at the first order when $\varepsilon \to 0$

$$\int_{\partial D_F} (\vec{V}^{\varepsilon} - i\,\omega\, \vec{u}'^{\varepsilon}).\vec{n}\, d\sigma = \int_D \operatorname{div} \left(\tilde{\vec{V}}^{o} - i\,\omega\,\pi\, u^{o}\right) dx \quad .$$

For the second term, we can write at the first order with respect to ε

$$\int_{D_F} \operatorname{div} \vec{u}'^{\varepsilon}\, dv = \int_{D_F} \left(\operatorname{div}_x \vec{u}^o + \operatorname{div}_y \vec{u}'^1\right) dv$$

$$= \int_D \pi \operatorname{div} \vec{u}^o\, dx + \int_D \left(\frac{1}{|Y|} \int_{Y_F} \operatorname{div}_y \vec{u}'^1\, dy\right) dx\ .$$

Since $\vec{u}'^1(x,y)$ is Y-periodic

$$\int_Y \operatorname{div}_y \vec{u}'^1\, dy = \int_{\partial Y} \vec{u}'^1.n\, d\sigma_y = 0$$

and

$$\int_{Y_F} \operatorname{div}_y \vec{u}'^1\, dy = -\int_{Y_S} \operatorname{div}_y \vec{u}'^1\, dy = -\int_{Y_S} \left(e_{\ell m x}(\vec{u}^o) \operatorname{div}_y \vec{w}^{\ell m} + P^o \operatorname{div}_y \vec{w}^o\right) dy$$

taking into account the expression (14) of $\vec{u}^1(x,y)$. Then, at the first order when $\varepsilon \to 0$:

$$\int_{D_F} \operatorname{div} \vec{u}'^{\varepsilon} dv = \int_D \left(\alpha_{\ell m}\, e_{\ell m}(\vec{u}^o) + \beta\, P^o\right) dx$$

with

$$(24) \begin{cases} \alpha_{\ell m} = \pi\ \delta_{\ell m} - \dfrac{1}{|Y|} \displaystyle\int_{Y_S} \operatorname{div}_y \vec{w}^{\ell m}\, dy \\ \beta = - \dfrac{1}{|Y|} \displaystyle\int_{Y_S} \operatorname{div}_y \vec{w}^o\, dy\ . \end{cases}$$

Finally the first approximation of the fluid mass conservation gives, according to the fact that D is arbitrary, the macroscopic equation

$$(25) \quad i\omega \frac{\pi}{a_o^2} P^o + \rho_o \operatorname{div}\left(\vec{V}^o - i\omega\pi\vec{u}^o\right) + i\omega\rho_o\left(\alpha_{\ell m}\, e_{\ell m}(\vec{u}^o) + \beta P^o\right) = 0\ ,$$

with the homogenized coefficients (24). It can be proved, considering the strain energy function that β is positive.

4. - Momentum conservation law for the mixture - At the first order when $\varepsilon \to 0$, the momentum conservation for the mixture contained in D is :

$$- \omega^2 \int_{D_S} \rho_S u_i^o \, dv + i \omega \rho_o \int_{D_F} V_i^o \, dv = \int_{\partial D} \sigma_{ij}^o n_j \, d\sigma$$

where $\sigma_{ij}^o(x,y)$ is the first approximation of the stress tensor component $\sigma_{ij}^{\varepsilon}$. In the fluid, by (3) :

$$\sigma_{ij}^{\varepsilon}(x) = \sigma_{ij}^{Fo}(x,y) + O(\varepsilon) = - P^o \delta_{ij} + O(\varepsilon) ,$$

in the solid, by (6) and taking (9) into account :

$$\sigma_{ij}^{\varepsilon}(x) = \sigma_{ij}^{So}(x,y) + O(\varepsilon) = a_{ijkh} \left(e_{khx}(\vec{u}^o) + e_{khy}(\vec{u}^1) \right) + O(\varepsilon)$$

From (10), (12) and (13) $\boldsymbol{\sigma}^o(x,y)$ is a divergence free tensor (in the sense of distributions) in Y, so, as we have already seen, we can write :

$$\int_{\partial D} \sigma_{ij}^o n_j \, d\sigma = \int_{\partial D} \tilde{\sigma}_{ij}^o n_j \, d\sigma_x$$

where the macroscopic tensor is given by

$$\tilde{\sigma}_{ij}^o = \frac{1}{|Y|} \int_Y \sigma_{ij}^o(x,y) \, dy = \frac{1}{|Y|} \left(\int_{Y_F} \sigma_{ij}^{Fo} \, dy + \int_{Y_S} \sigma_{ij}^{So} \, dy \right) .$$

Let us define :

$$\tilde{\rho}_S = \frac{1}{|Y|} \int_{Y_S} \rho_S(x,y) \, dy ,$$

then, at the first order when $\varepsilon \to 0$, the momentum conservation law for the mixture leads to the macroscopic equation :

$$(26) \quad -\omega^2 \tilde{\rho}_S u_i^o + i\omega\rho_o \tilde{V}_i^o = \frac{\partial \tilde{\sigma}_{ij}^o}{\partial x_j} .$$

5\. - **Macroscopic constitutive behavior** - It appears in the macroscopic law (26) a new macroscopic quantity : the stress tensor $\tilde{\sigma}^o$. Taking into account (11), (9) and (14), it can be calculated in terms of the macroscopic field quantities $P^o(x)$ and $e_{kh}(\vec{u}^o(x))$:

$$\tilde{\sigma}_{ij}^o = \frac{1}{|Y|} \Big\{ \int_{Y_F} - P^o \delta_{ij} dy$$
$$+ \int_{Y_S} a_{ijkh} \big(e_{khx}(\vec{u}^o) + e_{\ell mx}(\vec{u}^o) e_{khy}(\vec{w}^{\ell m}) + P^o e_{khy}(\vec{w}^o) \big) dy \Big\}$$

hence the macroscopic constitutive law :

$$(27) \quad \tilde{\sigma}_{ij}^o = -\gamma_{ij} P^o + c_{ij\ell m} e_{\ell m}(\vec{u}^o)$$

with the homogenized coefficients

$$(28) \begin{cases} \gamma_{ij} = \pi \delta_{ij} - \dfrac{1}{|Y|} \displaystyle\int_{Y_S} a_{ijkh} e_{khy}(\vec{w}^o) dy \\ c_{ij\ell m} = \dfrac{1}{|Y|} \displaystyle\int_{Y_S} a_{ijkh} \big(\delta_{k\ell} \delta_{hm} + e_{khy}(\vec{w}^{\ell m}) \big) dy . \end{cases}$$

We can prove by the same method we have used in the model problem of the first lecture (§III.4) that coefficients $c_{ij\ell m}$ satisfy the classical symmetry and positivity conditions of elasticity. Furthermore we can establish that coefficients $\alpha_{\ell m}$ in (24) and $\gamma_{\ell m}$ in (28) are the same and clearly are symmetric. By taking, in (16), $\vec{v} = \vec{w}^{\ell m}$ we obtain :

$$\int_{Y_S} a_{ijkh} e_{khy}(\vec{w}^o) e_{ijy}(\vec{w}^{\ell m}) dy = - \int_{Y_S} \mathrm{div}_y \vec{w}^{\ell m} dy ,$$

and by taking, in (15), $\vec{v} = \vec{w}^o$ we have :

$$\int_{Y_S} a_{ijkh} e_{khy}(\vec{w}^{\ell m}) e_{ijy}(\vec{w}^o) dy = - \int_{Y_S} a_{ij\ell m} e_{ijy}(\vec{w}^o) dy$$

then, the symmetry of coefficients a_{ijpq} implies :

$$\alpha_{\ell m} = \gamma_{\ell m}$$

In conclusion, the homogenized motion is described by the field quantities $\vec{V}^o$, P^o, ρ^o, $\vec{u}^o$ and $\tilde{\sigma}^o$ related by the macroscopic relations (21), (23), (25), (26), (27) with the homogenized coefficients given by (22), (24), (28) which are determined by microstructure. In fact the bulk motion may be described by the relative mean velocity of the fluid $\vec{W}^o = \vec{V}^o - i\omega\pi\, \vec{u}^o$ and the displacement $\vec{u}^o$ in the solid which satisfy the motion equations

$$(29) \quad \vec{W}^o = - \mathbf{K}(\omega) \left(\text{grad } P^o - \rho_o\, \omega^2\, \vec{u}^o\right)$$

from (21), and from (25)

$$(30) \quad - \omega^2 \tilde{\rho}\, u_i^o + i\,\omega\, \rho_o\, \tilde{W}_i^o = \frac{\partial\, \tilde{\sigma}_{ij}^o}{\partial\, x_j}$$

where $\tilde{\rho} = \frac{1}{|Y|} \int \rho(x,y)\, dy = \frac{1}{|Y|} \left(\int_{Y_S} \rho_S\, dy + \rho_o\, |Y_F|\right)$

and the corresponding constitutive relations, from (27) and (25) respectively :

$$(31) \quad \tilde{\sigma}_{ij}^o = - \alpha_{ij}\, P^o + c_{ij\ell m}\, e_{\ell m}(\vec{u}^o)$$

$$(32) \quad P^o = - \left(\frac{\pi}{a_o^2\, \rho_o} + \beta\right)^{-1} \left(\frac{1}{i\,\omega} \text{div } \vec{W}^o + \alpha_{\ell m}\, e_{\ell m}(\vec{u}^o)\right).$$

IV - CONCLUDING REMARKS

First, we note that this study deals with not high frequency problems, indeed wavelengths are very much greater than the dimensions of the periods.

Generally the global medium is inhomogeneous and anisotropic. When we assume that the geometry of the medium in the reference configuration and the components behavior are strictly periodic, the homogenized coefficients are independent of x, so they are constants. In this case we can compare our equations with those given by Biot[1,2] for macroscopically uniform media. We point out that our equations[3] have been found some years later, in a similar form, by Burridge and Keller.[4] Not only do our equations agree with Biot's but the homogenization method gives the explicit analytical construction of the macroscopic coefficents ; this construction requires, typically, the solutions of some boundary value problems within the basic period Y, which can be treated numerically. Furthermore the constitutive relation (32) expressing P^o in terms of the bulk medium strain, which is postulated by Biot, is found in our analysis as a consequence of fluid mass conservation law. Elsewhere the influence of time dependence more clearly appears than in Biot's formula. Our present study deals with time harmonic perturbations, in fact we have obtained the Fourier transforms of the homogenized equations for general time dependent macroscopic field quantities. The general time dependent form of (30), (31), (32) is straightforward (it is convenient to introduce instead of $\tilde{\vec{W}}^o$ the corresponding relative displacement as a field quantity), but particularly the dependence on ω of the permeability tensor $\mathbb{K}$ in our study implies that the bulk medium is with memory effect. In fact (29) must be replaced by

$$\tilde{W}_i^o(t) = -\int_o^t G_{ik}(t-s)\left(\frac{\partial P^o}{\partial x_k}(s) + \rho_o \frac{\partial^2 u_k^o}{\partial t^2}(s)\right) ds$$

where $G_{ik}(t)$ is the inverse Fourier transform of $K_{ij}(\omega)$. The relative fluid velocity $\tilde{\vec{W}}^o(t)$ is a functional of grad $P^o + \rho_o(\partial^2\vec{u}^o/\partial t^2)$ for the time preceeding t. Direct study of the problem with general time dependence can be found in Sanchez-Palencia book.[5]

V - SPECIAL CASES

1. - Macroscopic homogeneity and isotropy - Then the homogenized coefficients (22), (24), (28) are constant and isotropic :

$$K_{ij}(\omega) = k(\omega)\,\delta_{ij}, \quad \alpha_{ij} = \alpha\,\delta_{ij}, \quad c_{ij\ell m} = \lambda^{*}\delta_{ij}\delta_{\ell m} + \mu^{*}(\delta_{i\ell}\delta_{jm}+\delta_{im}\delta_{j\ell})$$

where $k(\omega)$ is a complex number depending on ω and α, λ^{*}, μ^{*} constants. Thus (29)-(32) give the following equations for acoustic propagation

$$k^{-1}(\omega)\,\tilde{\vec{W}}^{o} - \rho_o\,\omega^2\,\vec{u}^{o} = \left(\frac{\pi}{a_o^2\,\rho_o} + \beta\right)^{-1} \text{grad div}\left(\frac{\tilde{\vec{W}}^{o}}{i\,\omega} + \alpha\,\vec{u}^{o}\right)$$

$$-\,\omega^2\,\tilde{\rho}\,\vec{u}^{o} + i\,\omega\,\rho_o\,\tilde{\vec{W}}^{o} = \mu^{*}\Delta\,\vec{u}^{o} + \text{grad}\left(\left\{\lambda^{*} + \mu^{*} + \alpha^2\left(\frac{\pi}{a_o^2\,\rho_o} + \beta\right)^{-1}\right\}\text{div}\,\vec{u}^{o}\right.$$

$$\left. + \,\alpha\left(\frac{\pi}{a_o^2\rho_o} + \beta\right)^{-1} \text{div}\,\frac{\tilde{\vec{W}}^{o}}{i\,\omega}\right) .$$

The propagation of plane wave displays one rotational wave and two dilatational waves, all being attenuative and dispersive.

Other special cases may be derived from our general study corresponding to different orders of the characteristic parameters of the components.

2. - Case where $\rho_o \ll \rho_s$ - Then the solid motion is very much slow than the fluid motion ($\rho_o \tilde{\vec{V}}^{o} = O(\rho_s\,\omega\,\vec{u}^{o})$). Direct study is given in Lévy (1977)[6]. Equations can be derived from (29)-(32), they are :

$$\tilde{\vec{V}}^{o} = -\mathbf{K}(\omega)\text{ grad } P^{o}$$

$$-\,\omega^2\,\tilde{\rho}_S\,u_i^{o} + i\,\omega\,\rho_o\,\tilde{V}_i^{o} = \frac{\partial\,\tilde{\sigma}_{ij}^{o}}{\partial\,x_j}$$

$$\tilde{\sigma}_{ij}^{o} = -\,\alpha_{ij}\,P^{o} + c_{ij\ell m}\,e_{\ell m}(\vec{u}^{o})$$

$$P^o = - \frac{a_o^2 \rho_o}{\pi} \operatorname{div} \frac{\vec{V}^o}{i\omega} .$$

The vibration of the fluid is independent of the solid motion. The fluid and solid vibrations are not coupled as in the general case.

3. - **Porous rigid solid filled with fluid** - Then $\vec{u}^\varepsilon = 0$, so $\vec{u}^o = 0$, $\vec{W}^o = \vec{V}^o$ and the homogenized coefficient β is also zero. From equations (29)-(32) we obtain :

$$\vec{V}^o = - \mathbb{K}(\omega) \ \operatorname{grad} P^o$$

$$P^o = - \frac{a_o^2 \rho_o}{\pi} \operatorname{div} \frac{\vec{V}^o}{i\omega}$$

which leads to the wave equation

$$i\omega \frac{\pi}{a_o^2 \rho_o} P^o = \frac{\partial}{\partial x_i} \left(K_{ij} \frac{\partial P^o}{\partial x_j}\right) .$$

This problem has been studied by Lévy and Sanchez-Palencia[7] taking thermal effects into account.

4. - **Porous elastic solid without fluid** - In the absence of fluid the medium is an elastic skeleton with voids. In the cavities $\rho_o = 0$, $P^o = 0$, and equations (30) and (31) remain to give

$$- \omega^2 \tilde{\rho}_S u_i^o = \frac{\partial}{\partial x_j} \left(c_{ij\ell m} e_{\ell m}(\vec{u}^o)\right)$$

with $c_{ij\ell m}$ given by (28). These coefficients are the same as in the static case treated by Auriault and Sanchez-Palencia[8]. The macroscopic medium is an anisotropic elastic solid and the previous relation is the equation for acoustic propagation.

VI - FINAL COMMENTS

We insist on the great variety of problems arising in this domain according to the topological properties of the mixture or to the orders of the different constitutive constants.

In our study[3] (Lévy 1979), the both phases are connected ; the case where the fluid phase is not connected (closed pores) is more simple and is treated in the same paper, the homogenized medium is a linear anisotropic elastic solid. If the elastic phase is not connected, a law similar to the unstationary Darcy's law holds for the averaged velocity of the suspension ; in the case of macroscopic homogeneity and isotropy the propagation of a plane wave displays only one dilatational wave (Fleury[9]).

If a different hypothesis is made on the order of the fluid viscosity, conclusions are very different from ours, in the case of large viscosity the bulk medium is viscoelastic (Sanchez-Hubert[10]).

REFERENCES

1. **Biot, M.A.** - Mechanics of deformation and acoustic propagation in porous media, *Journ. Appl. Ph.*, 33, 1482, 1962.

2. **Biot, M.A** - Generalized theory of acoustic propagation in porous dissipative media, *Journ. Ac. Soc. Am.*, 34, 1254, 1962 .

3. **Lévy, T.** - Propagation of waves in a fluid-saturated porous elastic solid, *Int. J. Engng. Sc.*, 17, 1005, 1979.

4. **Burridge, R., Keller, J.B.** - Poroelasticity equations derived from microstructure, *Journ. Ac. Soc. Am.*, 70, 1140, 1981.

5. **Sanchez-Palencia, E.** - *Non homogeneous Media and Vibration Theory*. Lect. Notes in Physics 127, Springer-Verlag, Berlin, 1980, 184.

6. Lévy, T. - Acoustic phenomena in elastic porous media, *Mech. Res. Comm.*, 4, 253, 1977.

7. Lévy, T., Sanchez-Palencia, E. - Equations and interface conditions for acoustic phenomena in porous media, *J. Math. Ana. Appl.*, 61, 813, 1977.

8. Auriault, J.L., Sanchez-Palencia, E. - Etude du comportement macroscopique d'un milieu poreux saturé déformable, *Journal de Méca.*, 16, 575, 1977.

9. Fleury, F. - Propagation of waves in a suspension of solid particles, *Wave Motion*, 2, 39, 1980.

10. Sanchez-Hubert, J. - Asymptotic study of the macroscopic behaviour of a solid-liquid mixture, *Math. Meth. Appl. Sc.*, 2, 1, 1980.

CHAPTER 4

4. - SUSPENSION OF PARTICLES IN A VISCOUS FLUID

I - INTRODUCTION

We consider a suspension of solid particles immersed in a viscous fluid, on which a density of forces $\vec{F}$ may be imposed by external means. The problems in fluid mechanics that we have solved with homogenization methods concern geometric patterns which are given independently of the solution of the problem : either flow in porous rigid media or small vibration problems where the perturbed and unperturbed positions are supposed to be coincident. Very different situations appear if we deal with a medium with large deformations, as a suspension, then the geometry pattern evolves in time and depends on both its configuration at the initial time and the flow for subsequent time. If the particles at the initial time are locally periodically distributed in the fluid, we shall see that the locally periodic structure evolves in time by keeping the locally periodic character ; consequently at any instant in the evolution of the system the required conditions of the homogenization techniques are satisfied with a period which is that of the structure at the same instant. As usual, we denote by ε the small ratio ℓ/L of the two length scales of the problem, the microscopic one ℓ, which characterizes the period, and the macroscopic one L, which is that of the studied macroscopic phenomena. As we deal with an asymptotic analysis when $\varepsilon \to 0$, we must specify the order with respect to ε of dimensionless parameters that appear in the problem. In this study the parameter ε describes both the smallness of the particles and their mutual distances with respect to L, thus the (non-small) concentration of the suspension is independent of ε

and is taken of order unity. We assume that the Reynolds number of the macroscopic flow, $\mathcal{R} = \rho_o UL/\mu$ (where U is a reference velocity, for example a boundary condition), and the dimensionless number associated with the given forces, $FL^2/\mu U$, are of order unity. Furthermore, we consider a problem in which the solid particles density ρ_S is of the same order as the fluid density ρ_o. Generally speaking we deal with a non dilute suspension in which inertia and viscous macroscopic effects are taken into account. And the hypothesis on $\vec{F}$ means, if the force is the gravity $\rho\vec{g}$, that the limit velocity of one particle falling in the viscous fluid, $2(\rho_S - \rho_o)g\,\ell^2/9\,\mu$, is small as ε^2 with respect to the reference velocity of the study ; this is the reason why sedimentation, which is then a very slow phenomena, will not appear in this study as a macroscopic effect. The homogenization method displays the motion equations of the bulk medium, which is an anisotropic fluid with microstructure, and the equations of the microstructure evolution.

II - FORMULATION OF THE PROBLEM

We consider a suspension of a great number of rigid particles S in a viscous incompressible fluid. We assume that at any instant t the geometry of the suspension is locally periodic with a local period that we can consider homothetic with the ratio ε of a basic period Y(t,x) in the stretched coordinates. At this stage the period of the medium is not completely defined it may be shifted by a translation ; for the sake of simplicity we consider each particle S entirely contained in its period which is a parallelepiped centered in the mass center G of S.

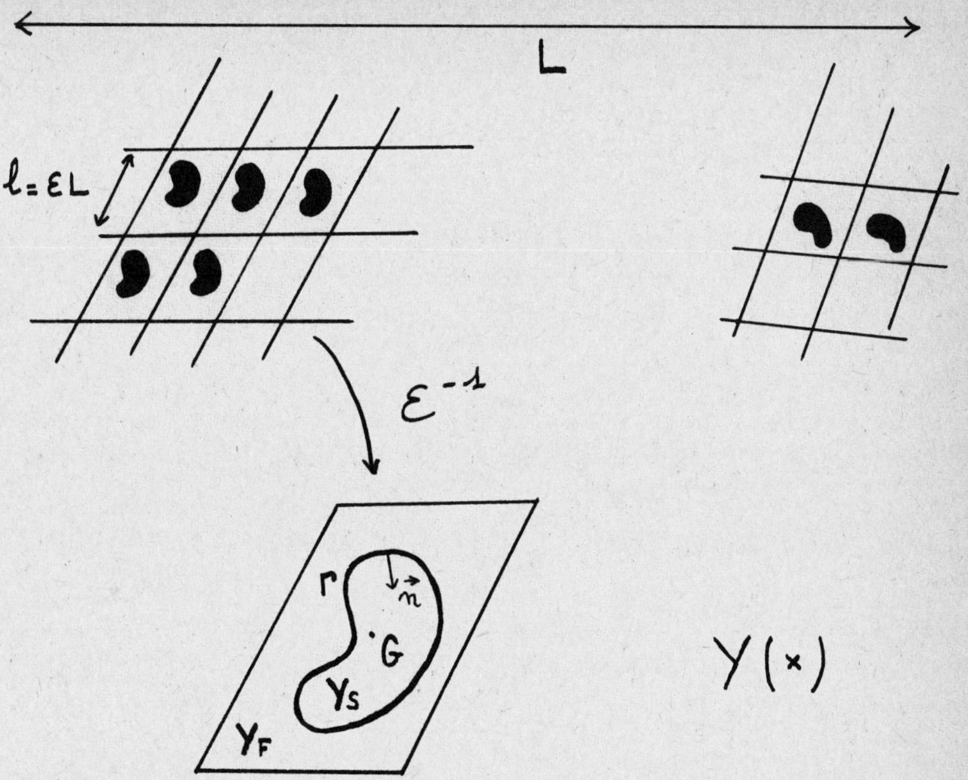

At a point of the fluid part of the medium, the velocity $\vec{V}^{\varepsilon}(t,x)$ and the pressure $P^{\varepsilon}(t,x)$ satisfy the equations :

(1) $\operatorname{div} \vec{V}^{\varepsilon} = 0$

(2) $\rho_o\left(\frac{\partial V_i^{\varepsilon}}{\partial t} + V_j^{\varepsilon}\frac{\partial V_i^{\varepsilon}}{\partial x_j}\right) = \frac{\partial \sigma_{ij}^{\varepsilon}}{\partial x_j} + F_i^{\varepsilon}$

with

(3) $\sigma_{ij}^{\varepsilon} = -P^{\varepsilon}\delta_{ij} + 2\mu D_{ij}(\vec{V}^{\varepsilon})$, $D_{ij}(\vec{V}) = \frac{1}{2}\left(\frac{\partial V_i}{\partial x_j} + \frac{\partial V_j}{\partial x_i}\right)$

In each particle S, $\vec{V}^{\varepsilon}(t,x)$ is a rigid solid velocity field, so :

(4) $D_{ij}(\vec{V}^{\varepsilon}) = 0$,

and because of the non-slip condition on S :

(5) $\vec{V}^{\varepsilon}$ is continuous on ∂S.

For each particle S, submitted to the density of forces $\vec{F}^{\varepsilon}$ and the fluid action, the dynamic laws imply :

$$(6) \quad \int_S \rho_S^{\varepsilon} \frac{d\vec{V}^{\varepsilon}}{dt}\, dv = \int_S \vec{F}^{\varepsilon}\, dv - \int_{\partial S} \sigma_{ij}^{\varepsilon}\, n_j\, \vec{e}_i\, d\sigma ,$$

$$(7) \quad \int_S \rho_S^{\varepsilon}(\vec{x}-\vec{x}_G) \wedge \frac{d\vec{V}^{\varepsilon}}{dt}\, dv = \int_S (\vec{x}-\vec{x}_G) \wedge \vec{F}^{\varepsilon}\, dv - \int_{\partial S} (\vec{x}-\vec{x}_G) \wedge \sigma_{ij}^{\varepsilon}\, n_j\, \vec{e}_i\, d\sigma$$

where $\vec{n}$ denotes the unit normal (outer to the fluid). According to the general features of homogenization method, we introduce the microscopic variable $y = (x - x_G)/\varepsilon$ defined in the basic period $Y = Y(t,x)$ and we assume that ρ_S^{ε} and $\vec{F}^{\varepsilon}$ are functions of the form $\rho_S(x,y)$ and $F(x,y)$, Y-periodic with respect to the y variable. We search asymptotic expansions of the solution of (1)-(7), when $\varepsilon \to 0$, in the form :

$$\vec{V}^{\varepsilon}(t,x) = \vec{V}^{o}(t,x,y) + \varepsilon\, \vec{V}^{1}(t,x,y) + \dots$$

$$P^{\varepsilon}(t,x) = P^{o}(t,x,y) + \dots$$

with functions $\vec{V}^{i}$, P^{i}, $i = 0,1,2,\dots$, Y-periodic with respect to y.

III - <u>ASYMPTOTIC EXPANSION OF THE SOLUTION</u>

As usual we insert in (1)-(7) the asymptotic expansions of $\vec{V}^{\varepsilon}$ and P^{ε} and identify powers of ε.

1. - <u>Determination of $\vec{V}^{o}$</u> - From (1) at $O(\varepsilon^{-1})$ and (2) at $O(\varepsilon^{-2})$:

$$\left.\begin{aligned} \operatorname{div}_y \vec{V}^{o} &= 0 \\ \frac{\partial}{\partial y_j}\left(D_{ijy}(\vec{V}^{o})\right) &= 0 \end{aligned}\right\} \text{ in } Y_F .$$

From (4) at $O(\varepsilon^{-1})$:

$$D_{ijy}(\vec{V}^o) = 0 \quad \text{in} \quad Y_S,$$

and from (5) at $O(\varepsilon^o)$:

$$\vec{V}^o \text{ is continuous on } \Gamma = \partial Y_S.$$

Moreover from (6) at $O(\varepsilon)$ and (7) at $O(\varepsilon^2)$:

$$\int_\Gamma D_{ijy}(\vec{V}^o) n_j \, d\sigma_y = 0$$

$$\int_\Gamma \varepsilon_{mki} \, D_{ijy}(\vec{V}^o) \, n_j \, y_k \, d\sigma_y = 0.$$

It is a differential problem for the Y-periodic function $\vec{V}^o(y)$, x being a parameter. As we shall see later on (§III.2), this problem admits a unique solution determined up to an additive vector $\vec{A}$ satisfying in the distributional sense $\vec{D}_{ijy}(\vec{A}) = 0$, $\forall$ i,j, in Y, that is to say taking the Y-periodicity into account : up to a constant vector which may depend on the parameters x and t. Hence the trivial solution of the present problem :

$$(8) \quad \vec{V}^o(t,x,y) = \vec{V}^o(t,x).$$

$\vec{V}^o$ does not depend on the microscopic variable. It is clear that this does not mean that the first term of the expansion of $\vec{V}^\varepsilon$ is the same as the one corresponding to fluid without particles. As we shall see in §IV, the equations satisfied by $\vec{V}^o(t,x)$ depend on the presence of particles.

2. - Determination of $\vec{V}^1$ and P^o - To begin, let us remark that from (1) and (4) at $O(\varepsilon^o)$:

$$\text{div}_x \vec{V}^o + \text{div}_y \vec{V}^1 = 0 \qquad \text{in } Y_F \text{ and } Y_S \ ;$$

so we can write

$$\int_Y (\mathrm{div}_x \vec{V}^o + \mathrm{div}_y \vec{V}^1)\, dy = 0,$$

taking into account the Y-periodicity of $\vec{V}^1$ and its continuity on Γ, the last term vanishes and we obtain a first macroscopic relation :

$$(9) \quad \mathrm{div}\, \vec{V}^o(t,x) = 0$$

(we omit the subindex x when there are only derivatives with respect to the macroscopic variable x).
Then the next approximation, with respect to ε, of equation (1)-(7), taking (9) into account, leads to :

$$\left.\begin{aligned} &(10) \quad \mathrm{div}_y \vec{V}^1 = 0 \\ &(11) \quad \frac{\partial \sigma^o_{ij}}{\partial y_j} = 0 \\ &(12) \quad \sigma^o_{ij} = - P^o \delta_{ij} + 2\mu \left(D_{ijx}(\vec{V}^o) + D_{ijy}(\vec{V}^1)\right) \end{aligned}\right\} \text{ in } Y_F$$

$$(13) \quad D_{ijx}(\vec{V}^o) + D_{ijy}(\vec{V}^1) = 0 \quad \text{in } Y_S$$

$$(14) \quad \vec{V}^1 \text{ continuous on } \Gamma$$

$$(15) \quad \int_\Gamma \sigma^o_{ij}\, n_j\, \vec{e}_i\, d\sigma_y = 0$$

$$(16) \quad \int_\Gamma \vec{y} \wedge \sigma^o_{ij}\, n_j\, \vec{e}_i\, d\sigma_y = 0 .$$

We are searching for a Y-periodic solution $\vec{V}^1$, P^o, of equations (10) (16). It is clear that the stress tensor σ^o_{ij} in Y_F is determined up to an additive pressure tensor $-\pi^o(t,x)\delta_{ij}$, and that the Y-periodic vector $\vec{V}^1$ is determined up to a vector $\vec{A}$ with $D_{ijy}(\vec{A}) = 0$, $\forall\, i,j$, that is to say a constant vector. In order to avoid difficulties associated with this indetermination of $\vec{V}^1$, we introduce the Hilbert space :

$$W = \{\vec{v}, \vec{v} \in (H^1(Y))^3, \ \vec{v} \text{ Y-periodic}, \ \int_Y \vec{v}(y)dy = 0\}$$

with the scalar product

$$(\vec{v},\vec{w})_W = \int_Y D_{ijy}(\vec{v}) \ D_{ijy}(\vec{w}) \, dy \ ,$$

and we defined the convex closed subset of W

$$K = \{\vec{v} \in W, \ \mathrm{div}_y \vec{v} = 0 \text{ in } Y_F, \ D_{ijy}(\vec{v}) = -D_{ijx}(\vec{V}^o) \ \forall \ i,j \text{ in } Y_S\}$$

The variational formulation of (10)-(16) is :

$$(17)\begin{cases} \text{Find } \vec{V}^1 \in K \text{ satisfying} \\ \\ \int_{Y_F} 2\mu \, D_{ijy}(\vec{V}^1) \left(D_{ijy}(\vec{v}) - D_{ijy}(\vec{V}^1)\right)dy = 0 \quad \forall \ \vec{v} \in K . \end{cases}$$

The existence and uniqueness of $\vec{V}^1 \in K$ (and thus defined up to an additive constant vector) are a standard result of the study of variational inequalities (Lions[1]). Furthermore as in study of Darcy's law, because $\mathrm{div}_y \vec{v} = 0$ in Y_F $\forall$ $\vec{v} \in K$ (see Temam[2]), there exists P^o of class L^2, Y-periodic, satisfying (11) with (12) in the distributional sense. Since $P^o(x,y)$ is defined up to $\pi^o(x)$ we can write :

(18) $P^o(x,y) = \pi^o(x) + p^o(x,y)$ with $\int_{Y_F} p^o(x,y)dy = 0$

and then $p^o(x,y)$ is completely defined by the $D_{ijx}(\vec{V}^o)$.

In order to express the linear dependence of the solution on $D_{ijx}(\vec{V}^o)$, we define the vectors $\vec{P}^{ij}(y)$ with components $P^{ij}_k = y_j \, \delta_{ik}$, then

$$D_{k\ell y}(\vec{P}^{ij}) = \frac{1}{2} (\delta_{ki} \ \delta_{\ell j} + \delta_{kj} \ \delta_{\ell i}).$$

We consider the set of admissible functions (analogous to K) :

$$U_{ad}(\vec{P}^{ij}) = \{\vec{\phi}, \vec{\phi} \in W, \ \mathrm{div}_y \vec{\phi} = 0 \ \text{ in } \ Y_F, \ D_{k\ell y}(\vec{\phi}) = D_{k\ell y}(\vec{P}^{ij}) \quad \forall \ k,\ell, \ \text{ in } \ Y_S\}$$

and the corresponding solutions $\vec{\chi}^{ij}$ of the variational problem (analogous to (17))

$$(19) \begin{cases} \text{Find } \vec{\chi}^{ij} \in U_{ad}(\vec{P}^{ij}) \text{ satisfying} \\ \int_{Y_F} 2\mu \, D_{k\ell y}(\vec{\chi}^{ij}) \, D_{k\ell y}(\vec{\phi} - \vec{\chi}^{ij}) dy = 0 \quad \forall \ \vec{\phi} \in U_{ad}(\vec{P}^{ij}) \ . \end{cases}$$

Note that $\vec{\chi}^{ij} = \vec{\chi}^{ji}$ and that for $y \in Y_S$:

$$(20) \quad \vec{\chi}^{ij}(y) = \frac{1}{2}\left(\vec{P}^{ij}(y) + \vec{P}^{ji}(y)\right) + \vec{\alpha}^{ij} + \vec{\beta}^{ij} \wedge \vec{y}$$

with $\vec{\alpha}^{ij} = \vec{\alpha}^{ji}$ and $\vec{\beta}^{ij} = \vec{\beta}^{ji}$ independent of y. Then we have :

$$(21) \quad \vec{V}^{1}(t,x,y) = - D_{ij}(\vec{V}^{o}) \, \vec{\chi}^{ij}(y) + \vec{A}(t,x)$$

and the asymptotic Y-periodic expansion of $\vec{V}^{\varepsilon}(t,x)$ is

$$\vec{V}^{\varepsilon}(t,x) = \vec{V}^{o}(t,x) + \varepsilon\left(-D_{ij}(\vec{V}^{o}) \, \vec{\chi}^{ij}(y) + \vec{A}(t,x)\right) + o(\varepsilon).$$

At this order, the microscopic variable y only appears in the functions $\vec{\chi}^{ij}$, which are solutions of variational problems (19) in the basic period Y taken at the considered time t and macroscopic point x. As the basic period Y, $\vec{\chi}^{ij}$ may be dependent on t and x.

IV - MACROSCOPIC MOMENTUM EQUATION

$\vec{V}^{o}(t,x)$ is a macroscopic field quantity and we know by (9) that the macroscopic flow is incompressible. We obtain the macroscopic momentum equation by applying the momentum conservation law at the mixture contained in an arbitrary macroscopic domain D consisting of a great number of periods :

$$\int_D \rho^\varepsilon \frac{\partial V_k^\varepsilon}{\partial t} dv + \int_D \rho^\varepsilon V_\ell^\varepsilon \frac{\partial V_k^\varepsilon}{\partial x_\ell} dv = \int_D F_k^\varepsilon dv + \int_{\partial D} \sigma_{k\ell}^\varepsilon n_\ell d\sigma .$$

At the first order ε^0 , when $\varepsilon \to 0$, we deduce the contribution of each term, using the asymptotic expansions of $\vec{V}^\varepsilon$ and $\boldsymbol{\sigma}^\varepsilon$ and the previous results. Let us note that the stress on ∂D is completely defined, in fact ∂D is made of faces of the neighbourhood periods of D and then lies in the fluid ; so the first approximation of $\boldsymbol{\sigma}^\varepsilon$ on ∂D is $\boldsymbol{\sigma}^0$ given by (12).

$$\int_D \rho^\varepsilon \frac{\partial V_k^\varepsilon}{\partial t} dv = \int_D \tilde{\rho} \frac{\partial V_k^0}{\partial t} dx + o(\varepsilon^0)$$

with the mean density $\tilde{\rho} = \frac{1}{|Y|} \left(\rho_o |Y_F| + \int_{Y_S} \rho_S dy\right)$,

if particles are homogeneous (ρ_S independent of y)

$$\tilde{\rho} = \rho_o + c(\rho_S - \rho_o)$$

where $c = |Y_S|/|Y|$ is the volumic concentration of the particles.

As for the non-linear terms, on account of (21), we have

$$\int_D \rho^\varepsilon V_\ell^\varepsilon \frac{\partial V_k^\varepsilon}{\partial x_\ell} dv = \int_D \tilde{\rho} V_\ell^0 \frac{\partial V_k^0}{\partial x_\ell} dx + \sum_{\text{periods} \subset D} \varepsilon^3 V_\ell^0 \int_Y \rho \frac{\partial V_k^1}{\partial y_\ell} dy + o(\varepsilon^0)$$

$$= \int_D \tilde{\rho} V_\ell^0 \frac{\partial V_k^0}{\partial x_\ell} dx - \int_D \beta_{kij\ell} D_{ij}(\vec{V}^0) V_\ell^0 dx + o(\varepsilon^0)$$

with the coefficients

$$\beta_{kij\ell} = \frac{1}{|Y|} \int_Y \rho \frac{\partial \chi_k^{ij}}{\partial y_\ell} dy = \frac{1}{|Y|} \int_{Y_S} (\rho_S - \rho_o) \frac{\partial \chi_k^{ij}}{\partial y_\ell} dy \quad ;$$

from the symmetry property of $\vec{\chi}^{ij}$, $\beta_{kij\ell} = \beta_{kji\ell}$, and if particles are homogeneous by (20) :

$$\beta_{kij\ell} = c(\rho_S - \rho_o)\left((\delta_{ik}\,\delta_{j\ell} + \delta_{i\ell}\,\delta_{jk})/2 + \varepsilon_{p\ell k}\beta_p^{ij}\right) .$$

We have also

$$\int_D F_k^\varepsilon \, dv = \int_D \tilde{F}_k \, dx + o(\varepsilon^o) \text{ with } \tilde{F}_k = \frac{1}{|Y|}\int_Y F_k(x,y)\, dy$$

and

$$\int_{\partial D} \sigma_{k\ell}^\varepsilon \, n_\ell \, d\sigma = \int_{\partial D} \sigma_{k\ell}^o \, n_\ell \, d\sigma + o(\varepsilon^o) .$$

Let us suppose (this will be proved later on, §V) that the Y-periodic stress tensor $\boldsymbol{\sigma}^o(x,y)$ defined in Y_F by (12), may be extended all over Y by a symmetric tensor $\hat{\boldsymbol{\sigma}}^o$ satisfying

(22) $$\frac{\partial \hat{\sigma}_{ij}}{\partial y_j} = 0 \qquad \text{in } Y_S$$

(23) $$\hat{\sigma}_{ij}^o \, n_j = \sigma_{ij}^o \, n_j \qquad \text{on } \Gamma.$$

Then we can define the averaged stress

(24) $$\tilde{\sigma}_{k\ell}^o = \frac{1}{|Y|}\int_Y \hat{\sigma}_{k\ell}(x,y)\, dy .$$

From (11) and (22), $\hat{\boldsymbol{\sigma}}^o(x,y)$ is a divergence free tensor in the y variable, then as we have already seen in preceeding lectures the integral of $\hat{\sigma}_{k\ell}^o \, n_\ell$ on a face of the basic period is the same as that of its mean value $\tilde{\sigma}_{k\ell}^o \, n_\ell$, and at the first order when $\varepsilon \to 0$

$$\int_{\partial D} \sigma_{k\ell}^o \, n_\ell \, d\sigma = \int_{\partial D} \tilde{\sigma}_{k\ell}^o \, n_\ell \, d\sigma_x .$$

Thus we obtain, at the first order when $\varepsilon \to 0$, the macroscopic momentum conservation law

$$(25)\quad \tilde{\rho}\,\frac{\partial V_k^o}{\partial t} + \tilde{\rho}\, V_\ell^o \frac{\partial V_k^o}{\partial x_\ell} - \beta_{kij\ell}\, V_\ell^o\, D_{ij}(\vec{V}^o) = \frac{\partial \tilde{\sigma}_{k\ell}^o}{\partial x_\ell} + \tilde{F}_k \,.$$

It involves homogenized stresses which are studied in the next section and the new term with coefficients $\beta_{kij\ell}$ which is a correction to the non-linear term.

V - AVERAGED STRESSES AND HOMOGENIZED CONSTITUTIVE EQUATION

The stress tensor $\boldsymbol{\sigma}^o(x,y)$ is defined by (12) in Y_F, but it is not defined in the solid part Y_S. The existence of a symmetric extension $\hat{\boldsymbol{\sigma}}^o$ of $\boldsymbol{\sigma}^o$ defined all over Y, satisfying (22) and (23) is a consequence of (15) and (16). In Y_S the $\hat{\sigma}_{k\ell}^o$ are defined, as $\sigma_{k\ell}^o$, up to $\left(-\pi^o(t,x)\,\delta_{k\ell}\right)$. Furthermore in Y_S, the existence of $\hat{\boldsymbol{\sigma}}^o(y)$ is known but $\hat{\boldsymbol{\sigma}}^o(y)$ is not uniquely determined, we prove in the following that the averaged value (24) is independent of the extension.

From (24)

$$\tilde{\sigma}_{k\ell}^o = \frac{1}{|Y|}\left(\int_{Y_F} \sigma_{k\ell}^o\, dy + \int_{Y_S} \hat{\sigma}_{k\ell}^o\, dy\right) \,.$$

and $\tilde{\sigma}_{k\ell}^o$ is, as $\sigma_{k\ell}^o$ in Y_F and $\hat{\sigma}_{k\ell}^o$ in Y_S, determined up to $\left(-\pi^o(t,x)\delta_{k\ell}\right)$. Up to the pressure π^o, we can write according to (12), (21) and (18)

$$(26)\quad \int_{Y_F} \sigma_{k\ell}^o\, dy = -\int_{Y_F} p^o\, \delta_{k\ell}\, dy + 2\mu\, D_{ijx}(\vec{V}^o) \int_{Y_F} D_{k\ell y}(\vec{P}^{ij} - \vec{\chi}^{ij})\, dy$$

$$= 2\mu\, D_{ijx}(\vec{V}^o) \int_{Y_F} D_{pqy}(\vec{P}^{ij} - \vec{\chi}^{ij})\, D_{pqy}(\vec{P}^{k\ell})\, dy \,.$$

Up to the pressure π^o

$$\int_{Y_S} \hat{\sigma}_{k\ell}^o\, dy = \int_{Y_S} \hat{\sigma}_{pq}^o\, D_{pqy}(\vec{\chi}^{k\ell})\, dy$$

since in Y_S $\quad D_{pqy}(\vec{\chi}^{k\ell}) = D_{pqy}(\vec{P}^{k\ell}) = \frac{1}{2}(\delta_{pk}\,\delta_{q\ell} + \delta_{p\ell}\,\sigma_{qk})$

so,

$$\int_{Y_S} \hat{\sigma}^o_{k\ell}\, dy = -\int_\Gamma \hat{\sigma}^o_{pq}\, \chi^{k\ell}_p\, n_q\, d\sigma_y = -\int_{\Gamma\cup\partial Y} \sigma^o_{pq}\, \chi^{k\ell}_p\, n_q\, d\sigma_y$$

in the last equality we have take into account the stress continuity (23) on Γ and the Y-peridicity of involved functions to replace Γ by $\Gamma \cup \partial Y$. Then

$$(27)\quad \int_{Y_S} \hat{\sigma}^o_{k\ell}\, dy = -\int_{Y_F} \frac{\partial}{\partial y_q}(\sigma^o_{pq}\; \chi^{k\ell}_p)dy = -\int_{Y_F} \sigma^o_{pq}\, D_{pqy}(\vec{\chi}^{k\ell})\, dy$$

$$= \int_{Y_F} p^o\, \delta_{pq}\, D_{pqy}(\vec{\chi}^{k\ell})\, dy$$

$$- 2\mu\, D_{ijx}(\vec{V}^o) \int_{Y_F} D_{pqy}(\vec{P}^{ij} - \vec{\chi}^{ij})\, D_{pqy}(\vec{\chi}^{k\ell})\, dy$$

and the first term vanishes because $\vec{\chi}^{k\ell}$ satisfies $\mathrm{div}_y\, \vec{\chi}^{k\ell} = 0$ in Y_F. Collecting (26) and (27) we find, up to $(-\pi^o\, \delta_{k\ell})$

$$\tilde{\sigma}^o_{k\ell} = \frac{1}{|Y|}\left\{2\mu\, D_{ijx}(\vec{V}^o) \int_{Y_F} D_{pqy}(\vec{P}^{ij} - \vec{\chi}^{ij}) D_{pqy}(\vec{P}^{k\ell} - \vec{\chi}^{k\ell})\, dy\right\}.$$

So the homogenized constitutive law is

$$(28)\begin{cases} \tilde{\sigma}^o_{k\ell} = -\pi^o(t,x)\,\delta_{k\ell} + a_{ijk\ell}\, D_{ij}(\vec{V}^o) \\ \text{with } a_{ijk\ell} = \dfrac{2\mu}{|Y|} \int_{Y_F} D_{pqy}(\vec{P}^{ij} - \vec{\chi}^{ij})\, D_{pqy}(\vec{P}^{k\ell} - \vec{\chi}^{k\ell})\, dy. \end{cases}$$

We can write, taking into account the properties (20) of the $\vec{\chi}^{mn}$ on Y_S and their Y-periodicity :

$$a_{ijk\ell} = \mu(\delta_{ik}\,\delta_{j\ell} + \delta_{i\ell}\,\delta_{jk}) + \frac{2\mu}{|Y|}\int_Y D_{pqy}(\vec{\chi}^{ij})\, D_{pqy}(\vec{\chi}^{k\ell})\, dy.$$

The coefficients $a_{ijk\ell}$ depend on the microstructure and it is easy to check that

$$a_{ijk\ell} = a_{jik\ell} = a_{ij\ell k} = a_{k\ell ij}$$

$$a_{ijk\ell} D_{ij} D_{k\ell} \geq a_o D_{ij} D_{ij} \quad \forall \; D_{ij} = D_{ji}, \; a_o > 0.$$

(28) generalizes to a suspension of rigid particles the constitutive law obtained by Bensoussan, Lions and Papanicolaou[3], for the mixture of slow viscous fluids. Let us remark that, since the bulk medium is incompressible, we may consider that the $a_{ijk\ell}$ are in fact determined up to coefficients of the form $a_{k\ell} \delta_{ij}$. For computation of the $a_{ijk\ell}$ when the suspension remains a cubic lattice of spheres at any instant see Nunan & Keller[4]. The bulk medium behaviour given by (28) is that of an incompressible anisotropic fluid with microstructure. Let us note that in the case of macroscopic isotropy the viscosity of the suspension is greater than that of the fluid.

VI - DEFORMATION OF THE STRUCTURE AND CONCLUSION

The macroscopic relations (25) and (28) contain homogenized coefficients $\beta_{kij\ell}$ and $a_{ijk\ell}$ depending on the microstructure of the medium by the averaging domain and the local solutions $\vec{\chi}^{ij}$ of the variational problems (19). These coefficients are calculated using the basic period $Y(t,x)$ at the considered time t and macroscopic point x. At the initial time t_o the structure of the medium is locally periodic with period $\varepsilon Y(t_o,x)$ and it is driven by the flow with the velocity $\vec{V}^{\varepsilon}$ found in §III

$$\vec{V}^{\varepsilon}(t,x) = \vec{V}^{o}(t,x) + \varepsilon \vec{V}^{1}(t,x,y) + o(\varepsilon)$$

with $\vec{V}^{1}$ Y-periodic with respect to the y variable (and given by (21)).

In order to describe the deformation of the structure we must identify appropriate variables $\mathcal{S}$, depending on t and x ; we can choose 3 vectors $\vec{d}^1$, $\vec{d}^2$, $\vec{d}^3$, of order ε, characterizing a period (for example $\vec{d}^1 = \vec{AB}$, or $\vec{d}^1 = \vec{G_1G_2}$ in a translated period...) and 3 angles for the orientation of the particle S.

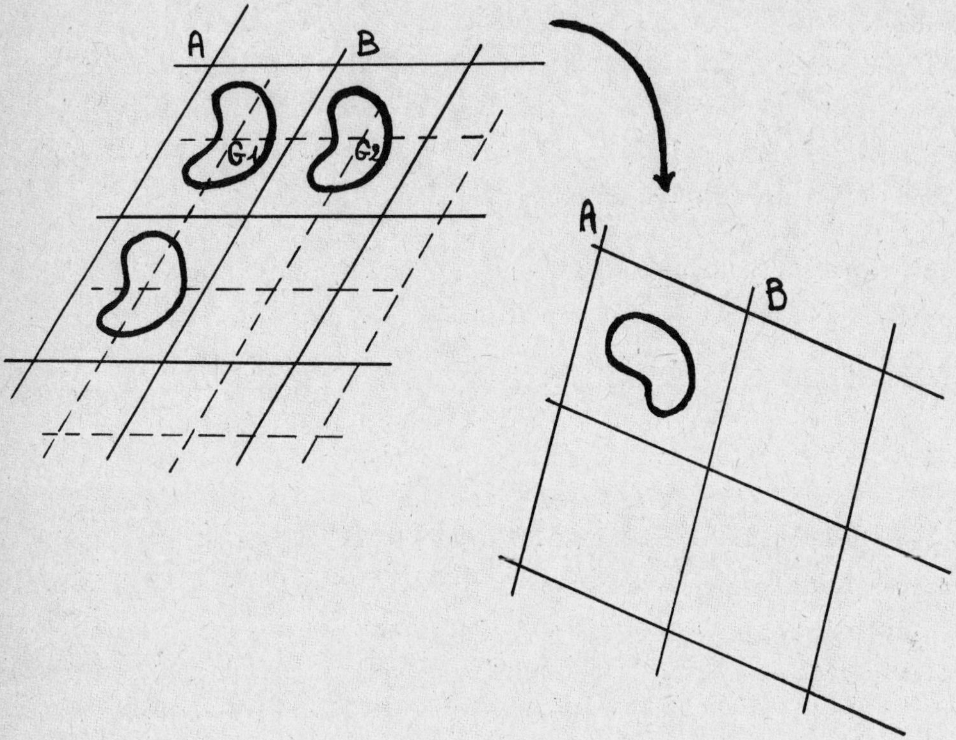

The deformation rate of the microstructure can be evaluated taking into account the expansion of $\vec{V}^\varepsilon$.

$$\frac{d\vec{d}^1}{d\,t} = \frac{d\,\vec{AB}}{d\,t} = \vec{V}^o(B) + \varepsilon\,\vec{V}^1(B) - \vec{V}^o(A) - \varepsilon\,\vec{V}^1(A) + o(\varepsilon),$$

the terms containing $\vec{V}^o$ give contributions of order ε (because $AB = O(\varepsilon)$) and the terms containing V^1 give contributions of order ε^2 because the y dependence is the same in A and B by Y-periodicity. Then we have at the leading order :

$$\frac{d\vec{d}^{1}}{d\,t} = \frac{\partial\,\vec{V}^{o}}{\partial\,x_i}\,d^{1}_{i}$$

and generally

$$(29)\quad \frac{d\vec{d}^{k}}{d\,t} = \nabla\,\vec{V}^{o}\,.\,\vec{d}^{k} \qquad\qquad k = 1,2,3.$$

Because of the Y-periodicity, the term $\varepsilon\,\vec{V}^{1}$ of the expansion of $\vec{V}^{\varepsilon}(t,x)$ does not modify, in first approximation, the evolution of $\vec{d}^{k}$ given by the term $\vec{V}^{o}(t,x)$ alone. In a similar manner, we can easily compute the angular velocity $\vec{\Omega}(S)$ of the solid particle S ; denoting by M and P two points of S we have :

$$\vec{\Omega}(S)\wedge\vec{PM} = \vec{V}^{\varepsilon}(M) - \vec{V}^{\varepsilon}(P) = \nabla\vec{V}^{o}\,.\,\vec{PM} - \varepsilon D_{ij}(\vec{V}^{o})\left(\vec{\chi}^{ij}(y_M) - \vec{\chi}^{ij}(y_P)\right) + o(\varepsilon)$$

taking (20) into account, this leads in first approximation to :

$$(30)\quad \vec{\Omega}(S) = \vec{\Omega}(t,x) = \frac{1}{2}\,\mathrm{rot}\,\vec{V}^{o} - D_{ij}(\vec{V}^{o})\,\vec{\beta}^{ij}\,.$$

The microstructure evolves in time according to (29) and (30) and then remains locally periodic. The deformation rate depends linearly on $\nabla\vec{V}^{o}$ and on the microstructure itself. When $\vec{V}^{o}$ is not small the deformation of the medium is not negligible, and the homogenized coefficients $\beta_{kij\ell}$ and $a_{ijk\ell}$ which depend on the microstructure, that is to say on the variables $\mathcal{S}$, are depending (by (29) and (30)) on $\nabla\vec{V}^{o}$. Particularly the $a_{ijk\ell}$ in the constitutive law (28) are depending on $\nabla\vec{V}^{o}$, which emphasizes the non-Newtonian behaviour of the bulk medium.

The macroscopic equations are (9), (25) with the constitutive relation (28), and the evolution equations of the microstructure are (29) and (30). We have elements to compute (at least in theory) the flow. Starting from the initial configuration at $t = t_o$, (9), (25) and (28) give the flow for a short interval of time Δt, (29) and (30) allow us to know the modified (locally periodic) configuration at time $t_1 = t_o + \Delta t$, and then we may proceed at t_1 as at t_o and so on.

VI - RELATED PROBLEMS

If the forcing term $\vec{F}$ has a different form than in this study (see also Lévy and Sanchez-Palencia[5]), the solution may be very different. Particularly the suspension of force free particles (with given couples) in a viscous incompressible fluid exhibits a macroscopic stress tensor which is not symmetric (Lévy[6,7]).

Otherwise the case of dilute suspensions ($c = |Y_S|/|Y| \ll 1$) may be obtained by an asymptotic analysis from the present study (Lévy and Sanchez-Palencia[8,9]). The equations of the bulk medium are obtained at the order of the concentration c, the macroscopic behaviour is generally that of an anisotropic incompressible fluid with non-Newtonian effects due to the evolution of the particles orientation. In the particular case of spherical identical solid particles the celebrated Einstein formula for the apparent viscosity of a dilute suspension of spheres is then obtained.

REFERENCES

1. **Lions, J.L.**, *Quelques méthodes de résolution des problèmes aux limites non linéaires*, Dunod, Paris, 1969, chap. 3.

2. **Temam, R.**, *Navier-Stokes equations*, North-Holland, Amsterdam, 1977, chap. 1.

3. **Bensoussan, A., Lions, J.L., Papanicolaou, G.**, *Asymptotic analysis for periodic structures*, North-Holland, Amsterdam, 1978, 129.

4. **Nunan, K.C., Keller, J.B.**, Effective viscosity of a periodic suspension, *J. Fluid Mech.*, 142, 269, 1984.

5. **Lévy, T., Sanchez-Palencia, E.**, Suspension of solid particles in a Newtonian fluid, *J. Non-Newt. Fl. Mech.*, 13, 63, 1983.

6. **Lévy, T.**, Application de l'homogénéisation à l'étude d'une suspension de particules soumises à des couples, *C.R.Acad.Sc. Paris*, 299 II, 597, 1984.

7. **Lévy, T.**, Suspension de particules soumises à des couples, *J. de Méca.*, to appear.

8. **Lévy, T., Sanchez-Palencia, E.**, Suspension diluée dans un fluide visqueux de particules solides ou de gouttes visqueuses, *C.R.Acad.Sc. Paris*, 297 II, 193, 1983.

9. **Lévy, T., Sanchez-Palencia, E.**, Einstein-like approximation for homogenization with small concentration, II Navier-Stokes equations. *J. Non Lin. Anal.*, to appear.

PART III

BOUNDARY LAYERS AND EDGE EFFECTS IN COMPOSITES

E. Sanchez-Palencia
Laboratoire de Mécanique Théorique
Université Pierre et Marie Curie
Tour 66 - 4 place Jussieu
F-75230 Paris Cédex 05, France

CHAPTER 1

1. - GENERAL INTRODUCTION TO ASYMPTOTIC METHODS

1. - GENERALITIES

In mechanics of continua we deal with the equations of motion (conservation of mass, balance of momentum, etc), the boundary conditions expressing the action of the outer regions on the medium, and the constitutive equations, associated with the own behavior of the material ; roughly speaking this is the law relating the forces acting upon a small piece of the material and the deformation of this piece, i.e. the strain-stress law.

Now, when considering a heterogeneous medium, formed by small parts of two (or more) constituents, we may adopt two different points of view. First, a microscopic or local one : we consider each portion of a constituent as a continuous medium, with its own equations, strain stress relation and boundary conditions at the contact surfaces with other portions. Of course, this is only accurate if the characteristic lengths of this portion satisfy the general hypotheses of continuum mechanics : they must be large with respect to molecular distances. On the other hand, we may adopt a macroscopic or global point of view : we disregard the lengths of the order of each portion of constituent and we only take into account macroscopic lengths. We then search for a homogeneous (or homogenized) material having overall mechanical properties analogous to that of a macroscopic sample of the composite medium. We can use experimental or phenomenological methods to study the global strain-stress laws of such a material. Moreover, we may search for laws relating the mechanical properties of each component and the geometrical form of the mixture with the overall properties of the homogenized material. To perform this, we study the behavior of a "*small piece*" of the macroscopic medium and this is nothing but a piece of the medium at the local level. This amounts to saying that the strain-stress law of the homogenized material is related to some solutions of the equations, strain-stress laws and boundary (= interface) conditions at the local level. The aim of the so called homogenization method is to go on with this program. Unfortunately, best results are only obtainable under hypotheses of periodicity of the local structure ; in this case, the method leads to a "*rigorous*" deduction of the macroscopic behavior. "*Rigorous*" is here understood in the sense of "*straight-forward from the preceeding considerations*" and moreover, in most problems, there is a mathematical proof of the convergence of solutions as the parameter ε (which measures the ratio of the micro- to macro lengths) tends to zero.

It should be noticed that the "*homogenized coefficients*" (or equivalently the strain-stress law for the homogenized medium) only depend on the local structure of the medium and may be obtained by numerical solution of some boundary value problems in a period of the structure, the boundary conditions being mostly of the periodic type.

Nevertheless, from our point of view, the interest of the method is mostly qualitative : it gives relevant information on the relation between the local and global behaviors. For most problems in mechanics the micro- and macro-processes are of very different nature, for instance, Darcy's law for fluid flow in porous media is the "*homogenized*" form of the Navier-Stokes equation, and viscoelasticity is in some cases the homogenized form of a mixture of an elastic solid and a viscous fluid.

The study is based on some knowledge of asymptotic methods (two-scale and matching asymptotic expansions) and boundary value problems for partial differential equations. In the sequel of this section we give some elements of these theories, but the reader not acquainted with them is advised to read the general reference books, for instance Van Dyke[1], Cole[2], Cole and Kevorkian[3] for asymptotic methods, and Necas[4], Lions Magenes[5] and Brézis[6] for partial differential equations and functional analysis, as well as Ciarlet[7,8] and Raviart[9] for numerical computations. On the other hand, general references for homogenization methods are Bensoussan, Lions and Papanicolaou[10], Lions[11] and SanchezPalencia[12].

In addition to specific notations, some general rules are used througout this series of lectures :

Symbols o and O are used for orders of magnitude associated with some parameter $\varepsilon \to 0$:

$$(1.1) \quad f(\varepsilon) = o(g(\varepsilon)) \iff \frac{f(\varepsilon)}{g(\varepsilon)} \to 0 \quad \text{as } \varepsilon \to 0$$

$$(1.2) \quad f(\varepsilon) = O(g(\varepsilon)) \iff \left|\frac{f(\varepsilon)}{g(\varepsilon)}\right| \leq \text{some constant as } \varepsilon \to 0$$

The classical convention of summation of repeated indexes is used :

$$(1.3) \quad \begin{cases} u_i v_i = \sum_1^3 u_i v_i \ ; \\ \delta_{ij} = 1 \quad \text{if} \quad i = j \ ; \ = 0 \ \text{otherwise.} \end{cases}$$

and sometimes, Greek indexes are used for sums on only two values (1,2 or 2,3).

Underlined symbols denote vectors of R^3 (or R^2), for instance

$$(1.4) \quad \underline{u} = (u_1, u_2, u_3)$$

but as an exception, the current point of R^3 is not underlined :

$$(1.5) \quad x = (x_1, x_2, x_3) \quad ; \quad y = (y_1, y_2, y_3)$$

Upper indexes as in u^1, u^2, are used to denote the terms of a sequence, (as lower indexes were used for the components of a vector in (1.4)).

When two scales are involved, e_{ijx}, e_{ijy} denote the strain tensor in the corresponding variables, for instance :

$$(1.6) \quad e_{ijy}(\underline{u}) = \frac{1}{2}\left(\frac{\partial u_i}{\partial y_j} + \frac{\partial u_j}{\partial y_i}\right) \qquad (\text{with } x = \text{parameter})$$

If Ω is a domain, $\partial\Omega$ denotes its boundary, and $\underline{n}$ its outer unit normal, $\underline{e}_i$ denotes the unit vectors of the axes.

The brackets $[u]$ denote the jump of a function u across a discontinuity.

The numbering of formulas, propositions, remarks, is by sections in each lecture, preceeded by the number of the lecture when it is not the present one (for instance, in this first lecture, (2.5) is formula 5 of section 2 ; but the same formula will be denoted by (1.2.5) when quoted in another lecture).

2. - TWO-SCALE ASYMPTOTIC EXPANSIONS AND LOCAL PERIODICITY

This method is classical in mechanics of vibrations, when a small perturbation modify a motion which should be otherwise periodic in time, for instance, the motion of a pendulum submitted to a small damping is such that each "*period*" is almost analogous to the preceeding one, but the cumulative effect of the damping provokes important differences (of the amplitude, for instance) of the motion of two "*far located in time*" periods. To study this one introduces, aside the ordinary time t, two variables (the so-called fast and slow times) $t^* = t$ and $\tau = \varepsilon t$ (with ε small

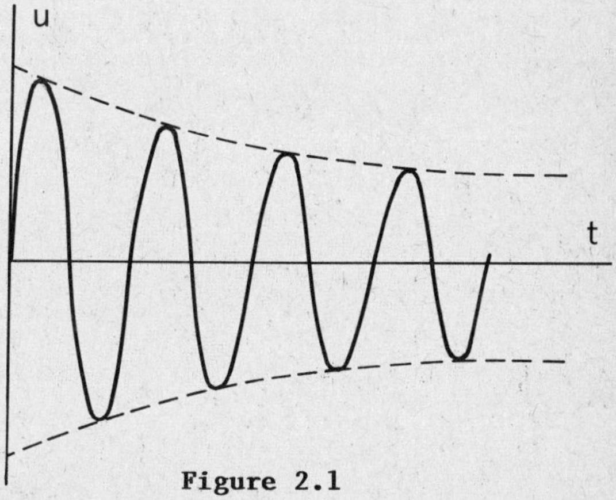

Figure 2.1

parameter) and we search for an asymptotic expansion of the solution $u^{\varepsilon}(t)$ under the form

$$(2.1)\quad u^{\varepsilon}(t) = u^{o}\left(t^{*}(t),\ \tau(t)\right) + \varepsilon u^{1}\ \left(t^{*}(t),\ \tau(t)\right) + \ldots$$

and we try to describe the local periodic phenomena by the dependence on t through t^{*}, and the slow modulation by the dependence on t through τ. We of course have

$$(2.2)\quad \frac{d}{d\,t} = \frac{\partial}{\partial t^{*}} + \varepsilon\,\frac{\partial}{\partial\,\tau}$$

Moreover, as a <u>convention</u> for the sake of simplicity, we drope the star in t^{*}, and write

$$(2.3)\quad u^{\varepsilon}(t) = u^{o}(t,\tau) + \varepsilon u^{1}(t,\tau) + \ldots\quad ;\quad t = \varepsilon t$$

$$(2.4)\quad \frac{d}{d\,t} = \frac{\partial}{\partial\,t} + \varepsilon\,\frac{\partial}{\partial\,\tau}$$

According to analogous considerations, let Ω be a body made of a composite material in the R^{3} space of the standard coordinates (x_1,x_2,x_3). Moreover, we assume that its mechanical properties are periodic with a small period, described with the aid of a small parameter ε as follows : In the auxiliar space of the variables (y_1,y_2,y_3) we consider a parallelepipedic period denoted by Y (with edges Y_1,Y_2,Y_3) as well as the parallelepipeds obtained by translations of an

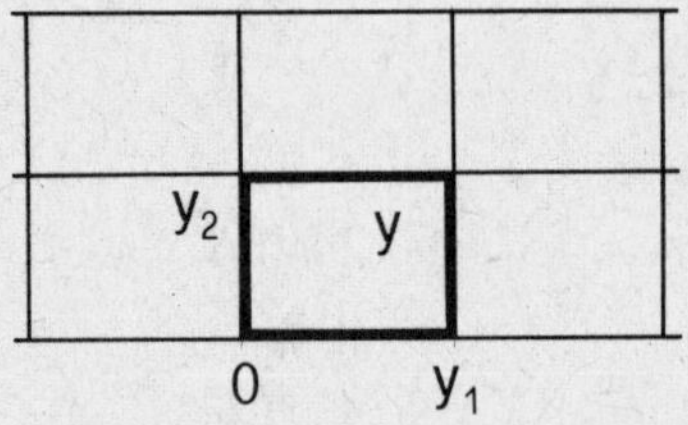

Figure 2.2

integer number of periods in the directions of the axes.

Let εY be the homothetic of Y with ratio ε. We consider the body Ω with the εY-periodic structure. Thus, some property $u^{\varepsilon}(x)$ (here u may denote displacement, stress or some other property of the mechanical process under consideration) is searched under the form of an asymptotic expansion

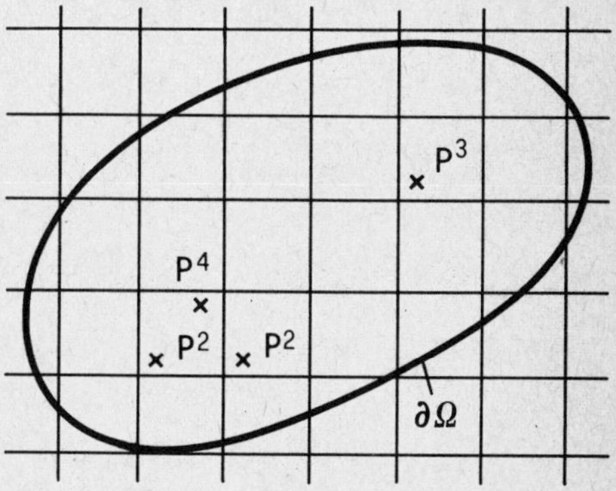

Figure 2.3

$$(2.5)\quad \begin{cases} u^{\varepsilon}(x) = u^{o}(z(x),y(x)) + \varepsilon u^{1}(z(x),y(x)) + \dots \\ z(x) = x \qquad ; \qquad y(x) = x/\varepsilon \end{cases}$$

or merely (with the preceeding convention)

$$(2.6)\quad \begin{cases} u^{\varepsilon}(x) = u^{o}(x,y) + \varepsilon u^{1}(x,y) + \dots \\ y = \dfrac{x}{\varepsilon} \quad ; \quad \dfrac{d}{d\, x_i} = \dfrac{\partial}{\partial\, x_i} + \dfrac{1}{\varepsilon}\dfrac{\partial}{\partial\, y_i} \end{cases}$$

and moreover, we intend to describe the influence of the periodic structure (resp. of the other non periodic causes of the phenomenon, as the boundary $\partial\Omega$ and so on) by the microscopic variable (resp. the macroscopic variable z or merely x) in (2.5), (2.6). To this end, we search for an expansion (2.5) or (2.6) with functions u^{i} Y-periodic with respect to the variable y and smooth with respect to x. Indeed, <u>each $u^{i}(x,y)$ is defined on $\Omega \times Y$</u> (or on $\Omega \times R^3$, which amounts to the same, as it is Y-periodic).

It is worhtwhile to see that each term $u^i(x,y)$ is a locally periodic function in the following sense. Let us compare the values of $u^i(x,y)$ at two points P^1, P^2 (Fig. 2,3) homologous by periodicity corresponding to two contiguous periods. By periodicity, the dependence on y is the same and the dependence on x is "*almost the same*" because the distance P^1P^2 is small and u^i is a smooth function of x. On the other hand, let P^3 be a point homologous to P^1 by periodicity, but located far from P^1. The dependence of u^i on y is the same, but the dependence on x is very different because P^1, P^3 are not near to each other. Finally we compare the values of u^i at two different points P^1, P^4 of the same period. The dependence on x is almost the same, but dependence on y is very different because P^1 and P^4 are not homologous by periodicity (in fact, the distance P^1, P^4 is "*large*" when measured with the variable y !).

It is evident that this locally periodic expansion is fit to describe the solution in regions of Ω far from its boundary, or from regions where the local effects are not εY-periodic, such as discontinuities of the microscopic structure. In such regions, the appropriate asymptotic expansions are almost periodic in the microscopic variable y only with respect to displacements which are tangential to the boundary, because the medium in fact is not periodic as for displacements normal to the boundary, and there is no reason for the solution to be "*almost invariant*" with respect to such displacements. As a consequence, near the boundary $\partial\Omega$ of the body (fig. 2.3) we must consider boundary layers where the solution is searched under the form (2.5) or (2.6), but now x runs in $\partial\Omega$ and y in the strip S (Fig. 2.4), and

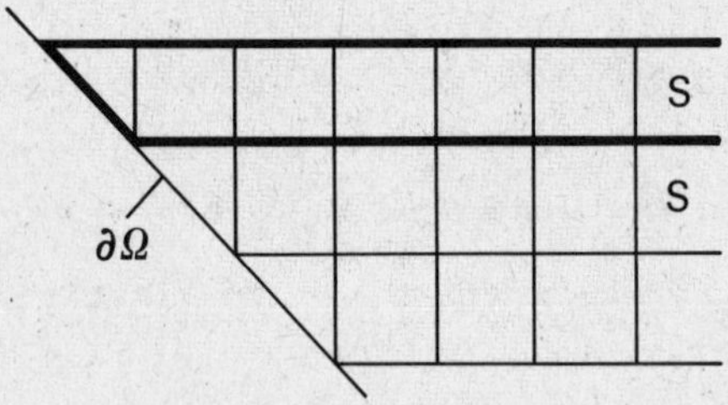

Figure 2.4

u^i is searched to be S-periodic. (Note, in Fig. 2.4 for instance, that S is a semi-infinite strip formed by Y-periods (plus perhaps "*parts*" of periods at the intersection with $\partial\Omega$). This situation is easily described for boundaries parallel to a coordinate plane, for instance x_3 = cost (Fig. 2.5).

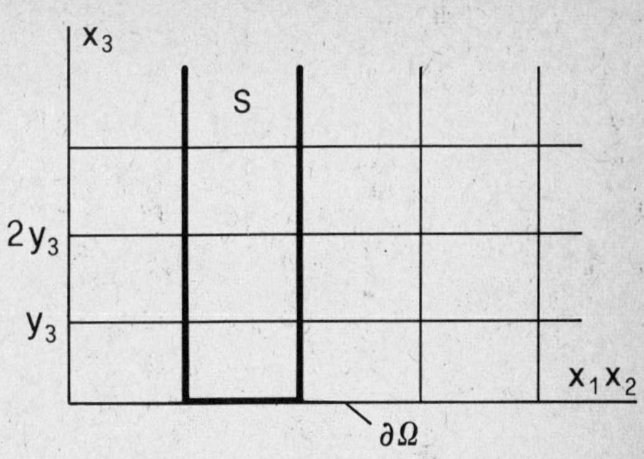

Figure 2.5

In this case, the solution in the boundary layer region takes the form : (the superscript BL is for "Boundary Layer") :

(2.7) $u^{\varepsilon}(x) = u^{oBL}(x,y) + \varepsilon u^{1BL}(x,y) + \ldots$

with

(2.8) $u^{iBL}(x,y) = u^{iBL}(x_1,x_2,y_1,y_2,y_3)$, y_1 and y_2 periodic with periods Y_1,Y_2, but not necessarily x_1,x_2,y_3 periodic.

Remark 2.1 - As for the expansion (2.6) far from the boundaries, the "*boundary conditions*" for the y variable amounts to the Y-periodicity. But in the boundary layer (2.7), the "*boundary conditions*" for y amounts to periodicity in y_1,y_2, genuine boundary conditions for $y_3 = 0$ and "*matching*" between (2.7) and (2.6) as $y_3 \to +\infty$ and $x_3 \to 0$. This amounts to saying that there is a "*transition region*" of the layer towards the bulk solution (2.6) far from the boundary.■

3. - MATCHED ASYMPTOTIC EXPANSIONS

We saw in the preceeding section that a function $u^{\varepsilon}(x)$ may have asymptotic expansions of different nature in different regions, for instance, in the boundary layer near $\partial\Omega$ and the bulk region at the interior of Ω. It is clear that two such expansions "*must agree*", i.e. the boundary layer contains a transition region between the genuine boundary layer and the "*outer*" region (outer to the boundary layer). As for this relation between the boundary layer and the bulk region, the tangential variables $(x_1, x_2, y_1, y_2$ in the case of (2.6), (2.7)) play the role of parameters, and x_3, y_3 are the relevant variables. We write

(3.1) $u^\varepsilon(x) = u^o(x_3) + \varepsilon u^1(x_3) + \ldots$ (outer or bulk expansion)

(3.2) $u^\varepsilon(x) = u^{oBL}(y_3) + \varepsilon u^{1BL}(y_3) + \ldots$
(inner or boundary layer expansion)

and we emphasize that the outer (resp. inner) expansion only depends on the outer or bulk variable x_3 (resp. on the inner or boundary layer variable $y_3 = x_3/\varepsilon$).

We now give some definitions. The outer (resp. inner) limit of a function $u^\varepsilon(x)$ is the limit as $\varepsilon \to 0$ for fixed outer variable x_3 (resp. inner variable y_3). In the same way, the m-term outer (resp. inner) expansion is the asymptotic expansion of m terms of u^ε for $\varepsilon \to 0$ with fixed outer variable x_3 (resp. inner variable y_3). For instance, $u^o(x_3)$ is the outer limit, and $u^{oBL}(y_3) + \varepsilon u^{1BL}(y_3)$ is the 2-term inner expansion. As sometimes we deal with expansions the first term of which are not of order $O(1)$, we also define the outer (resp. inner) representation as the first non-zero term of the outer (resp. inner expansion).

We now give the "*matching rules*" expressing that the outer and inner expansions (3.1), (3.2) agree in some intermediate transition region. Justification of these rules may be seen in the general references given in sect. 1.

The matching at order $O(1)$ is :

(3.3) Inner limit of (the outer limit) =

= Outer limit of (the inner limit)

Of course, the outer limit of u^{ε} in (3.1), (3.2) is $u^{o}(x_3)$; in order to compute its inner limit, we write it in the inner variable $y_3 = x_3/\varepsilon$, and we compute the limit as $\varepsilon \to 0$ for fixed y_3 ; this gives $\lim_{\varepsilon\to 0} u^{o}(\varepsilon y_3) = u^{o}(0)$ which is the left side of (3.3). Analogously, the right hand side is $u^{oBL}(+\infty)$. Thus, (3.3) amounts to

(3.4) $u^{o}(0) = u^{oBL}(\infty)$

or which amounts to the same, u^{o} at the boundary $\partial\Omega$ equals the boundary layer first term far from the wall (far in the small variable y_3). It is easily seen that (3.3) or (3.4) amounts to the existence of an "*intermediate variable*" z small (resp. large) with respect to x_3 (resp. y_3) such that (3.1) and (3.2) give at the first order, the same information for $z = O(1)$. We may take, for instance, $z = x/\varepsilon^{\frac{1}{2}}$.

A more general rule of matching is

Inner representation of (the outer representation) =

(3.5)

= Outer representation of (the inner representation)

The general matching rule of Kaplun and Lagerstrom is

The m-term inner expansion of (the n-term outer expansion) =

(3.6)

= the n-term outer expansion of (the m-term inner expansion)

where m and n are two integers ; m is often chosen to be either n or n + 1.

It should be noticed that sometimes the application of the matching rules is not straightforward, and some variants of them are used.

For instance, the matching of the gradient may be easy but that of the function itself may be difficult ; matching is often a matter of skill.

Remark 3.1 - Because of the matching, we may say that the thickness of the layer is $O(\varepsilon)$ (or in general the ratio of the inner variable to the outer one) as for large y_3 we are "*out of the layer*" (see (3.4)). But this is only the order ; the exact thickness as a length makes no sense.■

4. - SOME INDICATIONS ON SOBOLEV SPACES AND THE LAX-MILGRAM THEOREM

We recall here some results, but the reader is refered to the bibliography given in sect. 1 for proofs and other useful results.

Let Ω be an open set of R^n. We denote by $\mathcal{D}(\Omega)$ the space of the infinitely differentiable functions with compact support in Ω (i.e. which vanish out of a bounded set which do not approach $\partial\Omega$). If θ^i, $\theta \in \mathcal{D}(\Omega)$, we say that

$$(4.1) \qquad \theta^i \underset{i\to\infty}{\to} \theta \qquad \text{in } \mathcal{D}(\Omega)$$

iff the supports of all the θ^i are contained in a unique compact set of Ω and θ^i and all their derivatives tend uniformly to θ and its corresponding derivatives.

Let T be a continuous functional on $\mathcal{D}$, i.e. a law associating a number $\langle T,\theta\rangle$ to each $\theta \in \mathcal{D}$ such that it is is linear and

$$(4.2) \qquad \theta^i \to \theta \quad \Longrightarrow \quad \langle T,\theta^i\rangle \to \langle T,\theta\rangle$$

Such a functional is called a distribution and the set of such distributions is the space $\mathcal{D}'(\Omega)$. The concept of convergence is defined by :

$$(4.3) \qquad T^i \to T \qquad \text{iff } \langle T^i,\theta\rangle \to \langle T,\theta\rangle \quad \forall\, \theta \in \mathcal{D}(\Omega).$$

The derivatives of distributions are distributions defined by

(4.4) $\langle \partial T/\partial x_j, \theta\rangle = - \langle T, \partial\theta/\partial x_j\rangle \quad \forall\ \theta \in \mathscr{D}$

and it is seen that each distribution is indefinitely differentiable. Moreover, from (4.3) it is seen that <u>all</u> the derivatives of the sequence converge too.

Of course, if f is a function which is locally integrable on Ω, there is an associated distribution f^D defined by

(4.5) $\langle f^D, \theta\rangle = \int_\Omega f\ \theta\ dx$

$L^2(\Omega)$ is the space of all square integrable (in the Lebesque sense) functions. It is a Hilbert space (functions are defined a.e. = almost everywhere, i.e. up to the values in a set of vanishing measure) for the scalar product

(4.6) $(u,v)_{L^2} = \int_\Omega u\ v\ dx \quad ; \quad ||u||^2_{L^2} = \int_\Omega |u|^2\ dx$

$H^1(\Omega)$ is the space of the functions of $L^2(\Omega)$ such that their distributional derivatives of first order are associated (by (4.5)) with functions of $L^2(\Omega)$. It is a Hilbert space for the scalar product

(4.7) $(u,v)_{H^1} = (u,v)_{L^2} + \left(\frac{\partial u}{\partial x_j}, \frac{\partial v}{\partial x_j}\right)_{L^2(\Omega)}$

If $u \in H^1(\Omega)$, it is possible to define the <u>trace</u> of u on $\partial\Omega$ (or a part of $\partial\Omega$) and it is an element of $L^2(\partial\Omega)$ (and even of $H^{\frac{1}{2}}(\partial\Omega)$). The trace operator $u \to u|_{\partial\Omega}$ is continuous from $H^1(\Omega)$ to $L^2(\partial\Omega)$. In particular, the subspace of H^1 formed by the functions with trace 0 (i.e. functions of $H^1(\Omega)$ which vanish on $\partial\Omega$) is a Hilbert space denoted by $H_o^1(\Omega)$.

Of course, if $u \in H^1(\Omega)$, <u>grad</u> $u \in \underline{L}^2(\Omega)$; the converse is also true as follows from the Poincaré inequality (for bounded Ω)

(4.8) $\|u\|^2_{L^2} \leq C \left(\|\underline{\text{grad}}\, u\|^2_{L^2} + (\int_\Omega u\, dx)^2\right)$

by adding a constant to obtain $\int_\Omega u\, dx = 0$.

It follows that the space $H^1(\Omega)/R$ formed by <u>the equivalence classes of functions of H^1 defined up to an additive constant</u> may be equipped with the scalar product :

(4.9) $(u,v)_{H^1/R} = (\underline{\text{grad}}u, \underline{\text{grad}}v)_{\underline{L}^2}$

Another Poincaré inequality (for bounded Ω) is

(4.10) $\|u\|_{L^2} \leq C\, \|\underline{\text{grad}}\, u\|_{\underline{L}^2} \qquad \forall\, u \in H^1_o(\Omega)$,

and thus the right side of (4.9) may be taken as scalar product in H_o^1.

A useful tool in partial differential equations is the

<u>Theorem 4.1 (Lax-Milgram)</u>. <u>Let V be a Hilbert space, and $\ell(v)$, $a(u,v)$ a linear continuous and coercive forms on V, i.e. satisfying</u>

(4.11) $|\ell(v)| \leq C\, \|v\| \qquad \forall\, v \in V$

(4.12) $|a(u,v)| \leq C\, \|u\|\, \|v\| \qquad \forall\, u,v \in V$

(4.13) $|a(v,v)| \geq \alpha\, \|v\|^2 \qquad \forall\, v \in V$

Thus, there exists a unique $u \in V$ such that

(4.14) $a(u,v) = \ell(v) \qquad \forall\, v \in V$.

5. - THE NEUMANN PROBLEM FOR THE LAPLACIAN

As an example we consider an open, bounded domain $\Omega \subset R^3$, and we search for a function u such that

$$(5.1) \qquad - \Delta u = f \qquad (\Delta u \equiv \frac{\partial}{\partial x_i} \frac{\partial u}{\partial x_i}) \text{ in } \Omega$$

$$(5.2) \qquad \frac{\partial u}{\partial n} = 0 \qquad \text{on } \partial\Omega$$

where f is a given function of $L^2(\Omega)$.

Let us admit that this problem has a solution u ; then, integrating by parts (5.1) on Ω and using (5.2) we have :

$$(5.3) \qquad \boxed{\int_\Omega f \, dx} = - \int_\Omega \frac{\partial}{\partial x_i} \frac{\partial u}{\partial x_i} dx = - \int_{\partial\Omega} n_i \frac{\partial u}{\partial x_i} ds = \int_{\partial\Omega} - \frac{\partial u}{\partial n} ds \boxed{= 0}$$

i.e. the given function is not arbitrary ; it must satisfy (5.3). On the other hand, if this condition is satisfied, a solution u exists as we shall prove soon ; but it is not unique because if u is a solution, then $u + c$ for any $c \in R$ is also a solution. Moreover, if v is a solution of the homogeneous problem

$$(5.4) \qquad - \Delta v = 0 \quad ; \quad \frac{\partial v}{\partial n} = 0$$

by multiplying by v and integrating by parts (5.4) we have

$$\boxed{0} = \int_\Omega - \frac{\partial}{\partial x_i} (\frac{\partial v}{\partial x_i}) v \, dx = - \int_\Omega \frac{\partial}{\partial x_i} (\frac{\partial v}{\partial x_i} v) \, dx + \int_\Omega \frac{\partial v}{\partial x_i} \frac{\partial v}{\partial x_i} dx$$

$$= \int_{\partial\Omega} \frac{\partial v}{\partial n} v \, ds + \int_\Omega |\underline{\text{grad}}\ v|^2 \, dx = \boxed{\int_{\partial\Omega} |\underline{\text{grad}}\ v|^2 \, dx}$$

i.e. v is a constant. Thus, $V = H^1(\Omega)/R$ is the fit espace to search for the solution u.

In fact, (5.1), (5.2) is equivalent to the following abstract problem (or variational formulation).

Find $u \in H^1(\Omega)/R$ such that

(5.4)
$$\int_{\Omega} \frac{\partial u}{\partial x_i} \frac{\partial v}{\partial x_i} dx = \int_{\Omega} f v \, dx \qquad \forall \; v \in H^1(\Omega)/R .$$

where we note that, by virtue of the compatibility condition (5.3) satisfied by f, (5.4) takes the same value for any function of the equivalence class $v \in H^1(\Omega)/R$.

Indeed, by multiplying (5.1) by v and integrating by parts we obtain

(5.5) $$\int_{\Omega} f v \, dx = - \int_{\partial\Omega} \frac{\partial u}{\partial n} v \, ds + \int_{\Omega} \frac{\partial u}{\partial x_i} \frac{\partial v}{\partial x_i} dx$$

then, if u is a solution of (5.1), (5.2) it satisfies (5.4). Conversely, if u satisfies (5.4) by taking $v \in \mathcal{D}(\Omega)$ we see that u verify (5.1), and (5.5) becomes

$$\int_{\partial\Omega} \frac{\partial u}{\partial n} v \, ds = 0 \qquad \forall \; v \in H^1(\Omega)/R$$

and it follows that (5.2) is satisfied.

The existence and uniqueness of the solution $u \in H^1(\Omega)/R$ (note that, as a function, it is defined up to an additive constant) of (5.4) follows from the Lax-Milgram theorem thanks to the Poincaré inequality.

CHAPTER 2

2. - BOUNDARY LAYERS IN THERMAL CONDUCTION AND ELASTICITY

1. - A TWO-DIMENSIONAL PROBLEM IN THERMAL CONDUCTION

We consider a steady (i.e. independent of time) conduction problem in the rectangular domain D of Fig. 1, made of a heterogeneous layered medium, ε-periodic in the x_1 direction. The boundaries $x_2 = 0$, $x_2 = \ell_2$ are free, i.e. a homogeneous boundary condition is to be imposed there. On the boundaries $x_1 = 0$, $x_1 = \ell_1$ we prescribe a constant flux, i.e. a non homogeneous boundary condition.

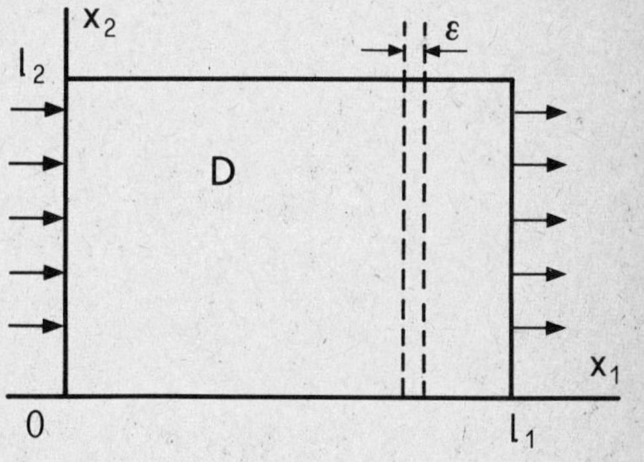

Figure 1.1

As the properties of the material are independent of x_2, we may consider that the period in the x_2 direction takes any value, for instance ε iftself, and we are of course in the classical homogenization framework with macroscopic variable $x = (x_1,x_2)$ and microscopic one $y = (y_1,y_2)$. The problem is

$$(1.1)\qquad -\frac{\partial}{\partial x_i}\left(a_{ij}\left(\frac{x_1}{\varepsilon}\right)\frac{\partial u^\varepsilon}{\partial x_j}\right) = 0 \qquad \text{on } D$$

$$(1.2)\qquad a_{2i}\frac{\partial u^\varepsilon}{\partial x_i} = 0 \qquad \text{for } x_2 = 0,\ x_2 = \ell_2$$

(1.3) $a_{1i} \frac{\partial u^\varepsilon}{\partial x_i} = 1$ for $x_1 = 0$, $x_1 = \ell_1$

where a_{ij} are the conductivity coefficients, which are symmetric (i.e., $a_{ij} = a_{ji}$) and positive (i.e. (1.1) is elliptic). Of course (1.1) is understood in the sense of distributions : if the coefficients are piecewise constant (case of a heterogeneous medium) we have the transmission condition (the symbol $[\ ]$ denotes the jump)

(1.4) $[u^\varepsilon] = 0 \quad ; \quad [a_{ij} \frac{\partial u^\varepsilon}{\partial x_j} n_i] = 0$

at the interfaces.

Far in the $y = x/\varepsilon$ variable from the boundaries, according to classical homogenization we have the expansion :

(1.5) $u^\varepsilon(x) = u^o(x) + \varepsilon u^1(x,y) + O(\varepsilon^2)$

where

(1.6) $u^1(x,y) = \frac{\partial u^o(x)}{\partial x_i} w^i(y)$

and $w^i(y)$, $i = 1,2$ are 1-periodic solutions of

(1.7) $- \frac{\partial}{\partial y_p} \left[a_{pm}(y_1)(\frac{\partial w^i}{\partial y_m} + \delta_{im})\right] = 0$

defined up to an additive constant. This gives :

(1.8) $$\begin{cases} w^1 \equiv \left(\int_o^1 \frac{d\xi}{a_{11}(\xi)}\right)^{-1} \int_o^{y_1} \frac{d\xi}{a_{11}(\xi)} - y_1 \\ w^2 \equiv 0 \end{cases}$$

which are of course independent of y_2 (i.e. periodic with any period). The homogenized coefficients are (the upper index h denotes "*homogenized*") :

$$a_{ik}^h = \int_o^1 \left(a_{ik}(y_1) + a_{ij}(y_1) \frac{\partial w^k(y_1)}{\partial y_j}\right) dy_1 \quad \Longrightarrow$$

$$(1.9) \quad a_{11}^h = \left(\int_o^1 \frac{d\xi}{a_{11}(\xi)}\right)^{-1} \; ; \; a_{12}^h = \int_o^1 a_{12}(\xi) d\xi \; ; \; a_{22}^h = \int_o^1 a_{22}(\xi)\, d\xi$$

The homogenized equation and boundary conditions (obtained by the flux method, for instance), are :

$$(1.10) \quad - a_{11}^h \frac{\partial^2 u^o}{\partial x_1^2} - 2a_{12}^h \frac{\partial^2 u^o}{\partial x_1 \partial x_2} - a_{22}^h \frac{\partial^2 u^o}{\partial x_2^2} = 0$$

$$(1.11) \quad a_{2j}^h \frac{\partial u^o}{\partial x_j} = 0 \qquad \text{for} \quad x_2 = 0 \; , \; x_2 = \ell_2$$

$$(1.12) \quad a_{1j}^h \frac{\partial u^o}{\partial x_j} = 1 \qquad \text{for} \quad x_1 = 0 \; , \; x_1 = \ell_1$$

thus, (1.10)-(1.12) defines $u^o(x)$ (up to an additive constant of course). The local structure is given by (1.5), (1.6), (1.8). We now write equation (1.1) in terms of the divergence of the flux

$$(1.13) \quad - \frac{\partial \sigma_i^\varepsilon}{\partial x_i} = 0 \quad , \quad \sigma_i^\varepsilon \equiv a_{ij}\left(\frac{x_1}{\varepsilon}\right) \frac{\partial u^\varepsilon}{\partial x_j}$$

we expand $\underline{\sigma}^\varepsilon$ according to (1.13), (1.4), (1.5), (1.6) :

$$\underline{\sigma}^\varepsilon = \underline{\sigma}^o(x,y) + \varepsilon \underline{\sigma}^1(x,y) + \ldots \quad \text{and we have :}$$

$$(1.14) \begin{cases} \sigma_1^o(x,y) \equiv a_{11}(y_1)\left(1 + \dfrac{\partial w^1(y_1)}{\partial y_1}\right) \dfrac{\partial u^o(x)}{\partial x_1} + a_{12}(y_1) \dfrac{\partial u^o(x)}{\partial x_2} \\ \sigma_2^o(x,y) \equiv a_{12}(y_1)\left(1 + \dfrac{\partial w^1(y_1)}{\partial y_1}\right) \dfrac{\partial u^o(x)}{\partial x_1} + a_{22}(y_1) \dfrac{\partial u^o(x)}{\partial x_2} \end{cases}$$

We see that u^o (resp. u^1) is defined up to an additive constant (resp. function of x) but the components of the flux (1.14) are uniquely defined. Let us consider the boundary condition (1.11) for instance. It amounts to

$$(1.15) \quad \int_o^1 \sigma_2^o(x,y_1)\, dy_1 = 0 \qquad \text{for} \quad x_2 = 0 \; , \; x_2 = \ell_2$$

but the exact boundary condition $\sigma_2^\varepsilon = 0$ should give at the first order

(1.16) $\sigma_2^o(x.y_1) = 0 \qquad \text{for } x_2 = 0 \;,\; x_2 = \ell_2$

Thus it is only satisfied in average for a period. This is the reason why a boundary layer appears. Of course, analogous considerations hold for the boundaries $x_1 = 0$, $x_1 = \ell_1$.

2. - BOUNDARY LAYERS FOR THE PRECEDING PROBLEM

Let us study as an example, the boundary layer in the vicinity of $x_2 = 0$. As we know $u^o(x_1,x_2)$ and $u^1(x_1,x_2,y_1, \;)$, we shall take into account the boundary layer by introducing a new complementary term in expansion (1.5) (see Remark 2.1 here after) which becomes :

(2.1) $u^\varepsilon(x) = u^o(x_1,x_2) + \varepsilon\left(u^1(x_1,x_2,y_1, \;) + u^{1c}(x_1,y_1,y_2)\right) + O(\varepsilon^2)$

in the layer, with u^{1c} S-periodic in the strip

(2.2) $S = \{(y_1,y_2) \;;\; y_1 \in (0,1) \;;\; y_2 \in (0,+\infty)\}$

(Fig. 2.1). This amounts to write $u^1 + u^{1c}$ instead of u^1 ; thus the equation for $u^1 + u^{1c}$ is the same as for u^1 out of the layer ; moreover, as u^1 satisfies this equation, u^{1c} satisfies it too ; this gives :

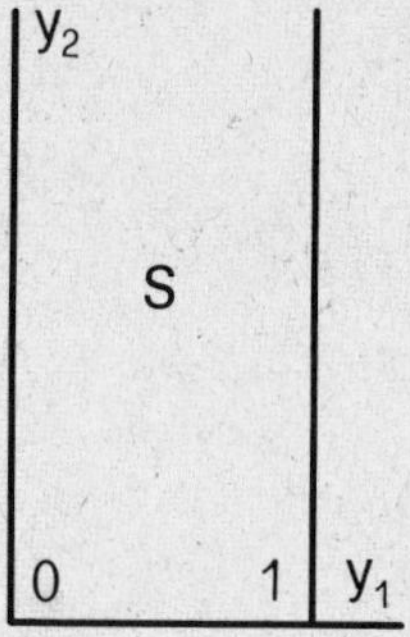

Figure 2.1

(2.3) $-\dfrac{\partial}{\partial y_j}\left(a_{ij}(y_1)\dfrac{\partial u^{1c}}{\partial y_j}\right) = 0$

As for the boundary conditions, in addition to the S-periodicity (i.e. 1-periodicity in y_1) we have

(2.4) $a_{2j}(y_1)\dfrac{\partial u^{1c}}{\partial y_j} = -\sigma_2^o(x_1,0,y_1) \qquad \text{for } y_2 = 0$

$$(2.5) \qquad \underline{\text{grad}}_y \; u^{1c} \xrightarrow[y_2 \to +\infty]{} 0$$

where the right side of (2.4) is precisely σ_2^o of (1.14), in order to obtain for the σ_2^o with $u^1 + u^{1c}$ instead of u^1, vanishing values at the boundary. As for (2.5), it is the matching condition for the gradient. It amounts to the vanishing of the gradient of the complementary term out of the layer.

Remark 2.1 - The preceeding framework is very easy but other forms of the expansion are also valid. For instance, if the genuine boundary layer theory framework is adopted, as in sect. 1.3, the expansion in the layer is

$$(2.6) \qquad u^\varepsilon(x) = u^{oBL}(x_1,y_1,y_2) + \varepsilon u^{1BL}(x_1,y_1,y_2) + O(\varepsilon^2)$$

In order to establish the relationship between (2.6) and (2.1), we note that the outer variable x_3 does not appear in (2.6). As (2.1) is an expansion of the two-scale type, with $y = x/\varepsilon$, we must write $x_2 = \varepsilon y_2$ in (2.1) with $y_2 = O(1)$ to obtain (2.6). This gives :

$$(2.7) \qquad u^o(x_1,x_2) = u^o(x_1,0) + \varepsilon \frac{\partial u^o}{\partial x_2}(x_1,0)y_2 + O(\varepsilon^2)$$

and by replacing it into (2.1) we obtain (2.6) with :

$$(2.8) \quad \begin{cases} u^{oBL}(x_1,y_1,y_2) = u^o(x_1,0) \\[2ex] u^{1BL}(x_1,y_1,y_2) = \dfrac{\partial u^o}{\partial x_2}(x_1,0)y_2 + u^1(x_1,0,y_1,y_2) + u^{1c}(x_1,y_1,y_2) \end{cases}$$

and it is easily seen that (2.4) and (2.5) amount to the boundary condition for u^{1BL} and the matching of the gradient, respectively. On the other hand, we note that, as the gradient out of the layer is oscillating, <u>the matching is easily written by using the complementary term which vanishes at infinity.</u> ■

The existence and uniqueness of u^{1c} satisfying (2.3)-(2.5) is easily proven, exactly as for the Neumann problem in sect. 1.5. In fact, this is a non homogeneous Neumann problem on $y_2 = 0$ (and on $y_2 = +\infty$) with periodicity conditions for y_1. Of course x_1 plays the role of a parameter and the solution is defined up to an additive constant (i.e. a function of x_1). Let V be the Hilbert space of the S-periodic functions defined up to an additive constant with gradient belonging to $L^2(S)$; then (2.3)-(2.5) amounts to find $u^{1c} \in V$ such that

$$(2.9) \qquad \int_S a_{ij}(y_1) \frac{\partial u^{1c}}{\partial y_j} \frac{\partial v}{\partial y_i} dy = \int_0^1 \sigma_2^0(x_1,0,y_1) v \, dy_1 \qquad \forall v \in V$$

where we note that, by virtue of condition (1.15), the right side of (2.9) takes the same value for v or v + constant, i.e. it is a functional on V. The existence and uniqueness then follow from the Lax-Milgram theorem (after applying the Poincaré inequality 1(4.8) on a bounded domain to prove that the functional is continuous).

Remark 2.2 - The preceeding proof only shows that $\underline{\text{grad}}_y \, u^{1c} \in \underline{L}^2(S)$; this amounts to saying that in some generalized sense it tends to 0 as $y_2 \to +\infty$; in fact this is true exponentially as was proved by Tartar (see Lions[11], sect. I.10.4).■

Exactly in the same way we may consider the case of a boundary which is ondulated ε-periodically in x_1. The functions u^o, u^1 are the same as before, but u^{1c} is modified (Fig. 2.2)

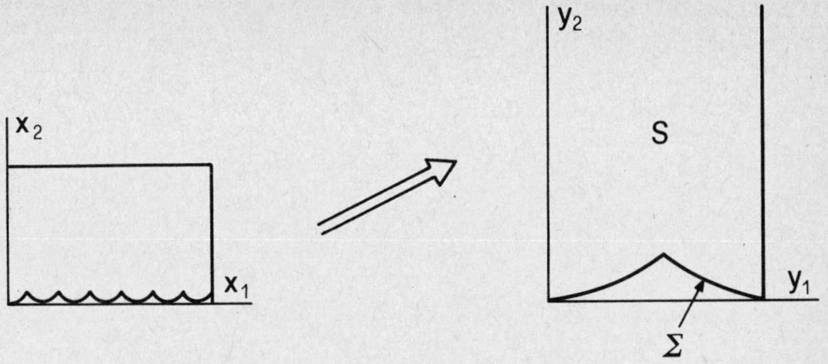

Figure 2.2

Remark 2.3 - In addition to the descriptions (2.1) and (2.2), we may define

$$(2.10)\quad u^*(x_1,y_1,y_2) \equiv \frac{\partial u^o}{\partial x_i}(x_1,0)y_i + u^1(x_1,0,y_1,y_2) + u^{1c}(x_1,y_1,y_2)$$

thus we have

$$(2.11)\quad \frac{\partial u^\varepsilon}{\partial x_i} = \frac{\partial u^*}{\partial y_i} + O(\varepsilon)$$

i.e. u^* represents in the microscopic variables the first term of the gradient. The equation and boundary conditions (2.3), (2.4) become :

$$(2.12)\quad -\frac{\partial}{\partial y_j}\left(a_{ij}(y_1)\frac{\partial u^*}{\partial y_i}\right) = 0 \qquad \text{in } S$$

$$(2.13)\quad a_{ij}\frac{\partial u^*}{\partial y_j}n_i = 0 \qquad \text{on } \Sigma$$

(which is also valid in the case of Fig. 2.2). We note that in this case both the equation and boundary condition are homogeneous and this will prove useful in the study of singularities (Lecture 5). ■

3. - LAYERS IN THREE-DIMENSIONAL ELASTICITY

Before starting our study we point out that in the preceeding section there was some mixture of Y-periodic (as u^1) and S-periodic (as u^{1c}), for instance in formula (2.1). In fact, our study is consistent only if Y-periodicity implies S-periodicity. This was the case in sect. 2, as well as in the situation of Fig. 3.1 (note, in particular that we may take Y' instead of Y as period of the structure). Moreover, this is always the case for layered media. In any case, we shall only consider in the present section the (three-dimensional) situation of Fig. 3.2 (or perhaps with ondulations of the boundary as in (Fig. 2.2).

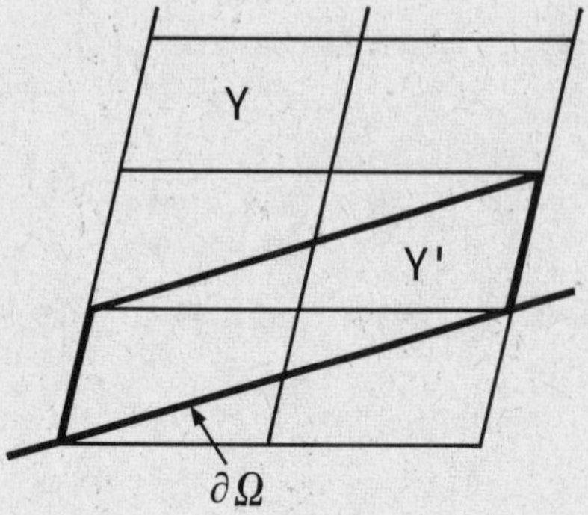

Figure 3.1

Using the general framework of homogenization in elasticity with εY-periodic coefficients $a_{ij\ell m}(y)$, (see for instance Sanchez[12], sect. 6.2, let

$$(3.1) \quad \underline{u}^{\varepsilon}(x) = \underline{u}^{o}(x) + \varepsilon \underline{u}^{1}(x,y) + \ldots \qquad y = x/\varepsilon$$

the asymptotic expansion far from the boundary. The corresponding expansions for strain and stress are

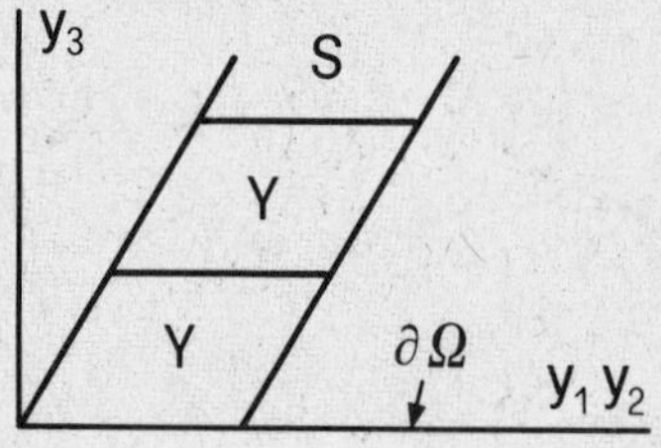

Figure 3.2

$$(3.2) \quad e^{\varepsilon}_{ij}(x) = e^{o}_{ij}(x,y) + \varepsilon e^{1}_{ij}(x,y) + \ldots$$

$$(3.3) \quad \sigma^{\varepsilon}_{ij}(x) = a_{ij\ell m}\, e^{\varepsilon}_{\ell m} = \sigma^{o}_{ij}(x,y) + \varepsilon\sigma^{1}_{ij}(x,y) + \ldots$$

where

$$(3.4)\quad \begin{cases} e^{o}_{ij}(x,y) \equiv e_{ijx}(\underline{u}^{o}) + e_{ijy}(\underline{u}^{1}) \\ \sigma^{o}_{ij} = a_{ij\ell m}\, e_{\ell m}(\underline{u}^{o}) \end{cases}$$

with

$$(3.5)\quad \underline{u}^{1} = e_{krx}(\underline{u}^{o})\, \underline{w}^{kr}(y) + \underline{\text{cost}}$$

where $\underline{w}^{kr}$ are the Y-periodic solutions of the local problems

$$(3.6)\quad -\frac{\partial}{\partial y_j}\left\{a_{ij\ell m}(y)\left(\delta_{k\ell}\,\delta_{mr} + e_{\ell my}(\underline{w}^{kr})\right)\right\} = 0$$

and the homogenized coefficients are

$$(3.7)\quad a^{h}_{ijkr} = \left\{a_{ij\ell m}\left(\delta_{k\ell}\,\delta_{mr} + e_{\ell my}(\underline{w}^{kr})\right)\right\}^{\sim}$$

where the tilde $\sim$ denote average on Y. Then $\underline{u}^{o}$ is the solution of the homogenized equation (3.8) and the boundary condition (which we write in (3.9) for the free boundary of Fig. 3.2) :

$$(3.8)\quad -\frac{\partial \tilde{\sigma}^{o}_{ij}}{\partial x_j} = f_i \quad ; \quad \tilde{\sigma}_{ij} = a^{h}_{ij\ell m}\, e_{\ell mx}(\underline{u}^{o})$$

$$(3.9)\quad \tilde{\sigma}^{o}_{3i} = 0 \qquad \text{on } \partial\Omega$$

Remark 3.1. - In (3.9), $\sim$ denotes the mean value on Y ; in fact it is also the mean value on the face Γ of the period (see Fig. 3.2), or on any section of the period $y_3 = c$ which is independent of c. Indeed, the local equation for u^1 is

$$-\frac{\partial\, \sigma^{o}_{ij}(x,y)}{\partial\, y_j} = 0$$

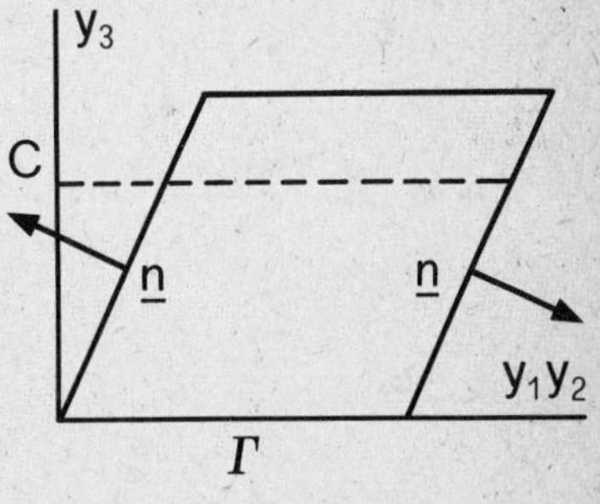

Figure 3.3

and integrating by part in the region of Y between $y_3 = 0$ and $y_3 = c$ (Fig. 3.3) we have

(3.10) $\int_{Y \cap \{y_3 = c\}} \sigma^o_{i3}\, dy_1\, dy_2 = \int_\Gamma \sigma^o_{i3}\, dy_1\, dy_2$

(not e that the integrals on the lateral faces cancel by Y-periodicity). ∎

Now the study of the boundary layer is exactly the same as in sect. 2. We introduce the complementary term $\underline{u}^{1c}$:

(3.11) $\underline{u}^\varepsilon(x) = \underline{u}^o(x) + \varepsilon\left(\underline{u}^1(x,y) + u^{1c}(x,y)\right) + O(\varepsilon^2)$

which satisfies (3.12)-(3.15) :

(3.12) $-\dfrac{\partial}{\partial y_j}\left(a_{ij\ell m}(y)\, e_{\ell my}(\underline{u}^{1c})\right) = 0$ in S

(3.13) $\underline{u}^{1c}(x,y)$ is S-periodic in y

(3.14) $\underline{\text{grad}}_y\, \underline{u}^{1c} \xrightarrow[y_3 \to +\infty]{} 0$

(3.15) $a_{3i\ell m}\, e_{\ell my}(\underline{u}^{1c}) = -\sigma^o_{3i} \equiv -a_{3i\ell m}\left(e_{\ell mx}(\underline{u}^o) + e_{\ell my}(\underline{u}^1)\right)$

The existence and uniqueness of $\underline{u}^{1c}$ (defined up to a constant (i.e. only depending on x_1) vector of R^3 follows from the Lax-Milgram theorem by noticing that it is equivalent to find $\underline{u}^{1c} \subset V$ such that

(3.16) $\int_S a_{ij\ell m}\, e_{\ell my}(\underline{u}^{1c})\, e_{ijy}(\underline{v})dy = \int_\Gamma \sigma^o_{3j}\, v_j\, dy_1\, dy_2 \qquad \forall\ \underline{v} \in V$

where V denote the Hilbert space of the S periodic vectors (defined up to a constant additive vector) with finite

$$\|\underline{v}\|^2_V = \int_S e_{ijy}(\underline{v})\, e_{ijy}(\underline{v})\, dy$$

We note, in particular that, by virtue of (3.9) and Remark 3.1, the right side of (3.16) takes the same value for $\underline{v}$ or $\underline{v}$ + constant.

Remark 3.2 - Dumontet[13] proved, for layered media, that (3.14) holds exponentially (see also Remark 2.2). ■

Remark 3.3. - It is clear that the fact that $\underline{u}^o$, $\underline{u}^1$, $\underline{u}^{1c}$ are only defined up to some additive constants (or functions of x) is immaterial as for the computation of the stress at first order :

$$(3.17) \quad \sigma^{\varepsilon}_{ij} = a_{ij\ell m} \left(e_{\ell m x}(\underline{u}^o) + e_{\ell m y}(\underline{u}^1 + \underline{u}^{1c}) \right) . \blacksquare$$

4. - INDICATIONS ON NUMERICAL COMPUTATION

The implementation of the S-periodic boundary value problems do not offer special difficulties when using the finite element discretization (MODULEF code for instance). The S-periodicity is imposed by numerical identification of the corresponding nodes. The fact that the solutions are only defined up to an additive vector is also classical in homogenization (as in Neumann problems) ; it suffices to prescribe a vanishing displacement at some point.

The only new feature is that S is not bounded. In fact, as the gradient is very rapidly decreasing, a truncation of S at $y_2 = L$ introduces no appreciable error, and the displacement may be prescribed to be zero there. Numerical experiments show that $\underline{\text{grad}}\ \underline{u}^{1c}$ vanish for a deep which is nearly a period. Fig. 4.1, taken from Dumontet[13] shows the complementary stress for an example of layered medium.

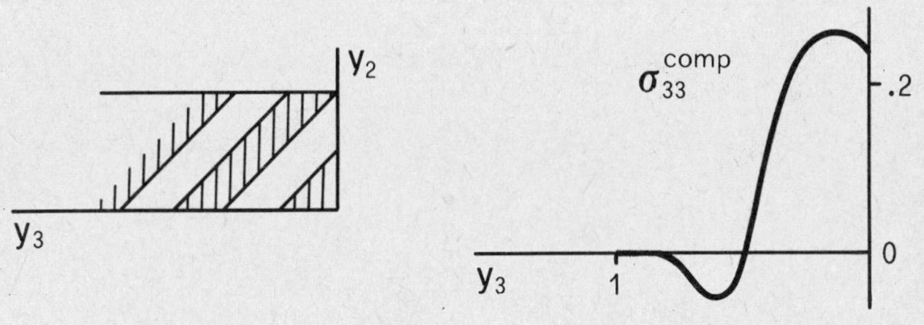

Figure 4.1

CHAPTER 3

3. - LAYERED PLATES IN TRACTION. BOUNDARY LAYERS

1. - PRELIMINARY REMARKS

Delamination phenomena often appear at the free edges of plates in traction. We shall see in Lecture 4 that singularities of the stress appear at the intersection at free edges with the interfaces between layers. In the present lecture we give a description of the stress and strain far from the edges and in the boundary layers close to them. As the study of layers and singularities is not very developped, we restrict ourselves to the simplest case of symmetric plates in traction ; thus, the considerations of sect. 2 are a particular case of the theory in the lectures by Caillerie.

We consider a plate of thickness 2ε occuping the region $x_3 \in (-\varepsilon,+\varepsilon)$, $(x_1,x_2) \in \omega$ (ε is a small parameter and ω denotes some domain of the x_1,x_2 plane).

The elasticity coefficients are functions of x_3/ε, (not of x_1,x_2), in particular they may be piecewise constant functions (case of the layers), satisfying the classical symmetry and positivity conditions :

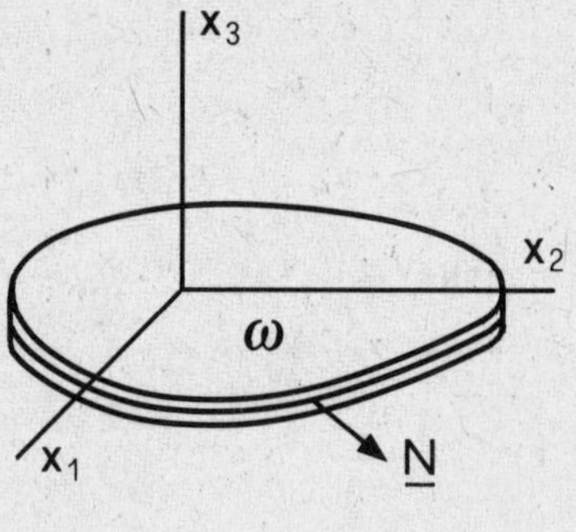

Figure 1.1

$$(1.1)\quad \begin{cases} a_{ijmn} = a_{mnij} = a_{mnji} \\ a_{ijmn}\, e_{mn}\, e_{ij} \geqq c\, e_{ij}\, e_{ij} \quad \forall\, e_{ij} \text{ (symmetric)} \end{cases}$$

In addition, we assume that the plate, as well as the given forces are symmetric with respect to $x_3 = 0$ (this symmetry is material, and concerns the elasticity coefficients).

According to classical homogenization, we introduce the local variable $y = x/\varepsilon$. In fact we are in the situation of a layer (as x_3 is small ; thus functions depend on x_1, x_2, y_1, y_2, y_3 ; moreover as the "period" of the coefficients in the 1 and 2 directions is any value, the classical periodicity in y_1, y_2 implies that functions are independent of y_1, y_2 :

$$(1.2)\quad f^\varepsilon(x) = f^o(x_1,x_2,y_3) + \varepsilon f^1(x_1,x_2,y_3) + \ldots \; ; \; y_3 = x_3/\varepsilon$$

and the classical formula

$$(1.3)\quad \frac{d}{d\,x_i} = \frac{1}{\varepsilon}\frac{\partial}{\partial\, y_i} + \frac{\partial}{\partial\, x_i}$$

will be used on account of the fact that the derivatives with respect to x_3, y_1, y_2 vanish. Then, this amounts to the method of dilatation in x_3 of Ciarlet and Destuynder[14].

2. - BEHAVIOR FAR FROM THE BOUNDARY OF ω

The exact equation and boundary conditions are :

$$(2.1)\quad \frac{\partial\, \sigma^\varepsilon_{ij}}{\partial\, x_j} = 0 \quad ; \quad \sigma^\varepsilon_{ij} = a_{ijmn}\, e_{mn}(\underline{u}^\varepsilon) \quad \text{in} \quad \omega \times (-\varepsilon,+\varepsilon)$$

$$(2.2)\quad \sigma^\varepsilon_{i3} = 0 \qquad \text{for} \quad x_3 = \pm\,\varepsilon$$

We perform the expansions :

$$(2.3)\quad \underline{u}^\varepsilon(x) = \underline{u}^o(x,y) + \varepsilon \underline{u}^1(x,y) + \ldots$$

$$(2.4)\quad \begin{cases} e_{ij}(u^{\varepsilon}) = \varepsilon^{-1} e_{ij}^{-1} + e_{ij}^{o} + \varepsilon e_{ij}^{1} + \dots \\ e_{ij}^{-1} = e_{ijy}(u^{o}) \quad ; \quad e_{ij}^{o} = e_{ijx}(\underline{u}^{o}) + e_{ijy}(\underline{u}^{1}) \quad ; \quad \dots \end{cases}$$

$$(2.5)\quad \sigma_{mn}(u^{\varepsilon}) = \varepsilon^{-1} \sigma_{mn}^{-1} + \sigma_{mn}^{o} + \varepsilon\, \sigma_{mn}^{1} + \dots$$

where the successive terms are the products of a_{mnij} by the corresponding terms of (2.4). Thus, (2.1), (2.2) become :

$$(2.6)\quad \begin{cases} \left(\frac{1}{\varepsilon} \frac{\partial}{\partial y_j} + \frac{\partial}{\partial x_j}\right) \left(\varepsilon^{-1} \sigma_{ij}^{-1} + \sigma_{ij}^{o} + \dots\right) = 0 \\ \sigma_{i3}^{p} = 0 \qquad \text{for} \qquad y_3 = \pm 1 \end{cases}$$

<u>It is easily seen that $\underline{u}^{o}$ does not depend on y_3</u>. Indeed, at the leading order ε^{-2}, (2.6) gives :

$$(2.7)\quad \begin{cases} \dfrac{\partial \sigma_{ij}^{-1}}{\partial y_j} = 0 \quad ; \quad \sigma_{ij}^{-1} \equiv a_{ijmn}\, e_{mny}(\underline{u}^{o}) \\ \sigma_{i3}^{-1} = 0 \quad \text{at} \quad y_3 = \pm 1 \end{cases}$$

we then multiply the first of (2.7) by u_i^{o} and we integrate by parts ; using the symmetry with respect to the indexes of σ_{ij} and e_{ij} and the boundary conditions of (2.7) we have :

$$0 = \int_{-1}^{+1} \frac{\sigma_{i3}^{-1}}{\partial y_3} u_i^{o}\, dy_3 = - \int_{-1}^{+1} \sigma_{i3}^{-1} \frac{\partial u_i^{o}}{\partial y_3}\, dy_3 =$$

$$= - \int_{-1}^{+1} \sigma_{ij}^{-1} \frac{\partial u_i^{o}}{\partial y_j}\, dy_3 = - \int_{-1}^{+1} \sigma_{ij}^{-1} e_{ijy}(\underline{u}^{o})\, dy_3 =$$

$$= - \int_{-1}^{+1} a_{ijmn}(y_3)\, e_{mny}(\underline{u}^{o})\, e_{ijy}(\underline{u}^{o})\, dy_3$$

which shows (see (1.1)) that all the $e_{ijy}(\underline{u}^{o})$ vanish ; then, taking x as a parameter, $\underline{u}^{o}$ is a solid displacement ; as it only depends on y_3, it is a translation, i.e. $\underline{u}^{o}$ is independent of y, Q.E.D.

As a consequence, (2.3)-(2.5) take the simpler form :

$$(2.8) \quad \underline{u}^{\varepsilon} = \underline{u}^{o}(x_1,x_2) + \varepsilon \underline{u}^{1}(x_1,x_2,y_3) + \ldots$$

$$(2.9) \quad \begin{cases} e^{\varepsilon}_{ij} = e^{o}_{ij} + \varepsilon e^{1}_{ij} + \ldots \\ e^{o}_{ij} = e_{ijx}(\underline{u}^{o}) + e_{ijy}(\underline{u}^{1}) \; ; \ldots \end{cases}$$

$$(2.10) \quad \sigma^{\varepsilon}_{ij} = a_{ijmn}\, e^{\varepsilon}_{mn} = \sigma^{o}_{ij} + \varepsilon\sigma^{1}_{ij} + \ldots$$

As we are searching for solutions symmetric with respect to $x_3 = 0$, the component 3 of $\underline{u}^{o}$ vanishes ; denoting by $\underline{e}_i$ the unit vectors of the direction of the axes, we have

$$(2.11) \quad \underline{u}^{o} = u^{o}_1\, \underline{e}_1 + u^{o}_2\, \underline{e}_2 \equiv u^{o}_{\alpha}(x_1,x_2)\, \underline{e}_{\alpha}$$

where the repeated Greek indexes sum from 1 to 2 (we shall use this notation all over this section). In order to study the local behavior we consider, as usual, $e_{ijx}(\underline{u}^{o})$ as given constants ; of course, only the terms where i,j take the values 1,2 are non-zero ; we denote them by $E_{\alpha\beta}$:

$$(2.12) \quad E_{\alpha\beta} \equiv \frac{1}{2}\left(\frac{\partial u^{o}_{\alpha}}{\partial x_{\beta}} + \frac{\partial u^{o}_{\beta}}{\partial x_{\alpha}}\right)$$

Equation (2.6) at order ε^{-1} gives (note that $\sigma^{-1} \equiv 0$) :

$$(2.13) \quad \begin{cases} \dfrac{\partial \sigma^{o}_{ij}}{\partial y_j} = 0 \quad ; \quad \sigma^{o}_{ij} \equiv a_{ijmn}\left(E_{mn} + e_{mny}(\underline{u}^{1})\right) \\ \sigma^{o}_{i3} = 0 \quad \text{on} \quad y_3 = \pm 1 \end{cases}$$

which is a boundary value problem for $u^1(x_1,y_3)$ (of course, x is taken as a parameter and E_{mn} as data). A variational formulation for this problem is easily obtained in the space

$$(2.14) \quad V = \underline{H}^1(-1,+1)/R^3 \equiv (H^1(-1,+1)/R)^3$$

(note that $\underline{u}^1$ will be defined up to an additive constant vector). Multiplying equation (1.13) by v_i (where $\underline{v}$ is a test function of V) and integrating exactly as in the computations after (2.7) we obtain

$$(2.15)\begin{cases} \underline{u}^1 \in V \quad \text{and,} \quad \forall\ \underline{v} \in V \quad \text{we have} \\ \int_{-1}^{+1} a_{ijmn}\ e_{mny}(\underline{u}^1)\ e_{ijy}(\underline{v})\ dy_3 = -\ E_{mn} \int_{-1}^{+1} a_{ijmn}\ e_{ijy}(\underline{v})\ dy_3 \end{cases}$$

which is the variational formulation of the local problem (1.13) (as (1.13) is easily obtained from (1.15)). The <u>existence and uniqueness</u> of $\underline{u}^1$ immediately follows from (2.15).

On account of the linearity with respect to the data E_{mn}, the solution of (1.15) writes :

$$(2.16)\quad \underline{u}^1 = E_{\alpha\beta}\ \underline{w}^{\alpha\beta}$$

where $\underline{w}^{\alpha\beta}$ is the solution of

$$(2.17)\begin{cases} \underline{w}^{\alpha\beta} \in V \quad \text{and} \quad \forall\ \underline{v} \in V \quad \text{we have} \\ \int_{-1}^{+1} a_{ijmn}\ e_{mn}(\underline{w}^{\alpha\beta})\ e_{ij}(\underline{v})\ dy_3 = -\int_{-1}^{+1} a_{ij\alpha\beta}\ e_{ij}(\underline{v})\ dy_3 \end{cases}$$

<u>When the functions $\underline{w}^{\alpha\beta}(y_3)$ are known, the stress state</u> (at the leading order) <u>is given by</u> :

$$(2.18)\quad \sigma^o_{ij}(x,y) = a_{ijmn}\left(E_{mn} + E_{\alpha\beta}\ e_{mny}(\underline{w}^{\alpha\beta})\right) \equiv$$

$$\equiv E_{\alpha\beta}\left(a_{ij\alpha\beta} + a_{ijpq}\ e_{pqy}(\underline{w}^{\alpha\beta})\right)$$

From here we may obtain the mean values of the stress (denoted by the tilde $\sim$) across the thickness :

$$(2.19)\quad \tilde{\sigma}^o_{\alpha\beta} = \frac{1}{2}\int_{-1}^{+1}\sigma^o_{\alpha\beta}\ (x_1,x_2,y_3)\ dy_3 = \frac{1}{2\ \varepsilon}\int_{-\varepsilon}^{+\varepsilon} \sigma^o_{\alpha\beta}\ (x_1,x_2,\frac{x_3}{\varepsilon})\ dx_3 =$$

$$= a^h_{\alpha\beta\gamma\delta}\ E_{\gamma\delta}$$

where a^h are the "homogenized coefficients" :

$$(2.20)\quad a^h_{\alpha\beta\gamma\delta} = \frac{1}{2}\int_{-1}^{+1} \left(a_{\alpha\beta\gamma\delta} + a_{\alpha\beta pq}\, e_{pqy}(\underline{w}^{\gamma\delta})\right) dy_3$$

The "macroscopic equations" are easily obtained from (2.6) at order O(1) :

$$(2.21)\quad \frac{\partial \sigma^1_{i3}}{\partial y_3} + \frac{\partial \sigma^o_{i\alpha}}{\partial x_\alpha} = 0$$

and integrating in y_3 from -1 to +1 and using the boundary conditions of (2.6), the term σ^1 vanishes ; the equation for $i = 1,2$ gives

$$(2.22)\quad \frac{\partial \tilde{\sigma}^o_{\beta\alpha}}{\partial x_\alpha} = 0$$

Consequently at the leading order we have a "plane elasticity problem" : the equations are (2.22), and on account of (2.19), (2.12),they are elasticity equations for $\underline{u}^o(x_1,x_2)$ with coefficients $a^h_{\alpha\beta\gamma\delta}$. The strain and stress fields are given by (2.9), (2.17), (2.18) and they depend on y_3.

Remark 2.1 - The coefficients $a^h_{\alpha\beta\gamma\delta}$ satisfy symmetry and positivity conditions analogous to (1.1). This is immediately proved by taking in the definition (2.17) of $\underline{w}^{\alpha\beta}$ (resp. $\underline{w}^{\gamma\delta}$) the test function $\underline{v} = \underline{w}^{\gamma\delta}$ (resp. $= \underline{w}^{\alpha\beta}$). ■

Remark 2.2 - We saw that the problem reduces to computing the functions $\underline{w}^{\alpha\beta} \in V$; each of them is a triplet of functions $w^{\alpha\beta}_i$ belonging to $H^1(-1,+1)$, defined up to an additive constant. Of course (2.17) amounts to

$$(2.23)\quad \frac{\partial}{\partial y_3}\left(a_{i3\alpha\beta} + a_{i3mn}\, e_{mny}(\underline{w}^{\alpha\beta})\right) = 0$$

in the sense of distributions (i.e. the jump of the bracket in (2.23) at the discontinuities of the coefficients vanishes) with boundary conditions at $y_3 = \pm 1$ which amounts to the vanishing of the bracket in (2.23), and this gives

$$(2.24) \quad a_{i3\alpha\beta} + a_{i3mn}\, e_{mn}(\underline{w}^{\alpha\beta}) = 0. \blacksquare$$

Remark 2.3 - In a layered plate, $\underline{w}^{\alpha\beta}(y_3)$ (and then $\underline{u}^1$) has a constant slope inside each layer ; consequently at the leading order the stress and strain fields are constant in each layer. This follows from (2.23) : in each region where the coefficients are constant, (2.23) becomes :

$$a_{i33m} \frac{d^2 w_m^{\alpha\beta}}{d\, y_3^2} = 0 \quad \Longrightarrow \quad \frac{d^2 w_m^{\alpha\beta}}{d\, y_3^2} = 0$$

because the matrix a_{i33m} is definite positive from (1.1). ■

3. - BEHAVIOR AT THE BOUNDARY OF ω

We refer to Fig. 1.1. Let $\partial\omega$ (or a part of it) be a free boundary, the outer stress is zero for $x \in \partial\omega \times (-\varepsilon,+\varepsilon)$. It is easily seen (by a method of flux, for instance) that the average stress $\tilde{\sigma}^o_{\alpha\beta}$ must satisfy

$$(3.1) \quad \tilde{\sigma}^o_{\alpha\beta}\, N_\beta = 0 \qquad \text{on } \partial\Omega, \ \alpha = 1,2.$$

where $\underline{N} = (N_1,N_2)$ denote the outer unit normal to $\partial\omega$ in its plane. Then, the stress tensor field $\sigma^o_{ij}(y_3)$ given by (2.18) enjoys the following property which will be useful in the study of the layers :

PROPOSITION 3.1 - Let condition (3.1) be fulfilled at some point (x_1,x_2) of $\partial\omega$. Then, the field of vectors with components $\sigma^o_{i\alpha}(y_3)N_\alpha$, $i = 1,2,3$ applied to the points $(x_1,x_2,\varepsilon y_3)$ of the vertical fiber associated with (x_1,x_2) have a resultant and moment equal to zero.

Proof - We first consider the resultant

$$(3.2) \quad \int_{-1}^{+1} \sigma^o_{i\alpha}\, N_\alpha \, dy_3 = 2N_\alpha \, \tilde{\sigma}^o_{i\alpha}$$

and (3.1) shows that the components 1 and 2 of this resultant vanish. As for the component $i = 3$, we see from (2.13) that $\sigma^0_{3j} \equiv 0$. We now consider the moment with respect to the point $(x_1, x_2, 0)$:

$$(3.3) \qquad \underline{M} = \int_{-1}^{+1} (\sigma^0_{i\alpha} N_\alpha \underline{e}_i) \wedge y_3 \underline{e}_3 \, dy_3 \quad ;$$

the component $i = 3$ vanish obviously ; for $i = 1,2$ it suffices to note that $\sigma^0_{\beta\alpha}$ take the same value at $+y_3$ and $-y_3$ by the symmetry of the problem. ■

In order to study the boundary layer at a point of $\partial\omega$ we take local axes y_1, y_2, y_3 as in Fig. 3.1 : y_1 is tangent to $\partial\omega$ and y_2 normal to $\partial\omega$, towards ω (then $\underline{N} = (0,-1,0)$). We note that they are associated with the local variables $y = x/\varepsilon$, as in the preceeding section, but now they have special directions ; in order to study the boundary layer we must (eventually) perform a change from the axes of sect. 2 to the present ones. Of course, $\sigma^0_{i3}(y_3)$ (and other analogous symbols) shall denote the expressions of sect. 2 for x_1, x_2 at the point of $\partial\omega$ which we study, and in the axes of Fig. 3.1. Moreover, under the assumption that $\partial\omega$ is a smooth curve in vicinity of the considered region, for small $|y|$ we may replace it by the axis Oy_1. According to the general consideration of lectures 1 and 2, we must replace

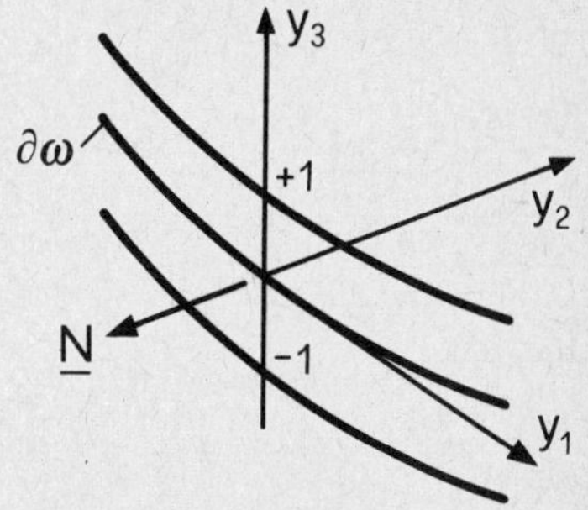

Figure 3.1

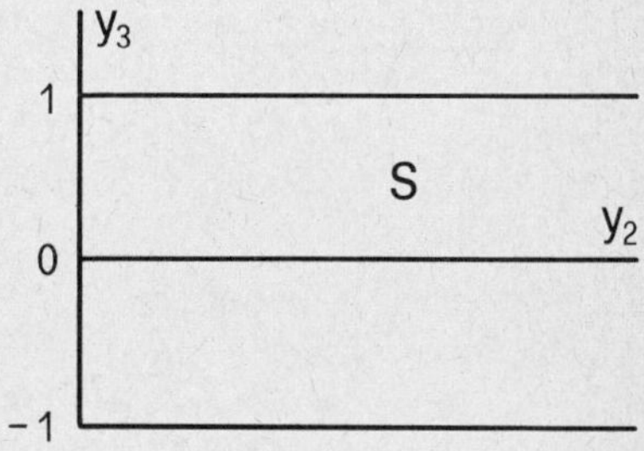

Figure 3.2

$u^1(y)$ by $u^1(y) + u^{1c}(y)$ where u^{1c} is a complementary term independent of y_1 (i.e. "periodic in y_1 of any period"), defined in the semi-infinite strip S of Fig. 3.2. The equations for $u^1 + u^{1c}$ are the same as for u^1, i.e. (2.13), as well as the boundary conditions, and in addition the corresponding stresses on the face $y_2 = 0$ must vanish. This gives the boundary layer problem for $u^{1c}(y_2, y_3)$ defined on S :

$$(3.4) \quad -\frac{\partial \sigma_{ij}(\underline{u}^{1c})}{\partial y_j} = 0 \quad ; \quad \sigma_{ij}(\underline{u}^{1c}) \equiv a_{ijmn}\, e_{mny}(\underline{u}^{1c}) \quad ; \quad \text{in } S$$

$$(3.5) \quad \sigma_{3j}(\underline{u}^{1c}) = 0 \qquad \text{on} \quad y_3 = \pm 1$$

$$(3.6) \quad \sigma_{2j}(\underline{u}^{1c}) = -\sigma^{o}_{2j}(y_3) \qquad \text{on} \quad y_2 = 0$$

$$(3.7) \quad e_{ijy}(\underline{u}^{1c}) \xrightarrow[y_2 \to +\infty]{} 0$$

where the last equation expresses that the strain (and consequently the stress) associated with $\underline{u}^{1c}$ vanishes far from $\partial\omega$. We note that $\underline{u}^{1c}$ is (at most) defined up to a rigid displacement. Of course, the corresponding stresses must be added to σ^{o}_{ij} in order to have the actual stresses near the boundary (at the leading order in ε).

We define the space W formed by the three dimensional vector fields $\underline{w}(y_2, y_3)$ defined on S (i.e. functions from S into R^3), defined up to a rigid displacement, with strain tensor $e_{ij}(\underline{v})$ belonging to $L^2(S)$, which is a Hilbert space for the scalar product

$$(3.8) \quad (\underline{w}, \underline{v})_W = \int_S e_{ij}(\underline{w})\, e_{ij}(\underline{v})\, dy_2\, dy_3$$

It is then easily seen that problem (3.4)-(3.7) amounts to the variational formulation :

Find $\underline{u}^{1c} \in W$ such that

$$(3.9) \qquad (\underline{u}^{1c},\underline{v})_W = \int_{-1}^{+1} \sigma^{o}_{2j}(y_3)\, v_j(0,y_3)\, dy_3 \qquad \forall\, \underline{v} \quad W .$$

The existence and uniqueness of the solution of (3.9) follows from the Lax-Milgram theorem. The fact that the right side of (3.9) defines a linear functional on W (i.e. takes the same value for vector fields the differences of which is a rigid displacement) follows from Proposition 3.1. This functional is also continuous (use the Korn inequality in a bounded domain).

Remark 3.2 - In this section (and in the preceeding one) we presented $\underline{u}^{1c}$ and $\underline{u}^{1}$ as local solutions in the framework of the asymptotic methods in the vicinity of some point (x_1,x_2). Nevertheless, they are also exact solutions for a semi-infinite plate with stresses independent of x_1,x_2. ■

Remark 3.3 - Numerical computations of the layers offer no difficulties, and truncation may be used, as in sect. 2.4. ■

4. - THE PARTICULAR CASE OF ISOTROPIC LAYERS

We consider here the particular case where the plate is made of an isotropic elastic body with Lamé constants $\lambda(y_3)$, $\mu(y_3)$; in the case of homogeneous layers, these functions are piecewise constant.

The local solutions $\underline{w}^{\alpha,\beta}(y_3)$ are immediately obtained from (2.24) :

$$(4.1)\begin{cases} \underline{w}^{11}(y_3) \equiv \underline{w}^{22}(y_3) = \displaystyle\int_{o}^{y_3} \frac{-\lambda(\xi)d\xi}{\lambda(\xi)+2\mu(\xi)}\, \underline{e}_3 \\ \underline{w}^{12}(y_3) \equiv 0 \end{cases}$$

The homogenized coefficients (2.20) are :

$$
(4.2)\quad \begin{cases} a^h_{1111} = a^h_{2222} = \dfrac{1}{2} \displaystyle\int_{-1}^{+1} \dfrac{4\mu(\lambda+\mu)}{\lambda + 2\mu}\, d\xi \\[2ex] a^h_{1212} = \dfrac{1}{2} \displaystyle\int_{+1}^{-1} \mu(\xi)\, d\xi = \tilde{\mu} \quad ; \quad a^h_{1122} = a^h_{1111} - 2\tilde{\mu} \end{cases}
$$

where, as usual $\tilde{\mu}$ denotes the mean value between -1 and +1. This amounts to a two-dimensional isotropic medium in plane deformation with the homogenized Lamé constants

$$
(4.3)\quad \lambda^h = \frac{1}{2} \int_{-1}^{+1} \frac{2\,\lambda\,\mu}{\lambda + 2\mu}\, d\xi \quad ; \quad \mu^h = \tilde{\mu}
$$

If the plate is submitted to a unit strain in the direction x_1 :

$$
(4.4)\quad E_{11} = 1 \quad ; \quad \text{other } E_{\alpha\beta} = 0 \quad (\alpha,\beta = 1,2)
$$

(where the notation (2.12) is used), the corresponding stresses (2.18) are :

$$
(4.5)\quad \begin{cases} \sigma^o_{i3} \equiv 0 \quad \text{for } i = 1,2,3 \quad ; \quad \sigma^o_{12} \equiv 0 \\[2ex] \sigma^o_{11}(y_3) \equiv \dfrac{4\mu(\lambda+\mu)}{\lambda + 2\mu}(y_3) \\[2ex] \sigma^o_{22}(y_3) \equiv \dfrac{2\,\mu\,\lambda}{\lambda + 2\mu}(y_3) \end{cases}
$$

and the corresponding mean values of the stress are :

$$
(4.6)\quad \tilde{\sigma}^o_{11} = a^h_{1111} \quad ; \quad \tilde{\sigma}^o_{22} = a^h_{1122} \quad ; \quad \tilde{\sigma}^o_{12} = 0
$$

If we consider a <u>unit mean stress in the direction x_1</u>, according to (4.6) this amounts to give the strains

$$
(4.7)\quad E_{11} = \frac{a^h_{1111}}{(a^h_{1111})^2 - (a^h_{1122})^2} \quad ; \quad E_{22} = \frac{-\,a^h_{1122}}{(a^h_{1111})^2 - (a^h_{1122})^2} \quad ; \quad E_{12} = 0
$$

and the corresponding stress field σ^o is immediately deduced from

(4.5) and from that obtained by permutation of the indexes 1,2. We note that they are constant in each layer (where λ , μ are constants) and $\sigma^o_{12} \equiv \sigma^o_{32} \equiv 0$.

The corresponding boundary layer is given by (3.4)-(3.7), and $\underline{u}^{1c}(y_2,y_3)$ is of course defined up to a rigid displacement. The data of the problem are (3.6), which amounts to the above mentioned stresses on $y_2 = 0$. They are normal stresses, constant on each layer (Fig. 4.1).

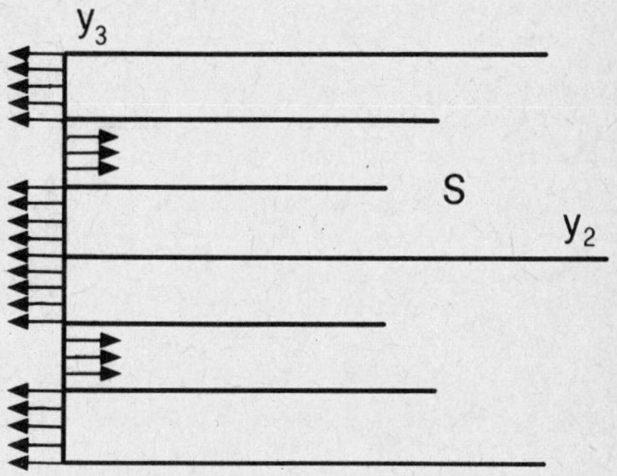

Figure 4.1

It follows from symmetry with respect to y_1 that we have (up to a rigid displacement) $u_1^{1c} \equiv 0$, i.e., $\underline{u}^{1c}$ may be taken to be a plane displacement in the plane y_2, y_3. For ulterior analysis of the singularities we note that the displacement vector may be written:

$$(4.8) \quad \underline{u}^{\varepsilon}(x) = \underline{u}^{o}(x) + \varepsilon\left(\underline{u}^{1}(x,y) + \underline{u}^{1c}(x,y)\right) + O(\varepsilon^2)$$

and the corresponding strain at the first order :

$$(4.9) \quad e^{\varepsilon}_{ij} = e_{ijx}(\underline{u}^{o}) + e_{ijy}(\underline{u}^{1} + \underline{u}^{1c})$$

where the terms $e_{ijx}(\underline{u}^o)$ are the E_{ij} of (4.7) (the notation (2.12) is used). This strain is the same as that of the vector field

$$(4.10) \quad \underline{u}^{*}(y) = E_{11}\, y_1\, \underline{e}_1 + E_{22}\, y_2\, \underline{e}_2 + \underline{u}^{1} + \underline{u}^{1c}$$

(incidentely we see that this is the asymptotic form of (4.8) for small x, written in y note the similitude with 2 (2.10)). Moreover, on account of the fact that $\underline{u}^1$ is a combination of the $\underline{w}^{\alpha\beta}$ given by (4.1), we see that $\underline{u}^1$ has the direction y_3 (and of course it depends only on y^3), $\underline{u}^*(y)$ may be written

(4.11) $\underline{u}^*(y) = E_{11}\, y_1\, \underline{e}_1 + \underline{v}(y_2, y_3)$ with $v_1 = 0$

(i.e. $\underline{v}$ is a displacement in the plane y_2, y_3. Of course, $\underline{u}^*(y)$ is defined in the strip S (Fig. 4.1) where it satisfies the elasticity system and the homogeneous boundary conditions

(4.12) $\sigma_{ij}(\underline{u}^*)n_j = 0$ on $y_2 = 0$ and $y_3 = \pm 1$.

C H A P T E R 4

4. - SINGULARITIES IN ELLIPTIC NON SMOOTH PROBLEMS

1. - INTRODUCTION

Most of the solutions of problems in mathematical physics are given by variational problems in spaces of the kind H^1 of Sobolev, i.e. they exist and are unique in spaces of functions having square integrable first order derivatives. This is a very poor regularity, and such solutions may be singular at some points ; more precisely, grad u (where u is the considered solution) may tend to infinity at some points.

Physically speaking, such solutions are meaningless at the vicinity of such singularities : in fact, the smallness hypotheses for linearization are not fulfilled. Then, such singularities show that new phenomena (non linearities, qualitative modification of the medium, etc) may appear. An example is the lightning rod. The singularities of grad u (u is a harmonic function, the electric potential) at the point O provokes ionization of the air, which becomes conducting near O. Another example is the rear edge of an airfoil (Fig. 1.2, a) and b)) :

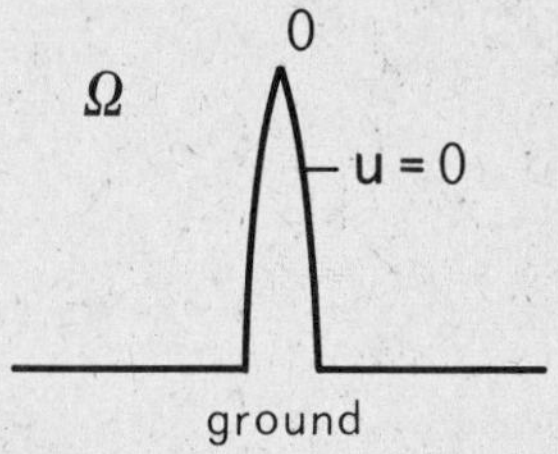

Figure 1.1

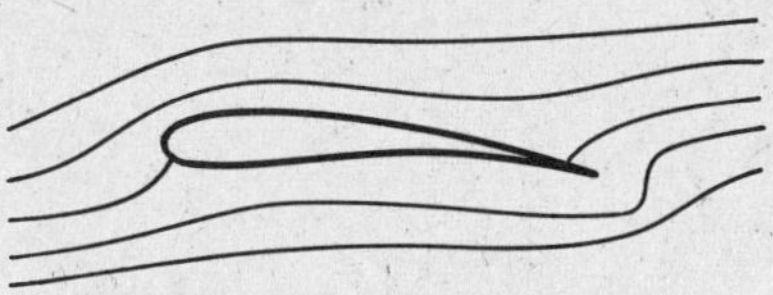

Figure 1.2.a)

Figure 1.2.b)

if the rear edge 0 of the airfoil is a turning point of the fluid flow, we have a singularity (in particular the pressure $\to -\infty$) : instead of this, there is a general modification of the flow, which leaves the airfoil precisely at 0, and a circulation around the airfoil appears ; this produces the lift force of the airfoil (Kutta-Joukovski). In elasticity theory, infinite values of grad u i.e. singularities of strain and stress provoke modifications of the elastic behavior : depending on the nature of the material, it may become plastic or a fracture may appear.

The study of singularities is well developped for second order elliptic equations in R^2 but there is much to do concerning problems in R^3 and elliptic systems, in particular the elasticity system. Fortunately, some problems in boundary layers are in fact in R^2, and we have at our disposal some (but not all) tools to study them. The principal references on these problems are Grisvard[15,16], Lemrabet[17,18], Kondratiev[19,20], Sanchez[21,22], Sovin[23] and for numerical computation, Lelièvre[23,24].

Let us consider an elliptic problem of the form

$$(1.1) \qquad - \frac{\partial}{\partial x_j} \left(a_{ij}(x) \frac{\partial U}{\partial x_i}\right) = f$$

in a domain Ω of R^2 with appropriate boundary conditions. Under suitable smoothness hypotheses about the coefficients and $\partial\Omega$, classical regularity theory holds. In particular, if f belongs locally (i.e. on a domain D which may go up to the boundary) to the H^m space (m real ≥ 0) and the boundary conditions are homogeneous (i.e. = 0) the solution

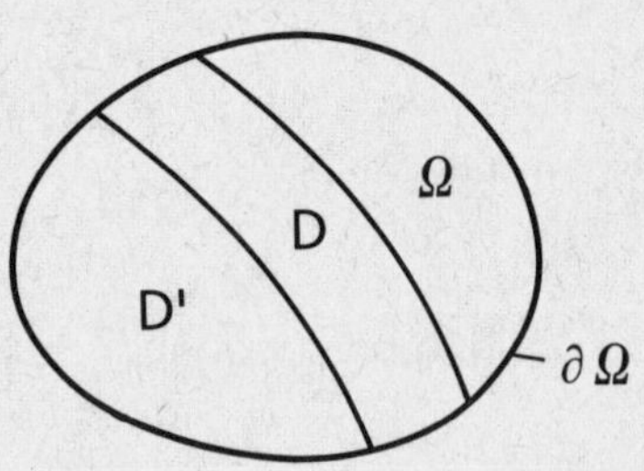

Figure 1.3

belongs to H^{m+2} in any subdomain D' included in D. Moreover, if Ω is bounded, an inequality of the type

$$(1.2) \qquad \|U\|_{H^{m+2}(\Omega)} \leq C\, (\|f\|_{H^m(\Omega)} + \|U\|_{H^m(\Omega)})$$

holds for $m \geqq 0$, with a constant C which depends only on Ω, m and the coefficients of the equation.

An analogous situation holds if the coefficients of (1.1) are piecewise smooth, having a discontinuity line Γ where (1.1) is considered in the distributtion sense :

$$(1.3) \qquad [U] = 0 \ , \quad \left[a_{ij} \frac{\partial U}{\partial x_j} n_i\right] = 0 \quad \text{on } \Gamma$$

then, if $f \in H^m$, the solution U belongs to H^{m+2} on each side of Γ (Fig. 1.4) (of course on Γ itself the solution is not of class H^n, $n \geqq 2$ as the first derivatives are not continuous across Γ) (see Ladyzhenskaya et Ouralceva[26], sect. III.16) in regions where Γ is smooth, buth not (as we shall see later) at points as A,B in Fig. 1.4. For instance, in problems with layers, as in sect. 2.1 and 3.2, far from the boundary, but singularities may appear at the intersection of layers with the boundary $\partial\Omega$.

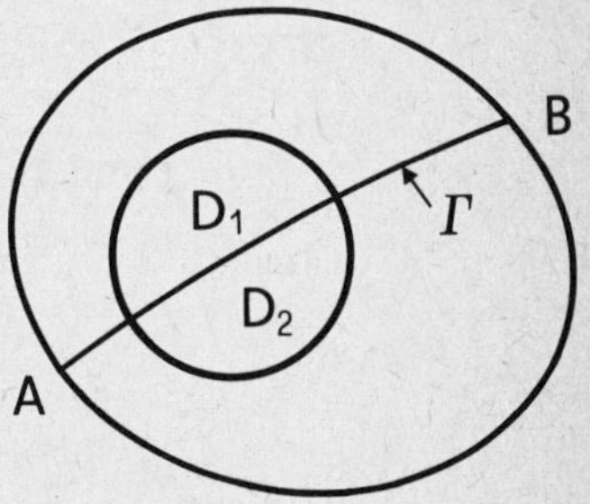

Figure 1.4

A different situation appears if the coefficients are smooth (constant, say) but the boundary $\partial\Omega$ does not, in particular if it has angular points. In such a case, the local regularity depends on the angle ϕ of the domain.

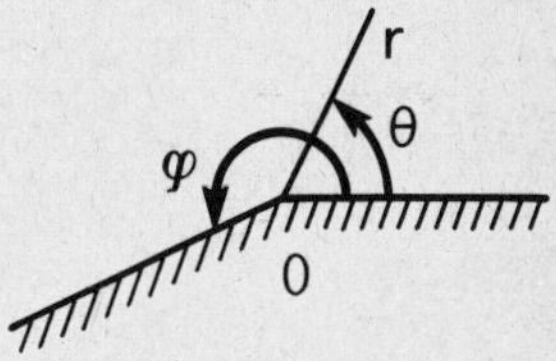

Figure 1.5

For instance, let consider the Laplace equation with Neumann condition :

$$(1.4) \quad -\Delta U = 0 \quad ; \quad \frac{\partial U}{\partial n} = 0$$

(in fact the right side is zero in a neighbourhood of the origin but may be $\neq 0$ elsewhere). We search for solutions of the form (r,θ = polar coordinates)

$$(1.5) \quad U(x_1,x_2) = r^{\alpha} u(\theta)$$

and we obtain for $u(\theta)$:

$$(1.6) \quad -u'' - \alpha^2 u = 0 \quad ; \quad u' = 0 \quad \text{for} \quad \theta = 0,\phi$$

with the solutions

$$(1.7) \quad u = A \cos \alpha\theta \quad \text{for} \quad \alpha = 0, \pm\frac{\pi}{\phi}, \quad \frac{2\pi}{\phi}, \ldots$$

Of course, grad U behaves as $r^{\alpha-1}$; we are interested by solutions exhibiting a singularity as $r \to 0$, i.e. grad $U \to \infty$ as $r \to 0$ and this amount to $\text{Re}\ \alpha - 1 < 0$. On the other hand, if the solution exists according to a variational problem in $H^1(\Omega)$, grad $u \in \underline{L}^2(\Omega)$ and this implies $\text{Re}\ \alpha > 0$. We see from (1.7) that such singular solutions exist if $\phi \in (\pi,2\pi)$, i.e. if the domain is not convex, but they do not exist if $\phi \in (0,\pi)$, i.e. if Ω is convex. A picture of the flux lines (i.e. lines tangent to grad U) furnishes some insight on the physical phenomenon : for a convex (resp. non-convex) domain the flux lines spread out (resp. push to each other) as shown in Fig. 1.6.a), b).

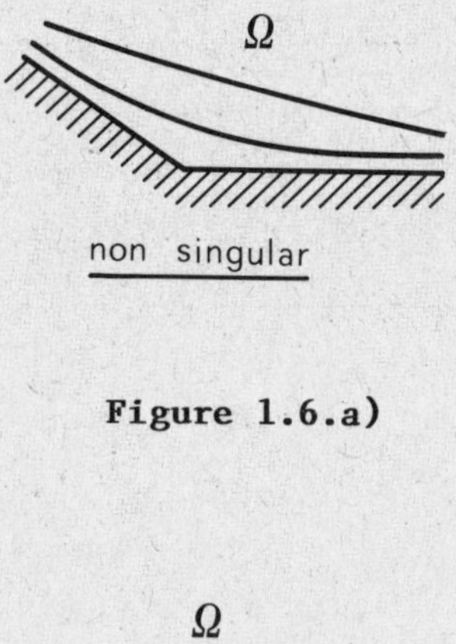

Figure 1.6.a)

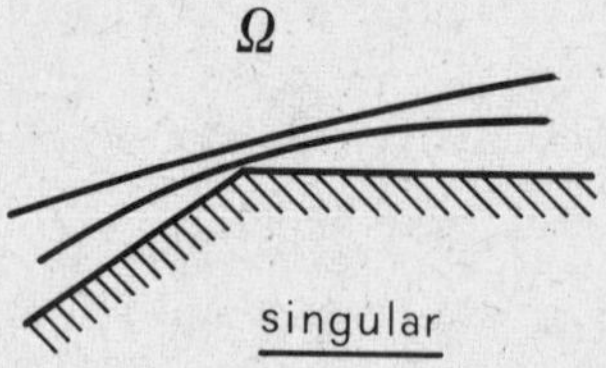

Figure 1.6.b)

Remark 1.1 - Singular solutions of the form (1.5) (i.e. with $0 < \text{Re}\,\alpha < 1$) furnish counter-examples to the regularity (1.2) but moreover, the theory (see Grisvard[15]) shows that they are the only singularities and in fact, if $u \in H^1(\Omega)$ is a solution, roughly speaking regularity holds for

$$U - \sum c_i\, r^{\alpha_i}\, u^i(\theta)$$

where Σ is extended to the singular solutions, and the coefficients c_i depend on U.

2. - SINGULARITIES AT THE BOUNDARY FOR TRANSMISSION PROBLEMS

We now consider the case where the interface Γ in the transmission problem (1.1), (1.3) touches $\partial\Omega$. We shall see that the convexity criterion for the Laplace equation (Fig. 1.6) becomes now a convexity with respect to the refracted fluxes.

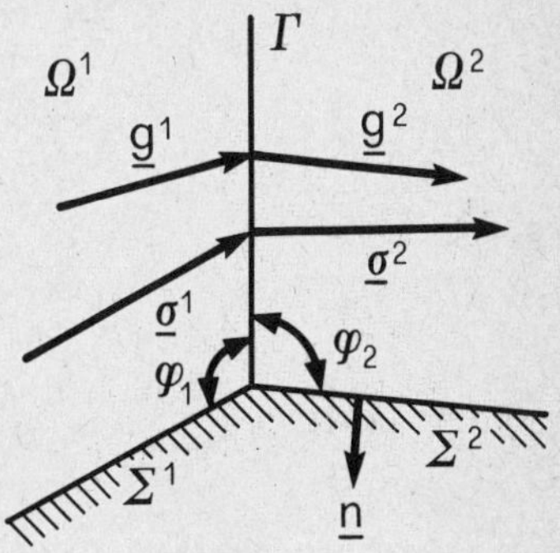

Figure 2.1

Let us consider to fix ideas, the transmission problem (1.1) with piecewise constant coefficients, the interface conditions across a line Γ of discontinuity of the coefficients being of course (1.3). Moreover, we consider Neumann boundary conditions

$$(2.1) \qquad a_{ij}\,\frac{\partial U}{\partial x_i}\,n_j = 0 \qquad \text{on } \partial\Omega$$

We are studying the vicinity of a point O where Γ intersects $\partial\Omega$ (Fig. 2.1). Let Ω^1 and Ω^2 be (in the vicinity of O) the two subdomains where the coefficients are constant. It will prove useful writing the equation and boundary conditions in terms of the vectors gradient $\underline{g}$ and the flux $\underline{\sigma}$, defined by :

(2.2) $g_i = \dfrac{\partial U}{\partial x_i}$; $\sigma_i = a_{ij} \dfrac{\partial U}{\partial x_j} \equiv a_{ij}\, g_j$

then, (1.1), (2.1) become :

$$(2.3)\quad \begin{cases} \operatorname{div} \underline{\sigma} = 0 \quad , \quad (\underline{\sigma} \equiv a\, \underline{g}) \text{ in } \Omega \\ \sigma_i\, n_i = 0 \quad \text{on } \partial\Omega \end{cases}$$

and of course the transmission conditions (1.3) become (the first is obtained by differentiation of the first (1.3) along Γ) :

(2.4) $[g_t] = 0$; $[\sigma_n] = 0$

where the indexes t,n denote "tangential" and "normal" components to Γ.

Solutions with constant gradient on each of the regions Ω^1, Ω^2 are associated with g and $\underline{\sigma}$ taking constant values $\underline{g}^i$, $\underline{\sigma}^i$ in Ω^i, $i = 1,2$. We shall say that $\underline{g}^2$, $\underline{\sigma}^2$ are the "refracted" of $\underline{g}^1$, $\underline{\sigma}^1$. To construct such solutions, we give arbitrarily either $\underline{g}^1$ or $\underline{\sigma}^1$ (the other is then obtained by

(2.5) $\sigma_i = a_{ij}\, g_j$

with the values of a_{ij} on Ω^1). Then, the two relations (2.4) and the two (2.5) with the values of a_{ij} on Ω^2 furnish uniquely the refracted vectors $\underline{g}^2$, $\underline{\sigma}^2$.

Now, coming back to Fig. 2.1, let us suppose that $\underline{\sigma}^1$ and the refracted $\underline{\sigma}^2$ are respectively parallel to the portions of $\partial\Omega$ in contact with Ω^1, Ω^2 (denoted by Σ^1, Σ^2). In this case, the Neumann boundary condition (i.e. the second of (2.3)) is satisfied. We then have the analogous, for equation (1.1), of the solution of constant gradient parallel to a straight boundary for the Laplace equation. We may gess (and we shall prove) that the presence of singularities is associated with non-convexity with respect to the line formed by $\underline{\sigma}^1$, $\underline{\sigma}^2$. Precisely :

PROPOSITION 2.1 - In the framework of this section (in particular Fig. 2.1), the Neumann problem (1.1), (2.1) has (resp. has not) a singularity at the point 0 of Fig. 2.1 (i.e. there exists a solution of the form (1.5) with $0 < \text{Re}\,\alpha < 1$) if when constructing a flux vector $\underline{\sigma}^1$ parallel to the portion of $\partial\Omega$ adjacent to Ω_1, pointing to 0 (see Fig. 2.2.a) and b)), the refracted vector $\underline{\sigma}^2$ is inside (respectively out of) Ω.

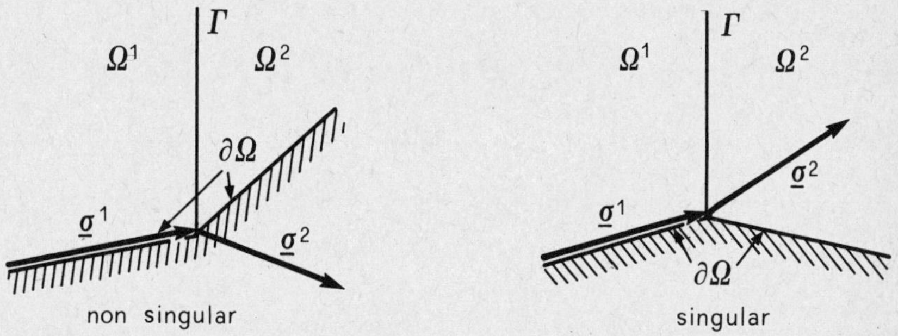

Figure 2.2

Remark 2.2 - In the singularity case, α is real and its value (as well as the corresponding function $u(\theta)$) may be computed by the method given in the forthcoming proof of Proposition 2.1. ■

Remark 2.3 - In the case of a Dirichlet (homogeneous) boundary condition $U = 0$ on $\partial\Omega$, Proposition 2.1 must be modified by taking instead of $\underline{\sigma}^1, \underline{\sigma}^2$ vectors normal to $\underline{g}^1$, $\underline{g}^2$. Of course in the case of Dirichlet and Neumann boundary condition in two adjacent portions of $\partial\Omega$ we shall take the normal to $\underline{g}^1$ and $\underline{\sigma}^2$. ■

Proof of Proposition 2.1 - Generally speaking the Neumann problem is associated with the bilinear form

(2.6) $\int_{\Omega} a_{ij} \frac{\partial u}{\partial x_j} \frac{\partial v}{\partial x_i} dx$

for $u,v \in H^1(\Omega)$. Performing a <u>linear</u> transformation $x \to x'$ on <u>each</u> of the subdomains Ω^1, Ω^2, the problem becomes another one with different angles and coefficients, but solutions with piecewise constant <u>g</u>, <u>σ</u> become solutions of the same class, and the Neumann and transmission conditions are preserved by the transformation (as they amount to belonging to H^1). Then, we take Γ as axis Ox_2 and we perform the transformation

$$(2.7)\quad \begin{cases} x'_1 = b_1 x_1 \quad ; \quad x'_2 = b_2 x_1 + x_2 \quad \text{with} \\ b_1 = a_{11}^{-1}(a_{11} a_{22} - a_{12}^2)^{\frac{1}{2}} \quad ; \quad b_2 = - a_{12} a_{11}^{-1} \end{cases}$$

on <u>each</u> of the regions Ω^1, Ω^2 (of course the coefficients a_{ij} are those of the corresponding region). This transformation preserves the Ox_2 axis and transforms (2.6) into the isotropic form :

$$(2.8)\quad \begin{cases} \int_{\Omega^1} c_1 \frac{\partial u}{\partial x_i} \frac{\partial v}{\partial x_i} dx + \int_{\Omega^2} c_2 \frac{\partial u}{\partial x_i} \frac{\partial v}{\partial x_i} dx \quad \text{with} \\ c = (a_{11} a_{22} - a_{12}^2)^{\frac{1}{2}} \end{cases}$$

Then, it suffices proving the proposition in the case (2.8), where the equation is the Laplacian on each Ω^i. Searching for solutions of the form (1.5) amounts to

$$u = \left\{ \begin{array}{ll} r^\alpha \cos \alpha\theta & \text{in } \Omega^1 \\ r^\alpha \cos \alpha(\phi_1 + \phi_2 - \theta) & \text{in } \Omega^2 \end{array} \right\} \text{ with}$$

$$(2.9)\quad c_1 \tan \alpha\ \phi_1 = - c_2 \tan \alpha\ \phi_2$$

where ϕ_1, ϕ_2 are the openings of Ω^1, Ω^2 (see Fig. 2.1) <u>after</u> transformation (2.7). The problem amounts to <u>discussing the existence of solutions $0 < \alpha < 1$ of (2.9) for given c_1, c_2, ϕ_1, ϕ_2</u>. We take $\phi_1 < \phi_2$

and we define $\omega = \alpha\,\phi_2$ ===>

(2.10) $\tan\omega = -\frac{c_1}{c_2}\tan\left(\frac{\phi_1}{\phi_2}\omega\right) \quad ; \quad \frac{\phi_1}{\phi_2} < 1$

the solutions ω are the abscissas of the intersection of the curves in the left and right sides of (2.10) (Fig. 2.3). In order to compare the solutions of (2.10) with the statement of the proposition, we define for the given values of c_1, c_2, ϕ_1 the angle $\phi_2 = \phi_2^r$ associated with the refraction across Γ, i.e. :

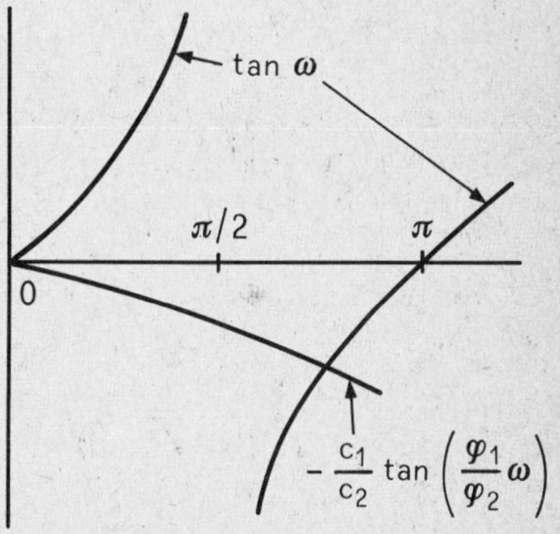

(2.11) $\tan\phi_2^r = -\frac{c_1}{c_2}\tan\phi_1$

Figure 2.3

which shows that for $\phi_2 = \phi_2^r$ we have $\alpha = 1$. Coming back to Fig. 2.3, we see that there is always a solution $\omega \in (\pi/2, \pi)$; if $\phi_2 > \pi$ this gives $\alpha \in (0,1)$, in agreement with the proposition. Moreover, if $\phi_2 < \pi/2$, we have $\alpha > 1$ (no singularity, also in agreement with the proposition). It only remains to study the case

(2.12) $\phi_1 < \phi_2 \quad ; \quad \pi/2 < \phi_2 < \pi.$

Let us consider (2.12) with $\phi_1 < \pi/2$. Then, for the given ϕ_1, ϕ_2 we construct c_1/c_2 according to the refraction law (2.11) with $\phi_2^r = \phi_2$; for this c_1/c_2 we have $\alpha = 1$. But from Fig. 2.3 we see that ω, and then α is a decreasing function of c_1/c_2, and we have or not singularity according to the statement of the proposition. In the case $\phi_1 > \pi/2$ we see (as a limit case of the preceeding one), there is a solution $\alpha \in (0,1)$. Then, for given c_1, c_2, ϕ_2, we see in fig. 2.3 that ω and then α is a decreasing function of ϕ_2, which remains > 0, and we have a singularity according to the proposition. ■

Remark 2.4 - In the preceeding proof only real solutions α where considered, but in fact there are not complex solutions of (2.9). To see this it suffices to verify that the solutions α are such that α^2 are the eigenvalues of the selfadjoint eigenvalue problem

$$(2.13)\begin{cases} \dfrac{d}{d\,\theta}\left(c(\theta)\dfrac{d\,w}{d\,\theta}\right) = \alpha^2\, c(\theta)\, w & \theta \in (0, \phi_1+\phi_2) \\ \dfrac{d\,w}{d\,\theta} = 0 \quad \text{for} \quad \theta = 0 \ , \quad \theta = \phi_1 + \phi_2 \ . \end{cases}$$

3. - SINGULARITIES IN PLANE ELASTICITY FOR A HOMOGENEOUS BODY

Elasticity problems are much more complex than the preceeding ones. Even in the case of an isotropic homogeneous body, complex solutions may appear , and a physical interpretation as in Fig. 1.6 is not known.

We consider the two-dimensional elasticity system in the plane x_1,x_2 with constant Lamé coefficients :

$$(3.1) \qquad \frac{\partial\,\sigma_{ij}}{\partial\,x_j} = 0 \qquad ; \qquad \sigma_{ij} = \delta_{ij}\,\lambda\,\mathrm{div}\,\underline{U} + 2\,\mu\, e_{ij}\,(\underline{U})$$

where of course the indexes run in 1,2. Here $\underline{U}$ has two components depending on x_1,x_2. This is the plane deformation problem, but it is known that the plane constraint problem has the same form, with modified coefficients. We denote by ϕ the angle of $\partial\Omega$ at some point O. Let Σ_1, Σ_2 be the parts of $\partial\Omega$ adjacent to O. The existence of singularities depends on the boundary condition (Dirichlet or Neumann) :

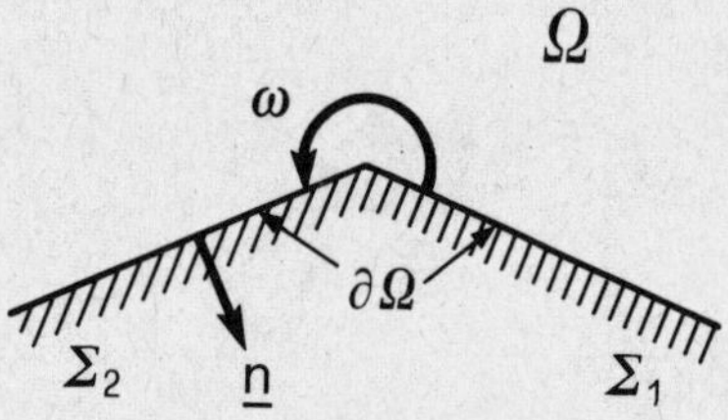

Figure 3.1

$$(3.2) \qquad \underline{U} = 0 \qquad\qquad \text{(Dirichlet)}$$

$$(3.3) \qquad \sigma_{ij}(\underline{U})\, n_j = 0 \qquad\qquad \text{(Neumann)}$$

on each Σ_i.

As in the preceeding sections, singularities of the strain or stress of solutions $\underline{U} \in \underline{H}^1$ are solutions of the form

(3.4) $$\underline{U}(x_1,x_2) = r^{\alpha} \underline{u}(\theta)$$

with $0 < \operatorname{Re} \alpha < 1$. The exponents α are the solutions of :

(3.5) For Dirichlet (3.2) condition on Σ_1 and Σ_2 :

$$\sin^2(\omega\alpha) = \left(\frac{\lambda + \mu}{\lambda + 3\mu}\right)^2 \alpha^2 \sin^2 \omega$$

(3.6) For Neumann (3.3) condition on Σ_1 and Σ_2 :

$$\sin^2(\omega\alpha) = \alpha^2 \sin^2 \omega$$

(3.7) For mixed conditions (i.e. Dirichlet (3.2) on Σ_1 and Neumann (3.3) on Σ_2) :

$$\sin^2(\omega\alpha) = \frac{(\lambda+2\mu)^2 - (\lambda+\mu)^2 \alpha^2 \sin^2 \omega}{(\lambda+\mu)(\lambda+3\mu)}$$

These equations are taken from Grisvard[16] where the corresponding form of the function $\underline{u}(\theta)$ in (3.4) is also given.

Remark 3.1 - From (3.5) and (3.6) we see that, for Dirichlet or Neumann conditions, there is (resp. is not) a singularity if $\omega > \pi$ (resp. $\omega < \pi$). But for mixed boundary conditions (3.7), there are singularities even for $\omega < \pi$. For instance, for $\omega = \pi/2$ and $\lambda = \mu$ (which amounts to Poisson coefficient $\nu = 0{,}25$), equation (3.7) has a solution $\alpha = 0{,}69$. ■

4. - A TRANSMISSION PROBLEM IN TWO-DIMENSIONAL ELASTICITY

We now consider the two-dimensional elasticity system (3.1) with piecewise constant Lamé coefficients λ,μ, with Neumann boundary conditions (i.e., (3.3) on $\theta = 0$ and $\theta = \phi_1 + \phi_2$, Fig. 4.1). The coefficients are

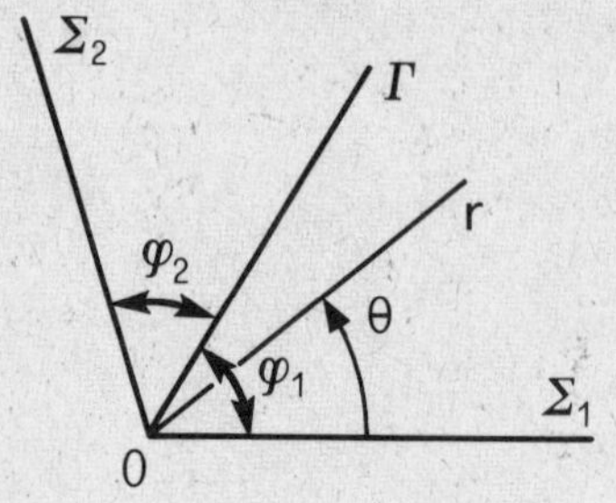

Figure 4.1

(4.1)
$$\lambda = \begin{cases} \lambda_o & \text{if } \theta \in (0,\phi_1) \\ \varepsilon^{-1}\lambda_1 & \text{if } \theta \in (\phi_1,\phi_2) \end{cases}$$
$$\mu = \begin{cases} \mu_o & \text{if } \theta \in (0,\phi_1) \\ \varepsilon^{-1}\mu_1 & \text{if } \theta \in (\phi_1,\phi_2) \end{cases}$$

where λ_o λ_1, μ_o, μ_1 are positive constants and ε is a positive parameter. As ε tends to zero, <u>the part between ϕ_1 and ϕ_2 becomes infinitely rigid.</u> Of course, (3.1) is considered in the distribution sense on the interface Γ ($\theta = \phi_1$) :

(4.2) $$[U] = 0 \quad ; \quad [\sigma_{ij}\, n_j] = 0 \qquad \text{on } \Gamma$$

The existence of a singularity at 0 for the different values of the elastic constants is considered in Bogy[27,28] and Dempsey and Sinclair[29,30]. We shall only give an asymptotic result :

<u>PROPOSITION 4.1</u> - <u>Under the hypotheses of the present section, searching for solutions of the form (3.4), the characteristic exponents $\alpha(\varepsilon)$ depend continuously on ε and converge, for $\varepsilon \to 0$ to the corresponding exponents of the problem in the region $\theta \in (0,\phi_1)$ with Dirichlet boundary condition on Γ and Neumann boundary condition on Σ_1 (Fig. 4.1).</u>

Remark 4.2 - Using Proposition 4.1 and Remark 3.1 we see, for instance, that for $\phi_1 = \phi_2 = \pi/2$, $\lambda_o = \mu_o$, and sufficiently small ε, there is a singularity $\alpha(\varepsilon)$ near the value 0,69. ■

The proof of Proposition 4.1 follows from the general relations of Dempsey and Sinclair[27]. It may also be obtained by a stiff perturbation in an implicite eigenvalue problem, starting with the method of next section.

5. - A GENERAL METHOD FOR COMPUTING SINGULARITIES

When singularities appear, the general form (roughly speaking) of the solution is

$$(5.1) \qquad U(x_1,x_2) = c\, r^{\alpha}\, u(\theta) + U^{regular}(x_1,x_2)$$

where α and $u(\theta)$ depend on the local geometry and coefficients of the problem, and the coefficient c and the regular part $U^{regular}(x_1,x_2)$ depend on the other data of the problem. The knowledge of α for a given problem shows if wether or not a singularity exists. Moreover, if $u(\theta)$ is known, the solution (5.1) may be computed in an accurate way by using a standard finite element discretization plus a special finite element in the vicinity of 0. This finite element is constructed to describe the singularity with not very important perturbation of the voids of the discretized matrix (see Lelièvre[23,24]).

The problem of finding α and $u(\theta)$ may be reduced to some implicit eigenvalue problem, and may be solved by numerical methods, (at least theoretically, for the real singular values α).

We now explain the method for an elliptic equation, but it is useful for general systems (with two independent variables x_1, x_2 of course). To fix ideas, we consider the problem of Fig. 2.1, where the domain Ω is an angle $\omega = \phi_1 + \phi_2$ in the vicinity of 0. Moreover, the boundary conditions are of the Neumann type and the coefficients depend only on θ in the vicinity of 0. The sesquilinear form associated with the problem is

$$(5.2)\qquad \int_\Omega a_{ij} \frac{\partial U}{\partial x_j} \frac{\partial V}{\partial x_i} dx$$

we take as Ω the angle

$$(5.3)\qquad \Omega = \{r,\theta \; ; \; r \in (0,\infty) \; ; \; \theta \in (0,\omega)\}$$

In order to search for solutions of the form $r^\alpha u(\theta)$ which do not belong to $H^1(\Omega)$ we take

$$(5.4)\qquad \begin{cases} U(x_1,x_2) = r^\alpha u(\theta) \; ; \quad u \in H^1(0,\omega) \\ V(x_1,x_2) = \phi(r)\, v(\phi) \; ; \quad v \in H^1(0,\omega) \; ; \quad \phi \in \mathscr{D}(0,\infty) \end{cases}$$

and the homogeneous equation with Neumann boundary conditions become

$$(5.5)\qquad 0 = \int_o^\infty rdr \int_o^\omega a_{ij} \frac{\partial(r^\alpha u)}{\partial x_j} \frac{\partial(\phi v)}{\partial x_i} d\theta$$

which after the change

$$\frac{\partial}{\partial x_1} = \cos\theta \frac{\partial}{\partial r} - \frac{\sin\Theta}{r} \frac{\partial}{\partial \theta}$$

$$\frac{\partial}{\partial x_2} = \sin\theta \frac{\partial}{\partial r} + \frac{\cos\theta}{r} \frac{\partial}{\partial \theta}$$

and after integrating with respect to θ, becomes :

$$0 = \int_o^\infty \left(F(\alpha,u,v)\, r^\alpha \phi' + \Phi(\alpha,u,v)\, r^{\alpha-1}\phi\right) dr, \quad \forall\, \phi \in \mathscr{D}(0,\infty)$$

or integrating by parts in r :

$$0 = \int_o^\infty \left(-\alpha F + \Phi\right) r^{\alpha-1} \phi \, dr$$

which amouts to

$$(5.6)\qquad 0 = -\alpha\, F(\alpha,u,v) + \Phi(\alpha,u,v) \equiv b(\alpha,u,v)$$

which defines a sesquilinear form b (depending on α) for $u,v \in H^1(0,\omega)$. The problem reduces to find the values of α such that a non zero $u \in H^1(0,\omega)$ exists satisfying

(5.7) $\quad b(\alpha;u,v) = 0 \qquad \forall\ v \in H^1(0,\omega)$

This is an implicite eigenvalue problem as it amounts to find the values of α for which zero is an eigenvalue of the operator $B(\alpha)$ associated with the form b.

In order to compute the singular values α, we discretize (by finite elements for instance) and use a finite dimensional basis $v^1,\ldots v^m$ of the discretized space $H^1(0,\omega)$. The searched values α are those for which the matrix with coefficients

(5.8) $\quad b_{st} = b(\alpha,u^s,u^t)$

is singular. For real α, as $0 < \alpha < 1$, it suffices to compute the determinant of the matrix for several α and to obtain by interpolation <u>the values for which it vanishes</u>. When the value α is known, the corresponding (discretized) $u(\theta)$ <u>is the corresponding eigenvector</u> which may be obtained by the inverse iteration method, (Ciarlet[6], sect. 6.4) for instance.

CHAPTER 5

5 - EXAMPLES OF SINGULARITIES IN THERMAL CONDUCTION AND ELASTICITY

1. - MISCELLANEOUS EXAMPLES

We know from Lecture 4 that boundary singularities (defined as points where grad u becomes infinity) are local phenomena associated with the form of the boundary, the coefficients and the boundary conditions. Then, strictly speaking, the presence of singularities has nothing to do with boundary layers. But of course, there is no singularities at regular points of the interfaces ; this is the reason why singularities usually appear at the boundary of the body ; but we shall see in Sect. 2 a case of singularity at an interior point.

Let us consider the homogenization problem of sect. 2.1, 2.2. The presence of boundary singularities may be studied without considering the boundary layers, using the criterion of Proposition 4.2.1. If there are no singularities, (i.e. infinite values of the gradient), the boundary layer implies modification of grad u with respect to the values out of the layer, but this modification only implies finite concentration factors of the gradient.

For instance, we consider a layered medium made of the different materials denoted by the superscripts ^{1,2}, i.e. the composite is made of alternating layers of the two media ; we consider the medium 1 (resp. 2) isotropic (resp. anisotropic) the matrix a_{ij} being

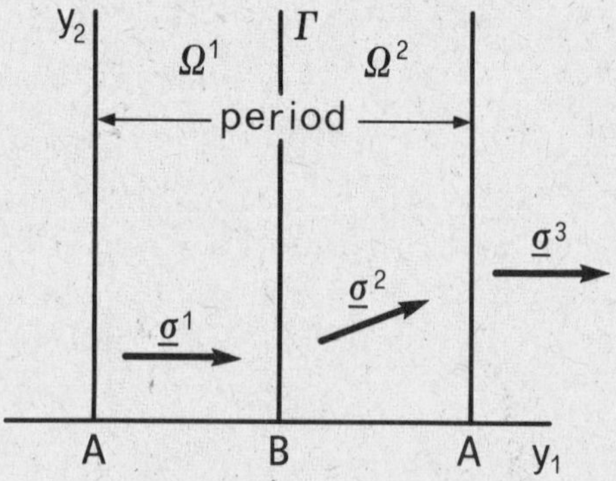

Figure 1.1

$$(1.1)\quad \begin{cases} a_{ij} = a^1\,\delta_{ij} & \text{in } \Omega^1 \\ a_{11} = a^2 \,,\ a_{22} = a^2 & \text{in } \Omega^2 \\ a_{12} = c & \text{in } \Omega^2 \end{cases}$$

where c is some value $\neq 0$. We consider the Neumann boundary condition on $y_2 = 0$. According to Proposition 4.2.1, we consider a horizontal vector $\underline{\sigma}^1$ in Ω^1. We construct the corresponding $\underline{g}^1$, $\underline{g}^2$, $\underline{\sigma}^2$ using of course the matrix a_{ij} in each region and the interface conditions across the boundary Γ. This gives a vector $\underline{\sigma}^2$ with component 2 of the sign of c ; thus, for $c > 0$ we have the picture of Fig. 1.1 and there are singularities at the points B. Of course, if $c < 0$, the singularities are at points A. If $c = 0$ (the two media are isotropic) there is no singularity. Of course, evident modifications allow us to consider 2 anisotropic media, interfaces Γ not normal to the boundary, and non-straight boundaries as in Fig. 2.2.2.

Let us now consider the analogous plane elasticity problem. The results of sect. 4.4 only concern the isotropic case. According to Remark 4.4.2, if one of materials, (Ω^1, for instance) is very hard with respect to the other (i.e. the ratio of the Lamé constants of Ω^2 to the ones of Ω^1 is sufficiently small) there are singularities at the points A and B of Fig. 1.1.

As a first example in anisotropic elasticity, which may be reduced to a scalar problem with an elliptic equation, we consider a cylindrical domain of R^3 (coordinates x_1, x_2, x_3) $\Omega \times R$ where Ω is some domain of the (x_1, x_2) plane and R is the x_3 axis. We consider the elasticity system for the displacement $\underline{U}$:

$$(1.2)\quad \frac{\partial}{\partial x_j}\left(b_{ijmn}\, e_{mn}(\underline{U})\right) = 0$$

with coefficients satisfying

$$(1.3)\qquad b_{ij3n} = 0 \quad \text{for } i,j,n \text{ running in } (1.2)$$

we may search for solutions of (1.2) with $U = (0,0,U_3(x_1,x_2))$ (shear solutions). The elasticity system becomes

$$(1.4)\qquad \frac{\partial}{\partial x_j}\left(a_{ij}\,\frac{\partial U_3}{\partial x_i}\right) = 0 \quad \text{in } \Omega, \qquad a_{ij} \equiv b_{3i3j}$$

The shear stresses are

$$(1.5)\qquad \sigma_{13} = a_{1i}\,\frac{\partial U_3}{\partial x_i} \quad ; \quad \sigma_{23} = a_{2i}\,\frac{\partial U_3}{\partial x_i}$$

and (1.4) amounts to

$$(1.6)\qquad \frac{\partial \sigma_{13}}{\partial x_1} + \frac{\partial \sigma_{23}}{\partial x_2} = 0$$

and consequently the Neumann condition

$$(1.7)\qquad n_j\, a_{ij}\,\frac{\partial U_3}{\partial x_i} = 0 \iff \sigma_{i3}\, n_i = 0$$

on the boundary amounts to the fact that the lateral surface of the cylinder is free. If the cylinder is made of two pasted pieces Ω^1, Ω^2, the application of Proposition 4.2.1 is straightforward. For instance, if Ω^1 is isotropic and Ω^2 is not (i.e. $b_{3132} \neq 0$) in the situation of Fig. 1.2.a), there is a singularity either at 0 or $0'$ (according to the sign of b_{3132}) ; but this singularity disappears in the case of Fig. 1.2.b).

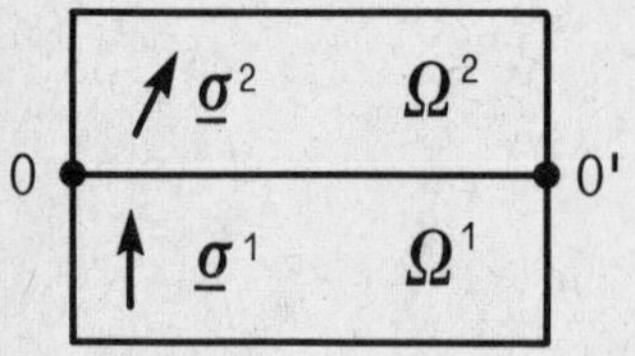

Figure 1.2.a

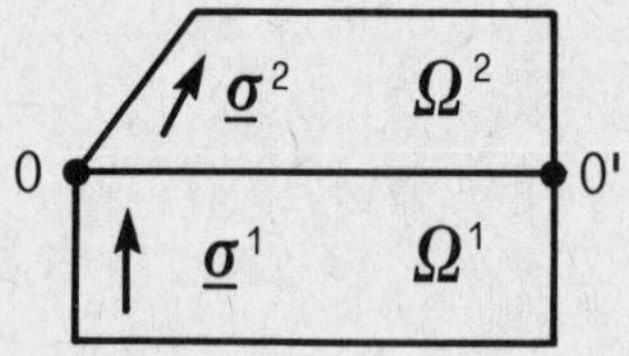

Figure 1.2.b

Numerical examples of this fact were obtained by Foura and Leguillon[31]. Moreover, it was proved in Sanchez[22] that this kind of singularities is preserved when (1.3) is not satisfied exactly.

2. - AN EXAMPLE OF SINGULARITY AT THE INTERIOR OF A COMPOSITE

We consider the elliptic equation

$$(2.1) \qquad \frac{\partial}{\partial x_i}\left(a \frac{\partial U}{\partial x_i}\right) = 0 \qquad \text{in } \Omega \subset R^2$$

with piecewise constant coefficient $a(x)$. Let Γ be a line dividing Ω into Ω_1 and Ω_2, where a takes the values a_1 and a_2 respectively. When searching for solutions of the form $U = r^\alpha u(\theta)$, we have symmetry with respect to Ox_2 and the solutions may be decomposed into symmetric and antisymmetric parts. As for the symmetric (resp. antisymmetric) gradient, Ox_2 may be considered as a Neumann (resp. Dirichlet) boundary. Each of these problems are easily studied with the criterion of Lecture 4, Proposition 2.1. It is seen that with the geometric disposition of Fig. 2.1, if $a_1 > a_2$ (resp. $a_2 > a_1$) the Dirichlet (resp. Neumann) problem is singular and the other one is not. As a whole, we have always a singularity.

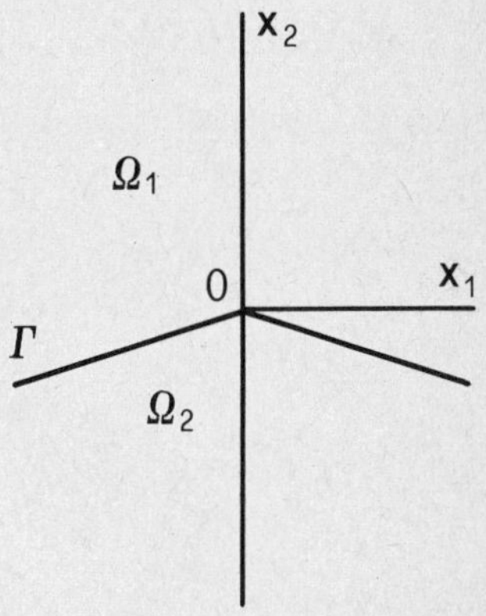

Figure 2.1

Anisotropic problems may also be considered in the same way, provided that the medium (i.e. Γ and the coefficients a_{ij}) is symmetric with respect to some axis Ox_2.

3. - EDGE SINGULARITIES FOR PLATES IN TRACTION

In the framework of Lecture 3, the eventual presence of singularities of the stress at the intersection of the interfaces and the boundary is an important question from both theoretical and practical point of view. In fact it involves singularites for the elasticity system in non-standard situations, and constitutes an open problem in the general case. In practice, the concentration of stress often implies delamination of the layers, and failure of the plate. Numerical results about this problem may be seen in Raju and Crews[32], Anquez and all[33], Barsoum and Freese[34], Wang and Chei[35], Zwiers and Ting[36] showing in almost all cases singularities. Destuynder[37] showed the existence of logarithmic singularities for some components of the stress in cross-fiber laminates. In fact, there are in general stronger (algebraic) singularities, in particular for the component σ_{33} which is most concerned with delamination. In this section we use the results of Sect.44 to prove such a fact in the case of a plate made of isotropic layers with very large ratio of rigidities (this means that the ratios λ_2/λ_1, μ_2/μ_1 of the Lamé constants are very large, as in sect. 4.4). The case of arbitrary rigidities may be investigated in the same way by using the results of Dempsey and Sinclair[21,22].

We turn back to sect. 3.4, in particular to (4.11), (4.12). We take the origin of coordinate 0 at the intersection of the interface Γ and the free boundary (up to a translation, the axes are the same as in lecture 3. The equations and boundary conditions are : (λ and μ are the Lamé coefficients, taking constant values λ_i, μ_i in Ω_i, i = 1,2) :

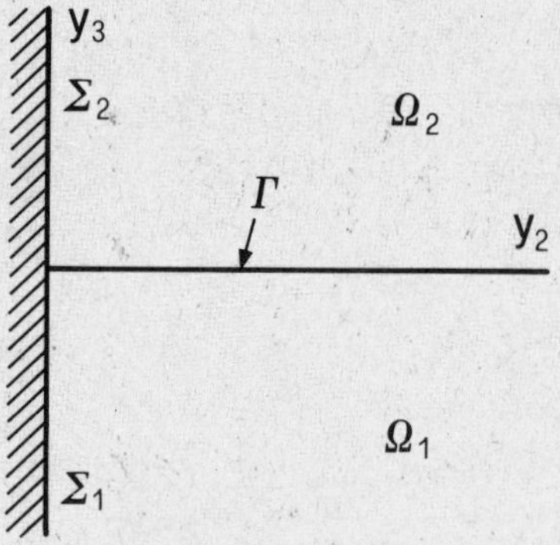

Figure 3.1

$$(3.1) \quad \underline{u}^*(y) = E_{11}\, y_1\, \underline{e}_1 + \underline{v}\,(y_2, y_3)$$

$$(3.2) \qquad \frac{\partial\, \sigma_{ij}(u^*)}{\partial\, y_j} = 0 \qquad \text{in } \Omega$$

$$(3.3) \qquad \sigma_{ij}\, n_j = 0 \qquad \text{on } \Sigma_1 \text{ and } \Sigma_2$$

where the component 1 of $\underline{v}$ vanishes identically. This gives (we take $E_{11} = 1$ without loss of generality) :

$$\sigma_{11}(\underline{u}^*) = \lambda + 2\mu$$

$$\sigma_{22}(\underline{u}^*) = \sigma_{22}(\underline{v}) + \lambda \quad ; \quad \sigma_{33}(\underline{u}^*) = \sigma_{33}(\underline{v}) + \lambda$$

$$\sigma_{12}(\underline{u}^*) = 0 \quad ; \quad \sigma_{13}(\underline{u}^*) = 0 \quad ; \quad \sigma_{23}(\underline{u}^*) = \sigma_{23}(\underline{v})$$

This amounts to (α, β run in (2.3)) :

$$(3.4) \qquad \frac{\partial\, \sigma_{\alpha\beta}(\underline{v})}{\partial\, y_\beta} = 0 \qquad \text{in } \Omega_1 \text{ and } \Omega_2$$

$$(3.5) \qquad \sigma_{22}(\underline{v}) = -\lambda \qquad \text{on } \Sigma_1 \text{ and } \Sigma_2$$

$$(3.6) \qquad \sigma_{23}(\underline{v}) = 0 \qquad \text{on } \Sigma_1 \text{ and } \Sigma_2$$

$$(3.7) \qquad \left(\sigma_{23}(\underline{v})\right) = 0 \qquad \text{on } \Gamma$$

$$(3.8) \qquad \left(\sigma_{33}(\underline{v}) + \lambda\right) = 0 \qquad \text{on } \Gamma$$

$$(3.9) \qquad \left(v\right) = 0 \qquad \text{on } \Gamma$$

We then have non homogeneous boundary conditions for the two-dimensional vector $\underline{v}$. As the results of sect. 4.4 only deal with homogeneous boundary conditions, we write $\underline{v} = \underline{w} + \underline{\hat{w}}$ and we search for a non singular vector $\underline{w}$ satisfying (3.4)-(3.9) ; we shall see that such a vector exists ; thus $\underline{\hat{w}}$ satisfies the homogeneous system and is singular according to Proposition 4.4.1. (for large ratio of rigidities of course).

We search for a solution $\underline{w}$ of (3.4)-(3.9) vanishing at the origin, with strains $e_{\alpha\beta}(\underline{w}) = \text{constant} = \varepsilon_{\alpha\beta}$ in each of the regions Ω_1, Ω_2. Differentiating (3.9) along Γ we obtain

$$(3.10) \quad \left(\frac{\partial v_2}{\partial y_2}\right) = 0$$

$$(3.11) \quad \left(\frac{\partial v_3}{\partial y_2}\right) = 0$$

which are equivalent to (3.9) for such kind of solutions. Moreover, we note that (3.11) may be satisfied by performing a rotation of Ω_1 with respect to Ω_2 around the axis y_1 ; consequently it is irrelevant. Relation (3.10) amounts to saying that ε_{22} takes the same value in Ω_1 and Ω_2. Denoting by $\varepsilon^i_{\alpha\beta}$ the values of $\varepsilon_{\alpha\beta}$ in Ω_i, we have for the time being, the unknowns ε^1_{22}, ε^2_{33}, ε^1_{33}, ε^1_{23}, ε^2_{23}. But (3.6) shows that the two later vanish. On the other hand, equation (3.4) is identically satisfied for the kind of solutions $\underline{w}$. Thus we have the unknowns ε_{22}, ε^1_{33}, ε^2_{33} and equations (3.5), (3.8) (note that (3.7) is now identically satisfied). This amounts to

$$(3.12) \quad \begin{cases} (\lambda_1 + 2\mu_1)\, \varepsilon_{22} + \lambda_1\, \varepsilon^1_{33} = -\lambda_1 \\ (\lambda_2 + 2\mu_2)\, \varepsilon_{22} + \lambda_2\, \varepsilon^2_{33} = -\lambda_2 \\ (\lambda_2 + 2\mu_2)\, \varepsilon^2_{33} + \lambda_2\, \varepsilon_{22} - (\lambda_1 + 2\mu_1)\, \varepsilon^1_{33} - \lambda_1\, \varepsilon_{22} = \lambda_1 - \lambda_2 \end{cases}$$

The vanishing of the determinant amounts to

$$(3.13) \quad \frac{(\lambda_1 + \mu_1)\mu_1}{\lambda_1} - \frac{(\lambda_2 + \mu_2)\mu_2}{\lambda_2} = 0$$

which does not vanish in the case of large ratio of rigidities. Then the non singular $\underline{w}$ exists and $\underline{v}$ <u>is singular</u> as $\hat{\underline{w}}$ (Proposition 4.4.1).

CHAPTER 6

6 - ELASTIC BODY WITH DEFECTS DISTRIBUTED NEAR A SURFACE

1. - SETTING OF THE PROBLEM

This lecture is based on Nguetseng and Sanchez[38] where details and further developments may be seen. The corresponding problem for the Laplace equation is in Nguetseng[39]. We deal with the elastic behavior of a body containing small cavities (holes) periodically distributed in the vicinity of a surface Σ interior to the body. As the holes may be in particular cracks, this problem is a model for the study of the concentration of stress for two pieces sticked together in an imperfect way. We use some homogenization method with respect to the variables which are tangential to Σ and matched asymptotic expansions for the normal variable.

Let Ω be a domain filled by an elastic body with constant elastic coefficients a_{ijmn} (but this is not essential). The body is clamped by its boundary and acted upon by a given force field $\underline{f}(x)$ but these asumptions are not essential . This is the "unperturbed problem", and we denote by $\underline{u}^o(x)$ the corresponding displacement which satisfies :

$$(1.1) \qquad - \frac{\partial \sigma_{ij}(\underline{u}^o)}{\partial x_j} = f_i \quad \text{in } \Omega \; ; \qquad \sigma_{ij} = a_{ij\ell m} \, e_{\ell m}(\underline{u}^o)$$

$$(1.2) \qquad \underline{u}^o = 0 \qquad \text{on } \partial\Omega$$

The "perturbed problem" (or "with holes") is defined as follows (Fig. 1.1 and 1.2). Let Σ be the section of Ω by the plane $x_3 = 0$, Ω^+ and Ω^- are the two parts of Ω. We consider the auxiliar space of the variable y_1 where the plane $y_3 = 0$ is divided into rectangles

ω which we consider as "periods". Let $\mathcal{H}$ (as "hole") be a bounded domain of R^3 intersecting a period ω. We also consider the holes obtained by ω-periodicity. The set of the holes will also be denoted by $\mathcal{H}$. We now define the "hollowed" or "perturbed" domain $\Omega_\varepsilon = \Omega - \varepsilon\mathcal{H}$, i.e., Ω less the $\varepsilon\mathcal{H}$ holes ($\varepsilon\mathcal{H}$ is the ε-homothetic of $\mathcal{H}$).

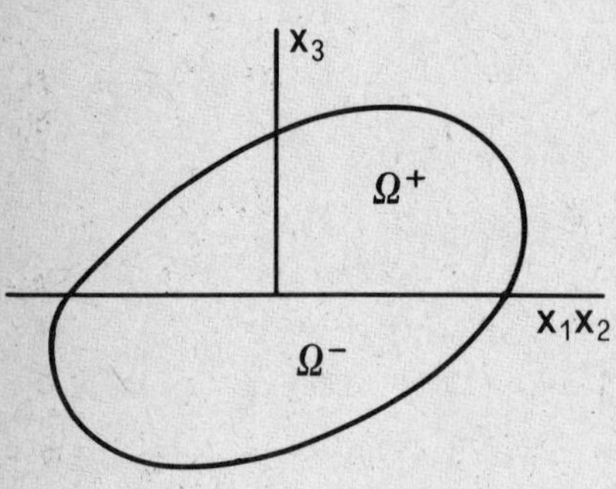

Figure 1.1

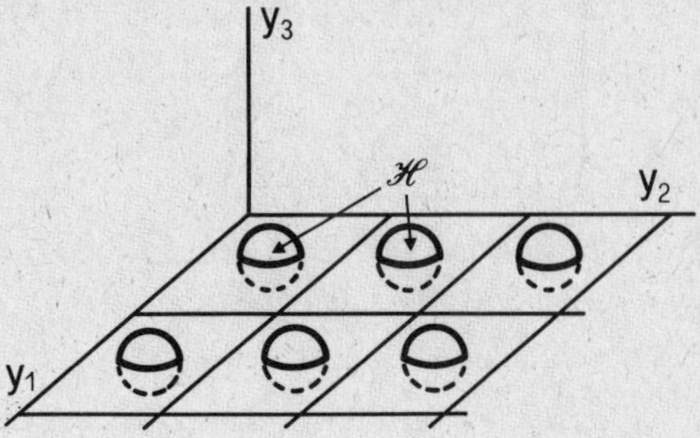

Figure 1.2

The perturbed problem is

$$(1.3) \qquad - \frac{\partial \sigma_{ij}(\underline{u}^\varepsilon)}{\partial x_j} = f_i \quad \text{in } \Omega_\varepsilon$$

$$(1.4) \qquad \underline{u}^o = 0 \quad \text{on } \partial\Omega$$

$$(1.5) \qquad \sigma_{ij}(u^\varepsilon)n_j = 0 \quad \text{on } \partial\Omega\mathcal{H} .$$

2. - ASYMPTOTIC EXPANSION

Here and in the sequel, Latin (resp. Greek) indexes run in 1,2,3 (resp. in 1,2).

To study the layer near Σ we consider the "inner variable" $y_3 = x_3/\varepsilon$. The inner and outer variables will be x_3 and y_3 respectively. On the other hand, each of the expansions (inner and outer) depends on the tangential variables x_1, x_2. The <u>outer expansion</u> is

$$(2.1) \qquad u^\varepsilon(x) = \underline{u}^o(x) + \varepsilon\, \underline{u}^1(x) + \varepsilon^2 \ \ldots \ ; \quad x = (x_1, x_2, x_3)$$

in Ω^+ and Ω^- (in fact two different expansions (2.1) in Ω^+ and Ω^-). We shall see that the first term $\underline{u}^o$ in (2.1) is in fact the solution of the unperturbed problem (1.1), (1.2). In the layer, according to two-scale general features for homogenization, the terms of the inner expansion will depend on x_α and $y_\alpha = x_\alpha/\varepsilon$. That is :

$$(2.2) \quad \begin{cases} \underline{u}^\varepsilon(x) = \underline{v}^o(x_\alpha, y_1, y_\alpha) + \varepsilon\, \underline{v}^1(x_\alpha, y_1, y_\alpha) + \varepsilon^2 \ldots \\ \text{for } y_\alpha = x_\alpha/\varepsilon \;;\; \underline{v}^j\text{-}\omega\text{-periodic in } y_\alpha \end{cases}$$

It will be useful considering the functions $\underline{v}^j$ of (2.2) as functions of $(x_1,x_2) \in \Sigma$ and (y_1,y_2,y_3) belonging to the "period" G

$$G = \{\omega \times (-\infty,+\infty)\} - \mathscr{H}$$

i.e. the "infinite prism" formed by the points such that y_α belongs to the period ω and y_3 takes any value, unless the hole $\mathscr{H}$. We shall say either ω-periodic or G-periodic.

It is easily seen (and is in agreement with the physical intuition) that the first term $\underline{u}^o$ of (2.1) is the solution of (1.1), (1.2).

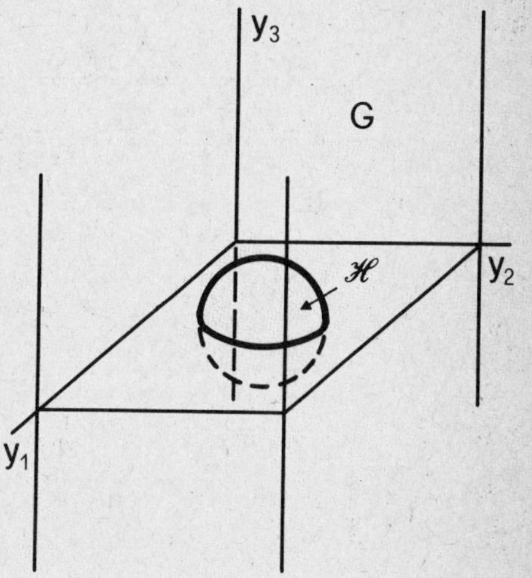

Figure 1.3

As for $\underline{v}^o$ in (2.2), putting the inner expansion into (1.3) we have at order ε^{-2} :

$$(2.3) \quad -\frac{\partial}{\partial y_j}\left(a_{ijmn}\, e_{mn}(\underline{v}^o)\right) = 0 \quad \text{in } G$$

which is the elasticity system in y, for x_α parameters. From (1.5) at order ε^{-1} :

(2.4) $a_{ijmn}\ e_{mny}(\underline{v}^o)n_j = 0$ on $\partial\mathcal{H}$

which is a homogeneous Neumann boundary condition on the boundary of the hole. Of course, $\underline{v}^o$ is G-periodic. The matching with (2.1) at the first order gives :

(2.5) $\underline{v}^o(x_\alpha, \pm\infty, x_\alpha/\varepsilon) = \underline{u}^o(x_\alpha, 0)$

i.e. taking x_α as parameters, $\underline{v}^o$ tends to a constant value at infinity. It follows that $\underline{v}^o$ is independent of y. Indeed, we multiply (2.3) by v_i^o and we integrate on the part of G with $|y_3| < C$ (denoted by G_c) :

(2.6) $$0 = \int_{G_c} \frac{\partial\ \sigma_{ij}(\underline{v}^o)}{\partial\ x_j} v_i^o\ dy = - \int_{\partial G_c} \sigma_{ij}(\underline{v}^o)\ n_j\ v_i^o\ ds + \\ + \int_{G_c} \sigma_{ij}(\underline{v}^o)\ e_{ij}(\underline{v}^o)\ dy$$

we note that the integral over ∂G_c has a zero contribution on $\partial\mathcal{H}$. because of (2.4) and on the lateral boundaries by G-periodicity ; (2.6) gives :

(2.7) $$\int_{G_c} a_{ijmn}\ e_{mn}(\underline{v}^o)\ e_{ij}(\underline{v}^o) = \int_{G\cap\phi_3=\pm c} a_{ijmn}\ e_{mn}(\underline{v}^o)\ n_j\ v_i^o\ ds$$

and from (2.5) the right side (and then the left) vanish and we have

(2.8) $\underline{v}^o(x_\alpha, y_1, y_\alpha) \equiv \underline{u}^o(x_\alpha, 0)$

3. - THE LOCAL PROBLEM IN THE VICINITY OF THE HOLES

At the present state, the outer and inner expansions for $\text{grad}\ u^\varepsilon$ are :

(3.1) $$\begin{cases} \underline{\text{grad}}\ u_i^\varepsilon(x) = \underline{\text{grad}}_x\ u_i^o(x) + \varepsilon\ \underline{\text{grad}}_x\ u_i^1 + \ldots \\ \underline{\text{grad}}\ u_i^\varepsilon(x) = \underline{\text{grad}}_{x_\alpha}\ u_i^o(x_\alpha, 0) + \underline{\text{grad}}_y\ v_i^1 + O(\varepsilon) \end{cases}$$

respectively where the symbol grad_{x_α} expresses the obvious fact that we consider the gradient of the function $u_i^o(x_1,x_2,0)$ which only depends on x_1,x_2. The matching rule for the gradients give (note that the component 3 gives an equation different from the others) :

$$(3.2)\quad \begin{cases} \dfrac{\partial u_i^o}{\partial x_\alpha}(x_\alpha,o) = \dfrac{\partial u_i^o}{\partial x_\alpha}(x_\alpha,o) + \lim\limits_{y_3\to\pm\infty} \dfrac{\partial v_i^1}{\partial y_\alpha}(x_\alpha,y) \quad ; \alpha = 1,2 \\ \dfrac{\partial u_i^o}{\partial x_3}(x_\alpha,o) = \lim\limits_{y_3\to\pm\infty} \dfrac{\partial v_i^1}{\partial y_3}(x_\alpha,y) \end{cases}$$

and some simplifications appear by considering instead of $\underline{v}^1$ the new unknown :

$$(3.3)\quad \underline{v}^*(x_\alpha,y) \equiv \underline{v}^1(x_\alpha,y) - y_3 \frac{\partial \underline{u}^o}{\partial x_3}(x_\alpha,o)$$

for which the matching becomes :

$$(3.4)\quad \lim_{y_3\to\pm\infty} \frac{\partial \underline{v}^*}{\partial y_j}(x_\alpha,y_\alpha,y_3) = 0$$

and the inner expansion for the gradient from (5.2) is :

$$(3.5)\quad \frac{\partial u_i^\varepsilon}{\partial x_j}(x) = \frac{\partial u_i^o}{\partial x_j}(x_\alpha,o) + \frac{\partial v_i^*}{\partial y_j}(x_\alpha,y) + O(\varepsilon) \quad ; \quad j = 1,2,3.$$

The equations and boundary conditions for $\underline{v}^*$ follow from (1.3) (1.5) and (3.5) :

$$(3.6)\quad \frac{\partial}{\partial y_j}\left(a_{ijmn}\, e_{mn}(\underline{v}^*)\right) = 0 \qquad \text{in } G$$

$$(3.7)\quad \sigma_{ijy}(\underline{v}^*)\, n_j = -\sigma_{ijx}(\underline{u}^o)\, n_j \qquad \text{on } \partial\mathscr{H}$$

that, with the matching condition (3.4) and the G-periodicity constitute the boundary value problem for $\underline{v}^*$. Of course, x_1,x_2 play the role of parameters and we have a problem in the domain G. The solution $\underline{v}^*$ is defined up to an additive constant, but the strain field at the leading order is well defined (cf. (3.5)). The data of the local problem are the $\sigma_{ijx}(u^o)$ appearing at the right side of (3.7). As the problem is linear we may write

$$(3.8) \quad \underline{v}^*(x_\alpha, y) = \sigma_{mnx}(\underline{u}^o)\, \underline{v}^{*mn}(y)$$

where the $\underline{v}^{*mn}$ only depend on the variable y. They are the solutions corresponding to

$$(3.9) \quad \sigma_{ijx}(\underline{u}^o) = \frac{1}{2}(\delta_{im}\,\delta_{jn} + \delta_{in}\,\delta_{jm}) .$$

Summing up, the term $\underline{v}^*(x_1,x_2,y)$ is defined by (3.8) where $\sigma_{mnx}(\underline{u}^o)$ denotes the value of the stress tensor corresponding to the unperturbed solution $\underline{u}^o$ taken at the points $(x_1,x_2,0)$ of Σ, and $\underline{v}^{*mn}(y)$ are the solutions of the local G-periodic problem (defined up to an additive constant vector)

$$(3.10) \quad -\frac{\partial}{\partial y_j}\left[a_{ijpq}\; e_{pqy}(\underline{v}^{*mn})\right] = 0 \quad \text{in } G$$

$$(3.11) \quad a_{ijpq}\; e_{pqy}(\underline{v}^{*mn})\, n_j = -\frac{1}{2}(\delta_{im}\, n_n + \delta_{in}\, n_m) \quad \text{on } \partial\mathscr{H}$$

$$(3.12) \quad \lim_{y_3\to\pm\infty} \underline{\text{grad}}_y\, \underline{v}^{mn}(y) = 0$$

Then, knowing the six local solutions $\underline{v}^{*mn}$ and the unperturbed solution, we may construct $\underline{v}^*$. The leading term (of order 1) of the stress tensor in any point of the vicinity of the surface Σ (defined by the coordinates (x_1,x_2,y_1,y_2,y_3) is

$$(3.13) \quad a_{ijmn}\left[e_{mnx}(\underline{u}^o) + e_{mny}(\underline{v}^*)\right]$$

or equivalently

$$(3.14) \quad \sigma_{ij}(\underline{u}^o) + \sigma_{mnx}(\underline{u}^o)\; a_{ijpq}\; e_{pq}(\underline{v}^{*mn})$$

Of course, the eventual presence of singularities of the stress depends on the form of the holes.

More developments may be seen in Nguetseng and Sanchez[38]. In particular, the proof of the fact (3.12) actually holds for the generalized solutions is made by Fourier transform methods, as in Sanchez[40].

REFERENCES

1 **Van Dyke, M.** - "Perturbation Methods in Fluid Mechanics". Academic Press, New York (1964).

2 **Cole, J.D.** - "Perturbation Methods in Applied Mathematics" Blaisdell, Toronto (1968).

3 **Cole, J.D. and Kevorkian** - "Perturbation Methods in Applied Mathematics". Springer, New York (1980).

4 **Necas, J.** - "Les méthodes directes en théorie des équations elliptiques". Masson, Paris (1967).

5 **Lions, J.L. et Magenes, E.** - "Problèmes aux limites non homogènes et applications". vol. I, Dunod, Paris (1967).

6 **Brezis, H.** - "Analyse fonctionnelle, théorie et applications". Masson, Paris (1983).

7 **Ciarlet, P.G.** - "The Finite Element Method for Elliptic Problems" North-Holland, Amsterdam (1978).

8 **Ciarlet, P.G.** - "Introduction à l'Analyse Numérique Matricielle". Masson, Paris, (1982).

9 **Raviart, P.A. et Thomas, J.M.** - "Introduction à l'Analyse Numérique des équations aux dérivées partielles". Masson, Paris, (1983).

10 **Bensoussan, A., Lions, J.L. and Papanicolaou, G.** - "Asymptotic Analysis for Periodic Structures". North-Holland, Amsterdam, (1978).

11 **Lions, J.L.** - "Some Methods in the Mathematical Analysis of Systems and their Control". Gordon and Breach, New York (1981).

12 **Sanchez-Palencia, E.** - "Non Homogeneous Media and Vibration Theory". Springer, Berlin, (1980).

13 **Dumontet, H.** - "Boundary layers stresses in elastic composites", in "Local Effects in the Analysis of Structures". Editor P. Ladevèze, Elsevier, Amsterdam (1986).

14 **Ciarlet, P.G. et Destuynder, P.** - "A justification of the two dimensional linear plate model". Jour. Méc., 18, (1979), p. 315-344.

15 **Grisvard, P.** - "Elliptic Problems in Non Smooth Domains". Pitman, London (1985).

16 **Grisvard, P.** - "Problèmes aux limites dans des polygones". Université de Nice, Pré-Publications Mathématiques, n° 45, (1984).

17 **Lemrabet, K.** - "Régularité de la solution d'un problème de transmission". Jour. Math. Pures Appl., 56 (1977), p. 1-38.

18 **Lemrabet, K.** - "An Interface Problem in a Domain of R^3". Jour. Math. Analys. Appl., 63 (1978), p. 549-562.

19 **Kondratiev, V.A.** - "Boundary Value Problems for Elliptic Equations in Domains with Conical or Angular Points". Trudy Moskovs. Mat. Obs., 16 (1967), p. 209-292 (= Transact. Moscow Math. Soc., 16, 1967, p. 227-313).

20 **Kondratiev, V.A.** - "The Smoothness of a solution of Dirichlet's Problem for 2nd Order Elliptic Equations in a Region with Piece-wise Smooth Boundary". Differ. Urav., 6 (1970), p. 1831-1843 (= Differ. Equat., 6, 1970), p. 1392-1401).

21 **Sanchez-Palencia, E.** - "Influence de l'anisotropie sur l'apparition de singularités de bord dans les problèmes aux limites relatifs aux matériaux composites". Compt. Rend. Acad. Sc. Paris, sér. I, 300, (1985), p. 27-30.

22 **Sanchez-Palencia, E.** - "On the Edge Singularities in composite media, Influence of the anisotropy". To be published in the Proceedings of the Stefan Banach Center, Warsaw, Semester on P.D.E., 1984, Editor, Prof. Bojarski.

23 **Sovin, J.A.** - "Elliptic boundary value problems for plane domains with angles and discontinuities reaching the boundary". Dokl. Akad. Nauk SSSR, 187 (1969) p. 995 (= Sov. Math. Dokl., 10, 1969, p. 985).

24 **Lelièvre, J.** - "Sur l'Utilisation de Fonctions Singulières dans la Méthode des Eléments Finis". Compt. Rend. Acad. Sc. Paris, 283 (1976) p. 863-865.

25 **Lelièvre, J.** - "Sur les éléments finis singuliers". Compt. rend. Acad. Sc. Paris, 283 (1976), p. 1029-1032.

26 **Ladyzenskaja, O.A. et Ouralceva, N.N.** - "Equation aux dérivées partielles de type elliptique". Dunod, Paris (1968).

27 **Bogy, D.B.** - "Edge dissimilar orthogonal elastic wadges under normal and shear loading". Transact. A.S.M.E., p. 460 (1968).

28 **Bogy, D.B.** - "On the problem of edge bo nded elastic quarter planes loaded at the boundary". Int. Jour. Solid. Struct., 6, p. 1287 (1970).

29 **Dempsey, J.P. and Sinclair, G.B.** - "On the stress singularities in the plane elasticity of the composite wedge". Jour. Elast., 9, p. 373, (1979).

30 **Dempsey, J.P. and Sinclair, G.B.** - "On the singular behavior of the vertex of a bi-material wedge". Jour. Elast., 11, p. 317, (1981).

31 **Foura, S. and Leguillon, D.** - Personal communication.

32 **Raju, I.J and Crews, J.H** - "Interlaminar stress singularities at a straight free edge in composite laminates". Comput. Struct., 14 (1981) p. 21-28.

33 **Anquez, L., Bern, A. et Renard, J.** - "Etude numérique des effets de bord libre dans les composites stratifiés". Rech. Aerosp., p. 29-40, (1985).

34 **Barsoum, R.S and Freese, C.E.** - "An Iterative Approach for the Evaluation of delamination stresses in laminated composites". Int. Jour. Num. Meth. Engng. 20, p. 1415-1431, (1984).

35 **Wang, S.S. and Choi, I.** - "Boundary layer effects in composite laminates, I. Free edge stress singularities and II, Free edge stress solutions and basic characteristics". Jour. Appl. Mech. 49 (1982), I, p. 541-548, II, p. 549-560.

36 **Zwiers, R.I., Ting, T.C. and Spalker, R.L.** - "On the logarithmic singularity of free edge stress in laminated composites". Jour. Appl. Mech., 49 (1982), p. 561-568.

37 **Destuynder, P.** - To appear, Jour. Mec. Theo. Appl. (1985).

38 **Nguetseng, G. and Sanchez-Palencia, E.** - "Stress Concentration for Defects distributed near a surface" in "Local Effects in

the Analysis of Structures". Editor P. Ladevèze, Elsevier, Amsterdam (1986).

39 **Nguetseng, G.** - "Problèmes d'écrans perforés pour l'équation de Laplace". Model. Math. Analyse Num. 19 (1985), p. 33-63.

40 **Sanchez-Palencia, E.** - "Un problème d'écoulement lent d'un fluide visqueux incompressible à travers une paroi finement perforée" in Ecole d'Eté E.D.F. - C.E.A. - I.N.R.I.A. sur l'homogénéisation. Eyrolles, Paris (1985).

PART IV

ELEMENTS OF HOMOGENIZATION FOR INELASTIC SOLID MECHANICS

Pierre M. Suquet
Laboratoire de Mécanique Générale
des Milieux Continus
Université des Sciences et Technique du Languedoc
Place Eugène Bataillon
F-34060 Montpellier Cédex

and

GRECO 47 "Grandes Déformations et Endommagement"

CHAPTER 1 : INTRODUCTION

These notes intend to give a brief summary of a few recent developments in the field of the behavior of heterogeneous materials with some emphasis on the dissipative or non-linear range. This topic has been widely discussed in the framework of polycrystals, and the main celebrated contributions by BUDIANSKI & al, HILL, HUTCHINSON, KRONER, MANDEL and others are recalled by A. ZAOUI in this volume. Less attention has been paid to plasticity of composite materials, mainly for two reasons. The first is that most of the composite materials develo ped in the past thirty years exhibit a brittle behavior rather than a ductile one. However, because of the importance of thermal loadings, we have been witnessing a significant development of metal matrix composites, with a highly non linear behavior. The second reason of the limited interest for the non-linear problems is the difficulty of the subject and almost no micromechanical problems have yet been solved in a closed form except simple ones. The following HILL's appreciation[1] (1967) is still valid twenty years later : "... As for non-linear systems, the computations needed to establish any complete constitutive law are formidable indeed, even with the piecewise linearization forced by the model". Indeed in most situations we shall limit ourselves to pointing out some simple qualitative facts, or elaborating models based on crude approximations, and we shall often turn to finite element computations to obtain more specific results.

A typical example

Recent experiments by LITEWKA & al [2] illustrate in an illuminating manner the main points in which the present work is interested, and some results of these authors are briefly outlined here. In order to model anisotropic damage they performed tension tests at various inclinations on thin perforated sheets (see figure 1a) . Figure 1b, borrowed from their work, reports the curves external stress/external strain that have been observed at various inclinations.

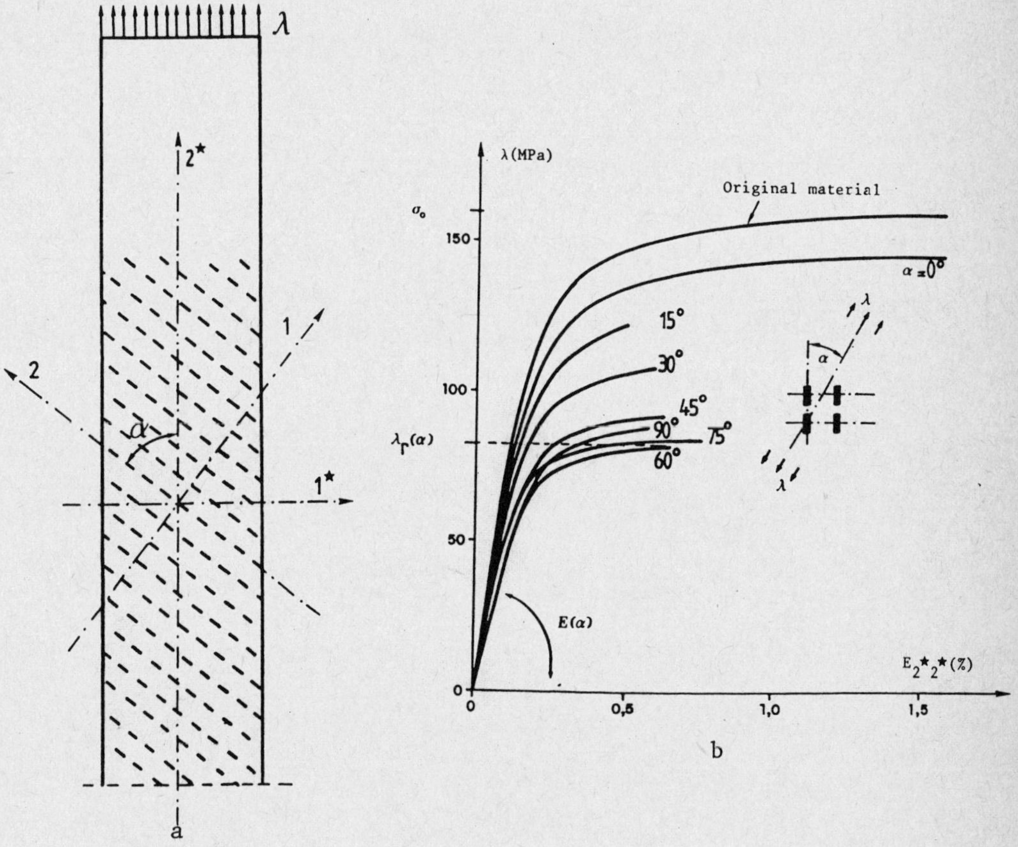

Tensile tests

stress-strain curves for various inclinations α

- Figure 1 -

Three different regimes in the behavior of this specific heterogeneous material are evidenced by this figure. For small external stresses and strains the material is in the linearly elastic range. For relatively large strains the external stress reaches a threshold which lead to rupture . A transient part is observed in which the hardening of the original material is affected by the perforations. The present work will devote one section to each of these three typical regimes : linear behavior, rupture of heterogeneous materials, overall elastic plastic hehavior of composites.

Contents

More specifically the paper is organized in the following way :

. Section 2 is devoted to general considerations on representative volume elements (r.v.e.), averaging and micromechanics. We pay a special attention to the boundary conditions imposed on the boundary of the r.v.e. which play an important role in non-linear problems. We set forth the importance and the generality of the so called HILL's macrohomogeneity equality which expresses the principle of virtual work between the microscopic and the macroscopic scales.

. Section 3 is devoted to linear problems. The concept of localization tensors introduced by HILL and MANDEL for heterogeneous elastic materials is exposed. We also consider Maxwell's viscoelastic bodies and we show that short range memory effects for the constituents give rise to long range memory effects for the composite.

. Section 4 is devoted to the failure of heterogeneous materials. We assume that the constituents possess an extremal yield locus which is the limiting set of all physical stress states. We propose a method of constructing the macroscopic extremal yield locus. The proposed set gives an overestimate of the actual set but this estimate turns out to be exact for rigid plastic or elastic plastic constituents.

. In section 5 we discuss the transient part of the stress strain curve of the composite, namely the influence of microscopic elasticity on macroscopic hardening. A large part of the qualitative analysis relies on HILL's and MANDEL's previous works [1,3]. Once the complexity of the exact law is recognized we turn to a few approximate models which yield more quantitative informations.

Notations

Throughout the following Einstein's convention of summation over repeated indices will be adopted. We shall avoid as far as possible the use of indices, denoting by a point or two points the summation over one or two indices. For instance

$\sigma . n$, $\sigma : \varepsilon$, $a : \varepsilon$, $\varepsilon' : a : \varepsilon$ stand for

$\sigma_{ij} n_j$, $\sigma_{ij} \varepsilon_{ji}$, $a_{ijkh} \varepsilon_{hk}$, $\varepsilon'_{ji} a_{ijkh} \varepsilon_{hk}$.

$\mathbb{R}^9_s$ is the space of 3×3 symmetric second order tensors.

ACKNOWLEDGEMENTS

Part of the work reported in this paper is taken from J.C. MICHEL's thesis, and from a joint study with O. DEBORDES, C. LICHT, J.J. MARIGO, P. MIALON and J.C. MICHEL. Many fruitful discussions with these persons are gratefully acknowledged.

CHAPTER 2

2. AVERAGES. BOUNDARY CONDITIONS

2.1. REPRESENTATIVE VOLUME ELEMENT

In the discussion of the overall properties of a highly heterogeneous medium two different scales are naturally involved :the macroscopic scale (termed x) on which the size of the heterogeneities is very small, and the so called "microscopic" scale (termed y) which is the scale of the heterogeneities. In order to derive a macroscopic (or homogenized) law for the composite one has to assume first that a "statiscally homogeneous specimen" or "representative volume element" can be defined in the composite. Experimentalists know that the assumption of statistical homogeneity can be a difficult matter illustrated for instance by the size effects encountered in the determination of the toughness of a composite. However we will disregard this difficulty and assume that at least one choice of the r.v.e. is possible. This choice of the r.v.e., or its modelling, determines a first difference between various theories of homogenization. In the model of spheres assembly (HASHIN [4]) the r.v.e. is filled with composite spheres of different sizes respecting the volumetric ratios of the phases ; in the self consistent scheme [5] the r.v.e. is successively modelled as an ellipsoidal inclusion of each phases in an infinite matrix endowed with the unknown macroscopic properties. In the homogenization theory of periodic media the r.v.e. is the unit cell, which generates by periodicity the entire structure of the composite. This unit cell is even sometimes modelled by an assembly of parallelepipedic blocks (ABOUDI [6]) . This variety of choices for the r.v.e. eventually results in different expressions of the macroscopic laws but the derivation of the latter follows, most of the time, the general procedure that has been settled by HASHIN [4] , HILL [7] , KRONER [8] and other pioneers of the subject of composite materials.

At a macroscopic point x we must consider two different families of variables : on the one hand macroscopic variables which stand in the homogeneous body the material properties of which we are looking for, on the other hand the microscopic variables which take place in the r.v.e. idealized by x at the macroscopic level.

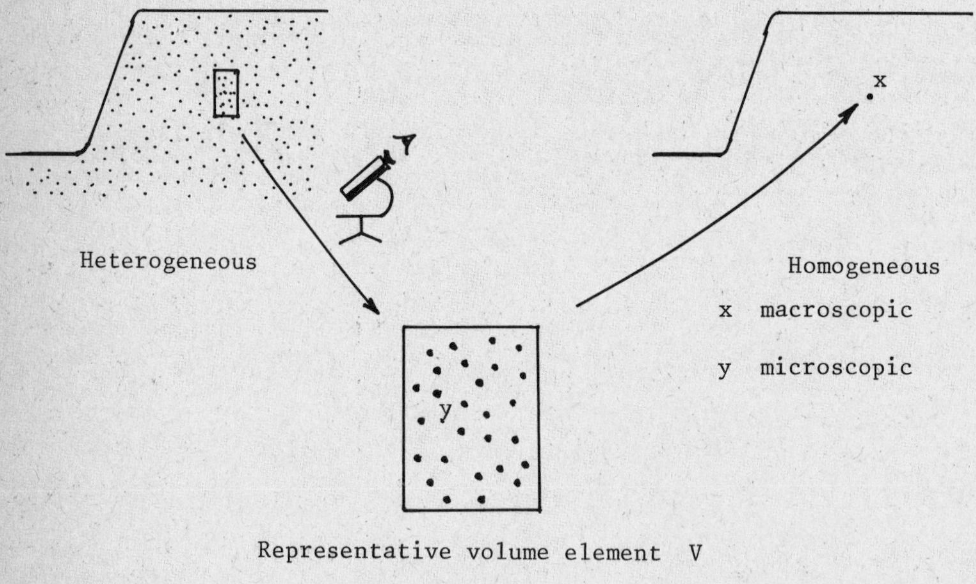

- Figure 2 -

For instance we shall distinguish

Σ E macroscopic stress and strain tensors

and

$\sigma(y)$ $\varepsilon(y)$ microscopic stress and strain tensors.

It results from classical arguments on oscillating functions that the macroscopic stress and strain tensors must be the averages of the microscopic corresponding quantities

$$\left.\begin{aligned} \Sigma_{ij} &= \frac{1}{|V|}\int_V \sigma_{ij}\,dy = \langle\sigma_{ij}\rangle^{(+)} \\ E_{ij} &= \frac{1}{|V|}\int_V \varepsilon_{ij}(u)\,dy = \langle\varepsilon_{ij}(u)\rangle \end{aligned}\right\} \quad (1)$$

where $\langle . \rangle$ stands for the averaging operator. However when the heterogeneities are voids or rigid inclusions, the stress or strain tensors remain to be defined in these heterogeneities, and more care is to be applied when considering the equality (1) (cf. § 3.2) .

Moreover, all the mechanical quantities which are usually assumed to be additive functions are averaged when proceeding from the microscopic level to the macroscopic one.

(+) $\langle\ \rangle$ is the average symbol

$$\left.\begin{array}{ll} \bar{\rho} = \langle \rho \rangle & \text{(additivity of mass)} \\ \bar{\rho}\mathcal{E} = \langle \rho\, e \rangle & \text{(additivity of internal energy)} \\ \bar{\rho}\, S = \langle \rho\, s \rangle & \text{(additivity of entropy)} \\ \mathcal{D} = \langle d \rangle & \text{(additivity of dissipation)} \end{array}\right\} \quad (2)$$

where capital letters refer to Macro quantities ; e , s , d respectively denote the specific internal energy, the specific entropy and the dissipation in the heterogeneous material.

2.2. LOCALIZATION

The procedure which relates Σ , E (and possibly their derivatives with respect to the time and other parameters), by means of (1)(2) , and of the micro constitutive laws, is termed *homogenization*. The inverse procedure, termed *localization*, amounts to a micromechanics problem which permits to determine microscopic quantities, for instance $\sigma(y)$ and $\varepsilon(y)$, from macroscopic ones, Σ and E . For this purpose the following system of equations, with data Σ or E , is to be solved for σ and $\varepsilon(u)$:

$$\left.\begin{array}{l} \text{microscopic constitutive law} \\ \operatorname{div} \sigma = 0 \quad \text{(micro equilibrium)} \\ \langle \sigma \rangle = \Sigma \quad \text{or} \quad \langle \varepsilon(u) \rangle = E \end{array}\right\} \quad (3)$$

This problem exhibits two noticeable differences with a classical problem :

i) the loading consists in the average value of one field (and not in surface or body forces)

ii) there is no boundary conditions.

Boundary conditions

Problem (3) turns out to be ill posed, due to the absence of boundary conditions, which are to be specified by a careful inspection of the status of the r.v.e. inside the heterogeneous medium. *These boundary conditions must reproduce, as closely as possible, the in situ state of the r.v.e. inside the material.* Therefore they strongly depend on the choice of the r.v.e. itself, and especially on its size. Although the attention will be focused on periodic media, we have to comment on two classical types of boundary conditions (HILL [1,7], HASHIN [4])

uniform stresses on ∂V : $\sigma.n = \Sigma.n$ on ∂V (4)

or

uniform strains on ∂V : $u = E.y$ on ∂V (5)

It is immediately seen that a displacement field which satisfies (5), and a divergence free field σ which satisfies (4), also satisfy

$$\langle \varepsilon(u) \rangle = E \quad , \quad \langle \sigma \rangle = \Sigma \quad .$$

In order to justify (4) or (5) (which are *not* equivalent boundary conditions) the r.v.e. must have a large size with respect of the heterogeneities size, so that the stress vector $\sigma.n$ or the displacement u on ∂V fluctuate about a mean with a wavelength small compared with the dimensions of the r.v.e.

However if periodic media are under consideration, and if the r.v.e. is chosen to be the unit cell the fluctuations of these fields about their average are large, and (4) or (5) are to be rejected. For a unit cell located at a sufficiently large distance from the boundary of the heterogeneous body, the strain and stress fields conform at the microscopic level to the periodicity of the geometry : σ and ε are "periodic fields", in a manner which will be specified soon. However it is already clear that the fields σ and ε, which depend on the two variables x (macro) and y (micro) are not exactly periodic throughout the composite : depending on the macrovariable they can vary from one place to the other, in a way similar to that of their averages $\Sigma(x)$ and $E(x)$. However their local variations, taken into account by their dependence on y, are supposed to be periodic. The precise meaning of these periodicity conditions is the following one :

<u>stress</u> : the stress vectors $\sigma.n$ are opposite on opposite sides of ∂V (where the external normal vectors n are also opposite) ;

<u>strain</u> : the local strain $\varepsilon(u)$ is split into its average and a fluctuating term

$$\varepsilon(u) = E + \varepsilon(u^\star) \quad , \quad \langle \varepsilon(u^\star) \rangle = 0 \quad ,$$

E is the Macro-strain, while $u^\star$ can be shown to be a periodic field, up to a rigid displacement that we disregard. The final form of the periodicity conditions on ∂V is :

$$\sigma.n \text{ anti periodic} \quad , \quad u = Ey + u^\star \quad u^\star \text{ periodic.} \qquad (6)$$

We term (4) (5) or (6) a set of "boundary conditions on ∂V" *for the pair* (u,σ) : (4) imposes stringent requirements on σ and none on

u , (5) imposes stringent requirements on u and none on σ , while (6) imposes requirements on both fields. For a specified set of boundary conditions, a displacement field u satisfying the boundary conditions will be said to be an *admissible displacement field*, while a divergence free stress field σ satisfying the boundary conditions will be said to be an *admissible stress field*. If moreover these fields satisfy

$$\langle \varepsilon(u) \rangle = 0 \quad , \quad \text{or} \quad \langle \sigma \rangle = 0$$

they are called "purely fluctuating fields" and a purely fluctuating stress field is a *self equilibrated* stress field.

Once the boundary conditions (4) (5) or (6) , are specified, the localization problem (3) is well posed (this assertion is to be checked in details for each constitutive law). In the variational discussion of this problem, the equality of virtual work plays obviously an important role, and can be expressed in simple terms.

Proposition 1. Let $\bar{\sigma}$ *and* $\bar{u}$ *be admissible fields of stress and displacements. Then the average of the microscopic work of* $\bar{\sigma}$ *in the strain field* $\varepsilon(\bar{u})$ *is equal to the macroscopic work* $\bar{\Sigma} : \bar{E}$

$$\langle \bar{\sigma} : \varepsilon(\bar{u}) \rangle = \bar{\Sigma} : \bar{E} \quad . \tag{7}$$

In order to prove (7) for the three sets of boundary conditions we introduce the purely fluctuating parts of $\bar{\sigma}$ and $\varepsilon(\bar{u})$

$$\bar{\sigma} = \bar{\Sigma} + \bar{\sigma}^{\star} \quad , \quad \text{where} \quad \langle \bar{\sigma}^{\star} \rangle = 0 \quad , \quad \operatorname{div} \bar{\sigma}^{\star} = 0$$

$$\varepsilon(\bar{u}) = \bar{E} + \varepsilon(\vec{u}^{\star}) \quad \text{where} \quad \langle \varepsilon(\bar{u}^{\star}) \rangle = 0 \quad .$$

An easy computation shows that

$$\langle \bar{\sigma} : \varepsilon(\bar{u}) \rangle = \langle (\bar{\Sigma} + \vec{\sigma}^{\star}) : \varepsilon(\bar{u}) \rangle = \bar{\Sigma} : \bar{E} + \langle \vec{\sigma}^{\star} : \varepsilon(\bar{u}) \rangle \quad ,$$

and

$$\langle \bar{\sigma} : \varepsilon(\bar{u}) \rangle = \langle \bar{\sigma} : (\bar{E} + \varepsilon(\vec{u}^{\star})) \rangle = \bar{\Sigma} : \bar{E} + \langle \bar{\sigma} : \varepsilon(\vec{u}^{\star}) \rangle \, .$$

A proper use of Green's theorem and of the equilibrium equations yields on the one hand

$$\langle \text{micro-work} \rangle = \bar{\Sigma} : \bar{E} + \frac{1}{|V|} \int_{\partial V} \vec{\sigma}^{\star} . n . \bar{u} \; ds \quad , \tag{8}$$

and on the other hand

$$\langle \text{micro-work} \rangle = \bar{\Sigma} : \bar{E} + \frac{1}{|V|} \int_{\partial V} \bar{\sigma} . n . \vec{u}^{\star} \; ds \quad . \tag{9}$$

If $\bar{\sigma}$ satisfies the boundary conditions (4) , i.e. if $\bar{\sigma}.n$ is uniform on ∂V , then $\vec{\sigma}^{\star}.n$ vanishes on ∂V and the equality (7) follows directly from (8) . If $\bar{u}$ satisfies (5) , then $\bar{u}^{\star}$ vanishes on ∂V and (7) follows directly from (9) . If $\bar{\sigma}$ and $\bar{u}$ satisfy (6) , the boundary integral in the second member of (9) vanishes since $\bar{\sigma}.n$ takes opposite values on opposite sides of ∂V , while $\bar{u}^{\star}$ takes equal values on these sets. This ends the proof of (7) which holds true for the three sets of boundary conditions (4) (5) or (6) . We shall term (7) the *equality of virtual work*(+) between the micro-

(+) (7) has sometimes been termed (7) HILL's Macrohomogeneity equality, or HILL's condition.

scopic scale and the macroscopic scale.

Remark. The equality (7) plays a central role in any homogenization theory. Up to a certain extent the boundary conditions on the boundary of the r.v.e. have a minor importance provided that they ensure the validity of (7). However in some statistical theories (7) is interpreted as an ergodic assumption, and deviations from it are sometimes considered (KRONER [8]).

Functional setting

The boundary conditions (4) (5) or (6) and the equilibrium equations can be expressed in a more compact manner, especially convenient for use of variational methods, namely

$$\left.\begin{array}{l} u = E.y + u^{\star} \quad , \qquad u^{\star} \in \mathcal{V}_o \\ \\ \sigma \in S_o = \varepsilon(\mathcal{V}_o)^{\perp} \end{array}\right\} \qquad (10)$$

where the space $\mathcal{V}_o$ of fluctuating displacements is one of the following ones, according to the type of selected boundary conditions :

case (4) $\quad \mathcal{V}_o = \hat{\mathcal{V}} = \{u^{\star} \in H^1(V)^3 \ ; \ \langle\varepsilon(u)\rangle = 0\}$

case (5) $\quad \mathcal{V}_o = \tilde{\mathcal{V}} = \{u^{\star} \in H^1(V)^3 \ ; \ u = 0 \ \text{ on } \ \partial V\}$

case (6) $\quad \mathcal{V}_o = \mathcal{V}_{per} = \{u^{\star} \in H^1(V)^3 \ ; \ u \ \text{ periodic on } \ \partial V\}$.

The macroscopic strain associated to a fluctuating $u^{\star}$ vanishes.

Therefore the equality of virtual work (7) yields

$$\sigma \text{ admissible} \Leftrightarrow \langle \sigma : \varepsilon(u^\star) \rangle = 0 \text{ for every } u^\star \text{ in } V_o$$

$$\text{i.e. } \sigma \in \varepsilon(V_o)^\perp .$$

The space of self equilibrated stress fields will be denoted by SE

$$SE = \{\sigma^\star \in \varepsilon(V_o)^\perp \; ; \; \langle \sigma^\star \rangle = 0\} .$$

In the next sections it will be understood that a choice of the boundary conditions, i.e. of the space V_o has been made among $\tilde{V}$, $\hat{V}$ or V_{per}.

CHAPTER 3

3. LINEAR PROBLEMS

We apply in this section the above considerations to linear constitutive laws, linear elasticity on the one hand, linear viscoelasticity on the other hand. It is to be understood that a choice of the boundary conditions on ∂V has been made, leading to a choice of V_o .

3.1 LINEAR ELASTICITY

Localization

The localization problem (3) in linear elasticity reads as

$$\left.\begin{aligned} &\sigma(y) = a(y) : \varepsilon(u(y)) = a(y) : (E + \varepsilon(u^\star(y)) \\ &\text{div}\ \sigma = 0 \quad , \quad \text{and boundary conditions} \ , \end{aligned}\right\} \quad (11)$$

where E *or* Σ is given. Since the material is heterogeneous the 4th order tensor of elastic stiffnesses a depends on the micro variable y . The fluctuating displacement $u^\star$ is therefore solution of the following Navier equations

$$\text{div}\ (a : \varepsilon(u^\star)) = - \text{div}\ (a : E) \quad , \quad \text{and boundary conditions} . \quad (12)$$

Assuming that the elasticity tensor a is constant on each constituent, it is readily seen that the second member of (12) reduces to body forces concentrated on the interface between the constituents :

$$\text{div}\,(a : E) = ([a] : E).n\,\delta_S$$

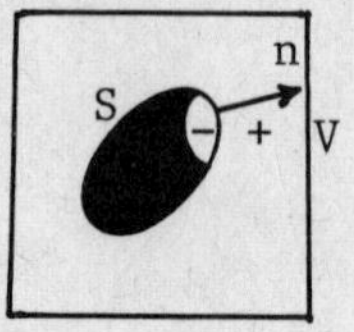

- Figure 3 -

where $[a] = a^+ - a^-$, and δ_S denotes the Dirac distribution on S .

It is worth noting that this concentrated loading is completely independent of the type of boundary conditions (4) (5) or (6) which have been selected in order to carry out the localization procedure.

It remains to prove that the problem (12) admits a solution, when Σ or E is given.

Proposition 2 . Under classical assumptions on the elastic tensor a *the problem* ((2) *admits a unique solution* $(\sigma,\varepsilon(u))$ *whatever is the set of boundary conditions* (4) (5) *or* (6) .

<u>E given</u>

Taking advantage of the fact that $<\sigma : \varepsilon(v)>$ vanishes for every v in V_o we obtain the following variational formulation of (12)

$$\left.\begin{aligned} & u^\star \in V_o \\ & <\varepsilon(v) : a : \varepsilon(u^\star)> = -<\varepsilon(v) : a : E> \quad \text{for every } v \text{ in } V_o \end{aligned}\right\} \quad (13)$$

It can be proved that $\varepsilon(V_o)$ is a Hilbert space (Problem 3.1), when endowed with the scalar product $<\varepsilon : \varepsilon'>$ ($\varepsilon, \varepsilon'$ belong to $\varepsilon(V_o)$). Then, under the classical assumptions of symmetry coercivity and boundedness of the elasticity tensor, the bilinear form

$$(\varepsilon,\varepsilon') \longrightarrow <\varepsilon : a : \varepsilon'>$$

is symmetric, continuous and coercive on the Hilbert space $\varepsilon(V_o)$. In a similar manner the following linear form is continuous on $\varepsilon(V_o)$

$$\varepsilon \longrightarrow <\varepsilon : a : E>$$

Thus, LAX-MILGRAM's theorem ensures existence and uniqueness of a solution $\varepsilon^\star = \varepsilon(u^\star)$, $u^\star$ in V_o, for (13). Existence and uniqueness of $\varepsilon(u) = \varepsilon^\star + E$ and of $\sigma = a : \varepsilon(u)$ follow directly.

Since the problem (13) is linear, its solution $\varepsilon(u^\star)$ depends linearly on the data E. More specifically let I_{ij} denotes the 2nd order tensor with components

$$(I_{ij})_{kh} = \frac{1}{2}(\delta_{ik}\,\delta_{jh} + \delta_{ih}\,\delta_{jk}) \quad ,$$

I the identity 4th order tensor has components

$$(I)_{ijkh} = (I_{ij})_{kh} \quad ,$$

and let $\varepsilon(\chi_{kh})$ denotes the solution of (13) when $E = I_{kh}$. $\varepsilon(\chi_{kh})$ is the field of fluctuating strains induced at the microscopic level by the 6 elementary states of macroscopic strain :

$$I_{11} = \begin{pmatrix} 1 & 0 & 0 \\ 0 & 0 & 0 \\ 0 & 0 & 0 \end{pmatrix} , \qquad 2\, I_{12} = \begin{pmatrix} 0 & 1 & 0 \\ 1 & 0 & 0 \\ 0 & 0 & 0 \end{pmatrix} \quad \ldots\ldots$$

(extension in direction 1) (shear strain between directions 1 and 2)

The solution $\varepsilon(u^{\star})$ of (13) for a general macro strain E is the superposition of the elementary solutions $\varepsilon(\chi_{kh})$

$$\varepsilon(u^{\star}) = E_{kh}\, \varepsilon(\chi_{kh}) \tag{14}$$

Finally the total field of microstrains amounts to

$$\varepsilon(u) = E_{kh}(I_{kh} + \varepsilon(\chi_{kh}))$$

$$\text{i.e.} \quad \varepsilon_{ij}(u) = D_{ijkh}\, E_{kh} = (D : E)_{ij} \tag{15}$$

where $D_{ijkh} = ((I_{kh})_{ij} + \varepsilon_{ij}(\chi_{kh}))$

D is 4th order tensor of *strain localization*(+) , since it yields the local strain $\varepsilon(u)$ in terms of the macroscopic strain E .

Homogenization

Once the localization procedure is known by (15) , the homogenization itself is straightforward :

(+) also termed "influence tensor" (HILL[1]), or "concentration tensor" (MANDEL[3])

$$\Sigma = \langle\sigma\rangle = \langle a : \varepsilon(u)\rangle = \langle a : D : E\rangle = \langle a : D\rangle : E$$

Then $\Sigma = a^{hom} : E$ where $a^{hom} = \langle a : D\rangle$.

In order to prove the symmetry of a^{hom} , which does not appear clearly on the above expression, we note further properties of the localization tensor D .

$$\langle D\rangle = I \quad , \quad \langle D^T\rangle = I \ ,$$

and for every admissible stress field $\overline{\sigma}$

$$\langle D^T : \overline{\sigma}\rangle_{ij} = \langle D^T_{ijkh}\ \overline{\sigma}_{kh}\rangle = \langle [(I_{ij})_{kh} + \varepsilon_{kh}(\chi_{ij})]\overline{\sigma}_{kh}\rangle = \overline{\Sigma}_{ij}$$

i.e. $\langle D^T : \overline{\sigma}\rangle = \overline{\Sigma}$ for every admissible $\overline{\sigma}$ (16)

The equality (16) allows to derive an equivalent expression of a^{hom} :

$$\Sigma = \langle D^T : \sigma\rangle = \langle D^T : a : \varepsilon(u)\rangle = \langle D^T : a : D\rangle : E$$

i.e. $a^{hom} = \langle D^T : a : D\rangle$ (17)

This last equality, which clearly shows the symmetry of a^{hom} , can be derived by energy considerations. Let us apply the averaging process to internal energies (cf. (2)) :

$$\overline{\rho}\mathcal{E} = \frac{1}{2} E : a^{hom} : E = \langle\rho e\rangle = \langle\frac{1}{2}\varepsilon(u) : a : \varepsilon(u)\rangle$$

$$= \frac{1}{2} E : \langle D^T : a : D\rangle : E \ ,$$

which yields the expression (17) for a^{hom} .

Σ given

We now consider the localization problem (11) when Σ is given :

$$\left.\begin{array}{ll} \varepsilon(u) = \varepsilon(u^{\star}) + E = A : \sigma & (E \text{ is unknown}) \\ \operatorname{div} \sigma = 0 \text{ , and boundary conditions} & \\ \langle\sigma\rangle = \Sigma & (\Sigma \text{ is known}) \end{array}\right\} \quad (18)$$

where A is the 4th order tensor of elastic compliances, inverse of a (A depends on the microscopic variable y) . Admit for a moment that (18) has a unique solution σ . Since the problem is linear its solution depends linearly on the data Σ . More specifically if C_{kh} denotes the solution of (18) with $\Sigma = I_{kh}$, we have

$$\sigma(y) = \Sigma_{kh} C_{kh}(y)$$

Let us define the 4th order tensor of *stress localization* C , by

$$C_{ijkh} = (C_{kh})_{ij} ,$$

then $\sigma = C : \Sigma$ (19)

This last equality allows to compute the tensor of macroscopic elastic compliances

$$E = \langle\varepsilon(u)\rangle = \langle A : \sigma\rangle = \langle A : C\rangle : \Sigma$$

$$\text{i.e.} \quad A^{hom} = \langle A : C \rangle \quad . \tag{20}$$

Further properties of C^T help to prove the symmetry of A^{hom} : $\langle C^T \rangle = I$, and for every admissible strain field $\varepsilon(\bar{u})$

$$\langle C^T : \varepsilon(\bar{u}) \rangle_{ij} = \langle C^T_{ijkh} \, \varepsilon_{kh}(\bar{u}) \rangle = \langle (C_{ij})_{kh} \, \varepsilon_{kh}(\bar{u}) \rangle$$

by the equality of virtual work

$$= \langle (C_{ij})_{kh} \rangle \langle \varepsilon_{kh}(u) \rangle = \bar{E}_{ij} .$$

Therefore

$$E = \langle C^T : \varepsilon(u) \rangle = \langle C^T : A : \sigma \rangle = \langle C^T : A : C \rangle : \Sigma$$

$$\text{and} \quad A^{hom} = \langle C^T : A : C \rangle \tag{21}$$

It remains to prove the existence and uniqueness of a solution of (18) . Taking advantage of the fact that $\langle \tau : \varepsilon(u) \rangle$ vanishes for every self equilibrated stress field τ , we obtain the following variational formulation of (18)

$$\left. \begin{array}{l} \sigma \in K \\ \\ \langle \bar{\sigma} - \sigma : A : \sigma \rangle = 0 \quad \text{for every } \bar{\sigma} \text{ in } K \end{array} \right\} \tag{22}$$

where $K = \{\bar{\sigma} \in \varepsilon(V_o)^{\perp} \; ; \; \langle \bar{\sigma} \rangle = \Sigma\}$.

Under the classical assumptions of symmetry, coercivity and boundedness of A , the bilinear form

$$(\tau,\sigma) \rightarrow \langle \tau : A : \sigma \rangle$$

is symmetric, continuous and coercive on $L^2(V)^9_s$. K is clearly a non empty closed convex set of this space and LAX-MILGRAM's theorem ensures the existence and uniqueness of a solution σ of (22) . Existence and uniqueness of $\varepsilon(u) = A : \sigma$ follows directly. This ends the proof of proposition 2 .

Note that we can write (22) in an equivalent form

$$\left. \begin{array}{l} \sigma \in K \\ \langle \tau : A : \sigma \rangle = 0 \quad \text{for every } \tau \text{ in SE} \end{array} \right\} \qquad (23)$$

<u>Equivalence between imposed strains and imposed stresses</u>

The tensors a^{hom} and A^{hom} constructed by imposing either a given E , or a given Σ are inverse tensors, provided that the *same* boundary conditions (4) (5) or (6) have been chosen to solve the localization problems. Indeed using the symmetry of a^{hom} we get

$$a^{hom} : A^{hom} = a^{hom\ T} : A^{hom} = \langle D^T : a \rangle : \langle A : C \rangle$$

But we notice from the true definition of D and C that $D^T : a$ is an admissible stress field

$$(D^T : a)_{ijkh} = a_{pqkh} [(I_{ij})_{pq} + \varepsilon_{pq}(\chi_{ij})] \quad ,$$

while $A : C$ is an admissible strain field

$$(A : C)_{khlm} = A_{khrs}(C_{lm})_{rs} .$$

Then by the equality of virtual work

$$\langle D^T : a\rangle : \langle A : C\rangle = \langle D^T : a : A : C\rangle = \langle D^T : C\rangle$$

$$= \langle D^T\rangle : \langle C\rangle = I .$$

Once the boundary conditions have been chosen among (4) (5) or (6) or any other type (see Problems), we can compute several pairs of elasticity tensors

$$\begin{array}{lll} \mathcal{V}_o = \hat{\mathcal{V}} & \hat{a}^{hom} , \hat{A}^{hom} & \\ \mathcal{V} = \mathcal{V}_{per} & a^{hom}_{per} , A^{hom}_{per} & \quad (24) \\ \mathcal{V} = \tilde{\mathcal{V}} & \tilde{a}^{hom} , \tilde{A}^{hom} & . \end{array}$$

As it has been proved previously the compliances tensors and the stiffnesses tensors computed by the same type of boundary conditions are inverse. However we point out that constructing the stiffnesses tensor by the assumption of uniform strain on ∂V , and the compliances tensor by the assumption of uniform stress on ∂V leads to an approximate theory, since these two tensors are not rigourously inverse. As pointed out by HILL [1] and MANDEL [3]

$$\tilde{a}^{hom} : \hat{A}^{hom} - I = O((\frac{d}{\ell})^3)$$

where d is the typical size of the heterogeneities, and ℓ is the typical size of the r.v.e. . For large r.v.e., containing of large number of heterogeneities the ratio d/ℓ is small, and the choice of the boundary conditions is unimportant. However for periodic media, when the r.v.e. is taken to be a unit cell, d and ℓ are of the same order, the different boundary conditions lead to substantial differences.

It can be proved (cf. Problems) that the strain energies defined by the 3 tensors (24) are ordered in the following way

$$E : \hat{a}^{hom} : E \leqslant E : a^{hom}_{per} : E \leqslant E : \tilde{a}^{hom} : E$$

(reverse inequalities for compliances).

For a periodic medium, the assumption of uniform strains on ∂V overestimates the stiffnesses, while the assumption of uniform stresses on ∂V underestimates it.

3.2 LINEAR ELASTICITY. COMPARISON EXPERIMENTS/COMPUTATIONS

We turn back to the tensile tests on perforated thin sheets, performed by LITEWKA & al [2] as described in the introduction. In view of the periodicity of the structure the r.v.e. is chosen to be the unit cell. The boundary conditions are of the periodic type (6) and $\mathcal{V}_o$ equals $\mathcal{V}_{per}$. Since a solution in a closed form seems to be unattainable, we solve the localization problem (13) (imposed macroscopic strain) by a finite element method. The computations are performed under the plane stress assumption and the only elementary macroscopic strains which are of interest here, are

$$I_{11} = \begin{pmatrix} 1 & 0 \\ 0 & 0 \end{pmatrix} \quad , \quad I_{22} = \begin{pmatrix} 0 & 0 \\ 0 & 1 \end{pmatrix} \quad , \quad 2I_{12} = \begin{pmatrix} 0 & 1 \\ 1 & 0 \end{pmatrix}$$

Since the heterogeneities considered in this example are voids we need to comment the definition of the macroscopic strain and stress. Indeed the expressions Σ and E are ambiguous since σ and ε are not defined in the hole. However assuming that the void consists in a infinitely soft heterogeneity, we can extend the fields σ and u , into $\bar{\sigma}$ (+) and $\bar{u}$ everywhere defined on the r.v.e. . Thus

---

(+) $\bar{\sigma}$ clearly vanishes in the void

$$\Sigma_{ij} = \langle \bar{\sigma}_{ij} \rangle = \frac{1}{|V|} \int_V \bar{\sigma}_{ij}\, dy = \frac{1}{|V|} \int_{V^\star} \sigma_{ij}\, dy$$

$$E_{ij} = \langle \varepsilon_{ij}(\bar{u}) \rangle = \frac{1}{|V|} \int_V \varepsilon_{ij}(\bar{u})\, dy = \frac{1}{|V|} \int_{\partial V} \frac{1}{2}(\bar{u}_i n_j + \bar{u}_j n_i)\, ds$$

$$= \frac{1}{|V|} \int_{\partial V} \frac{1}{2}(u_i n_j + u_j n_i)\, ds$$

where V denotes the r.v.e., including the hole T, $V^\star$ denotes the material part of V, $V^\star = V - T$, and where we have assumed that T does not intersect ∂V (hence $\bar{u} = u$ on ∂V). With this modified definition of Σ and E, the whole preceeding section remains valid.

Periodicity conditions

The finite element computations required to solve (13) with $E = I_{11}, I_{22}, I_{12}$ are standard, except for the periodicity conditions. However these periodicity conditions reduce to ordinary ones if the unit cell admits two orthogonal axis of symmetry. In the examples under consideration here, the unit cells are symmetric with respect to the lines $y_1 = 0$ and $y_2 = 0$. It is easily shown that the periodicity boundary conditions reduce to usual ones indicated on figure 4 below, and that the computations can be carried out on a quarter cell

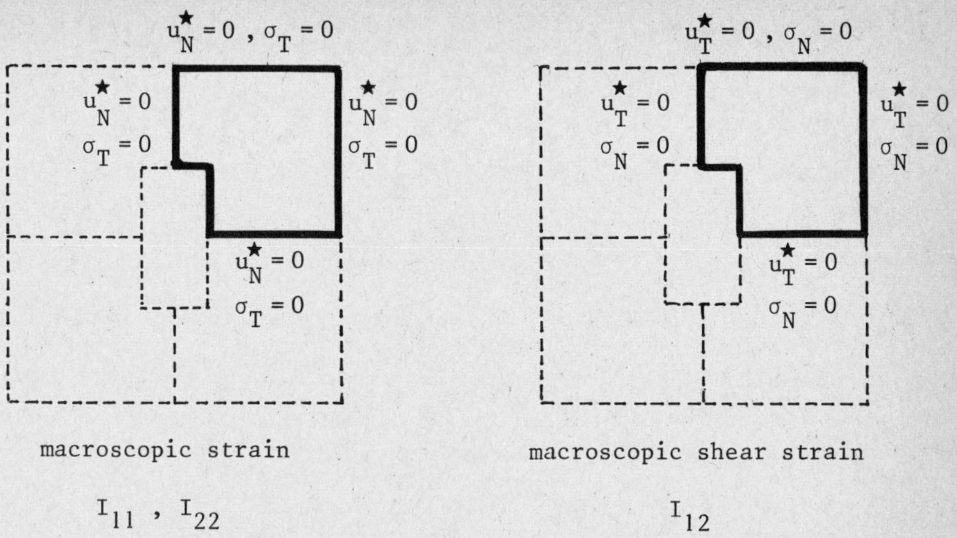

\- Figure 4 -

However in two important situations we cannot get rid of these periodicity conditions :

a) when the unit cell has no axis of symmetry. In this connection DUVAUT [9] and coworkers studied the influence of the shape of fibers cross section on the macroscopic stiffness of unidirectional composites, and the reader is referred to this work for more details.

b) when non linear materials are under consideration, the superposition principle (14) no more applies and it is not possible to separate the extension part and the shear part of a general macroscopic strain E , in order to carry out the numerical computations.

A survey of a few direct numerical procedures accounting for the periodicity conditions is given in DEBORDES & al [10] and MARIGO & al [11] .

Diagonal perforation pattern

Figures 5 & 6 show a few numerically computed microscopic strain, or stress states. The deformation of the perforation, and the stress concentration on its boundary are evidenced. A few noticeable facts deserve brief comments :

a) the deformation of the perforation is mainly due to the fluctuating strain $\varepsilon(\chi)$ which accounts for the heterogeneity of the material.

b) the deformed state corresponding to a shear macroscopic strain I_{12} clearly shows that the strain is *not uniform* on ∂V since straight lines do not remain straight lines. Indeed in most of the computations that the author has performed, the most significant differences between the various types of boundary conditions were observed on macroscopic strains of *shear* (with respect to the axis of the unit cell), and these differences resulted in significant variations of the shear moduli.

c) the agreement between the computations performed with the periodicity conditions and experiments is quite satisfactory.

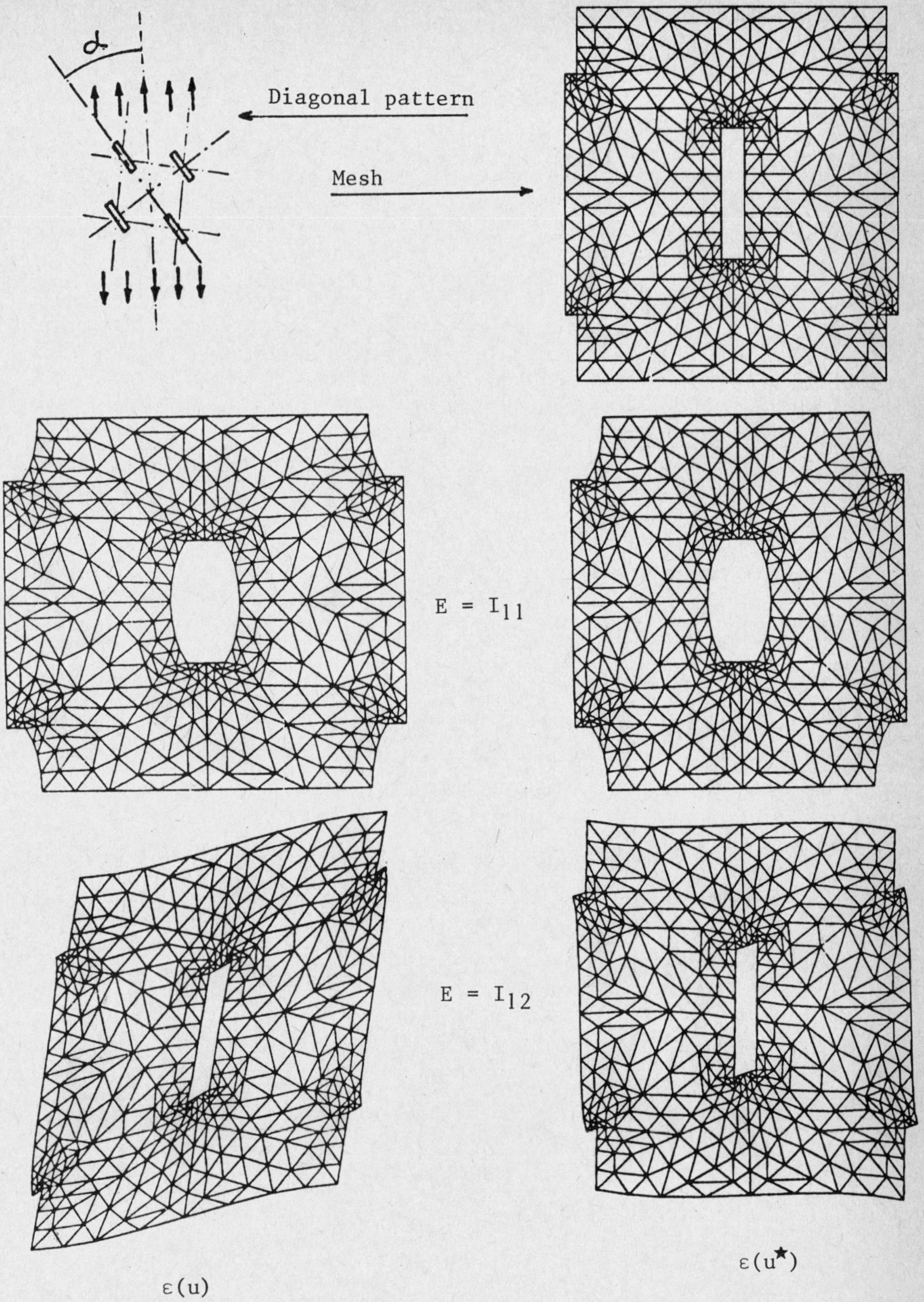

α
Diagonal pattern
Mesh
E = I11
E = I12
ε(u)
ε(u*)

- Figure 5 -

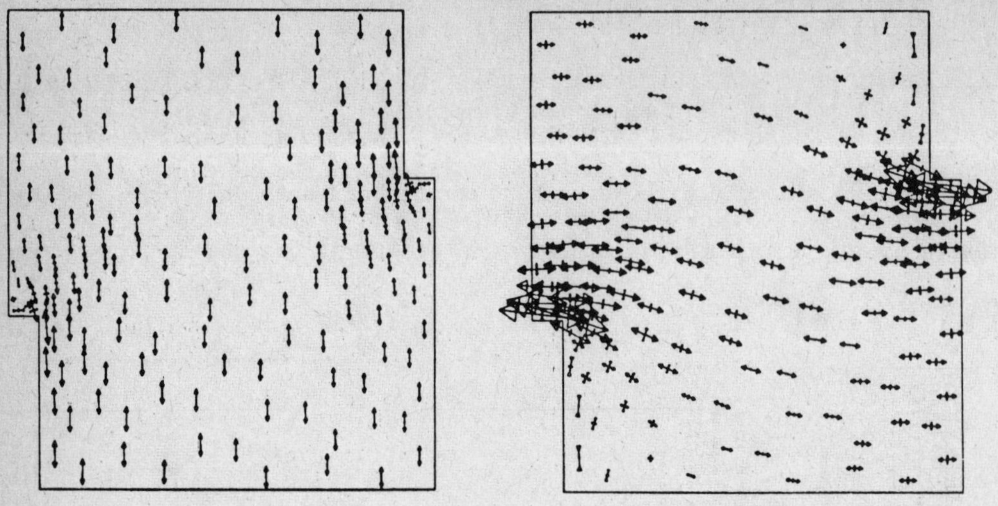

$$\Sigma = \begin{matrix} 0 & 0 \\ 0 & 1 \end{matrix} \qquad \Sigma = \begin{matrix} 1 & 0 \\ 0 & 0 \end{matrix}$$

microscopic stresses on a quarter cell

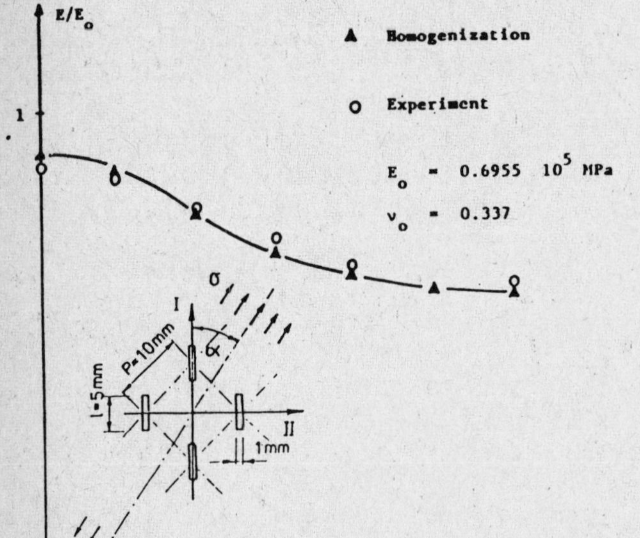

- Figure 6 -

Square perforation pattern(+)

The three types of boundary conditions (4) (5) or (6) are compared on the square perforation pattern. The stiffnesses predicted by the 3 theories are ordered in the way that energy arguments indicate : overestimate by the uniform strain theory, underestimate by the uniform stress theory.

Mesh

Square pattern

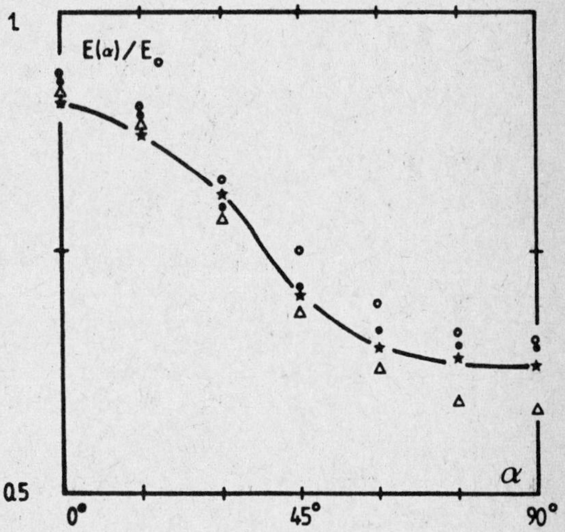

o Uniform strains on ∂V

● Periodicity conditions

★ Experiment [2]

Δ Uniform stresses on ∂V

Stiffnesses/orientation

- Figure 7 -

(+) The numerical computations for the square perforation pattern have been performed by J.C. MICHEL [12].

3.3 LINEAR VISCOELASTICITY

We consider an assembly of viscoelastic constituents obeying the MAXWELL's law :

$$\varepsilon(\dot{u}(y)) = A(y) : \dot{\sigma}(y) + B(y) : \sigma(y) \tag{25}$$

where A and B are 4th order tensors endowed with all the recommended properties (symmetry, boundedness, coercivity). Maxwell's law (25) has a "short memory" since it does not involve any creep function accounting for long range memory effects. The main result of this section is the following one.

Proposition 3. The macroscopic law derived from (25) *is*

$$\dot{E} = A^{hom} : \dot{\Sigma} + \int_0^t J(t-s) : \dot{\Sigma}(s)ds + B^{hom} : \Sigma \tag{26}$$

where the kernel J *will be specified in the text.*

Therefore the homogenization procedure gives rise to *long memory effects*, characterized by J. A similar result for KELVIN-VOIGT's materials was pointed out earlier by SANCHEZ & al [13] and FRANCFORT & al [14].

Localization

As in the elastic case the main point of the homogenization procedure is the localization problem. We assume that the macroscopic stress tensor follows a given path $\Sigma(t)$ and we search for the microscopic state $\sigma(t)$, $\varepsilon(u(t))$ thereby induced. The localization problem amounts to

$$\left.\begin{array}{l} A : \dot{\sigma} + B : \sigma = \varepsilon(\dot{u}) = \dot{E} + \varepsilon(\dot{u}^{\star}) \\ \operatorname{div} \sigma = 0 \quad , \quad \text{and boundary conditions} \\ \langle \sigma(t) \rangle = \Sigma(t) \quad . \end{array}\right\} \tag{26}$$

Applying the Laplace transform to (26) yields

$$\left.\begin{array}{l} (\lambda A + B)\hat{\sigma}(\lambda) = \hat{\dot{E}}(\lambda) + \varepsilon(\hat{\dot{u}}^{\star}(\lambda)) \\ \operatorname{div} \hat{\sigma}(\lambda) = 0 \quad , \quad \text{and boundary conditions} \\ \langle \hat{\sigma}(\lambda) \rangle = \hat{\Sigma}(\lambda) \quad , \end{array}\right\} \tag{27}$$

where $\sigma(t)$ is taken as vanishing in $t = 0$. Therefore we are let with a localization problem for a fictitious elastic material with elastic compliance $\lambda A + B$. Let us denote by $C(\lambda)$ the stress localization for the latter operator, solution of the following variational problem

$$\left.\begin{array}{l} \langle \tau : (A + \frac{1}{\lambda} B) : C(\lambda) \rangle = 0 \quad \text{for every } \tau \text{ in SE} \\ \\ \langle C(\lambda) \rangle = I \quad , \quad C(\lambda) \in \varepsilon(V_o)^{\perp} \quad . \end{array}\right\} \tag{28}$$

Let us denote by C^A and C^B the elastic stress localization tensors associated with A and B respectively, solutions of the following variational problem

$$\left.\begin{aligned} &\langle C^B \rangle = I \quad , \quad C^B \in \varepsilon(V_o)^{\perp} \\ &\langle \tau : B : C^B \rangle = 0 \quad \text{for every } \tau \text{ in } SE \end{aligned}\right\} \tag{29}$$

(similar problem for C^A) .

We set $C^{\star}(\lambda) = (C(\lambda) - C^B)/\lambda$ and we note that $C^{\star}(\lambda)$ satisfies

$$C^{\star}(\lambda) \in SE$$

$$\langle \tau : \lambda A + B : C^{\star}(\lambda) \rangle + \langle \tau : A : C^B \rangle = 0 \quad \text{for every } \tau \in SE \quad .$$

Using (29) for C^A yields another expression of the second term of the above equality :

$$\langle \tau : A : C^B \rangle = \langle \tau : A : C^B - C^A \rangle \quad ,$$

and $C^{\star}(\lambda)$ satisfies

$$C^{\star}(\lambda) \in SE \text{ , and for every } \tau \text{ in } SE$$

$$\langle \tau : A : (\lambda C^{\star}(\lambda) - C^A + C^B) \rangle + \langle \tau : B : C^{\star}(\lambda) \rangle = 0 \quad .$$

It is readily seen that $C^{\star}(\lambda)$ is the Laplace transform of the fourth order tensor $\mathcal{C}(t)$ solution of the following evolution equation

$$\left.\begin{aligned} &\mathcal{C}(t) \in SE \quad , \quad \mathcal{C}(0) = C^A - C^B \\ &\langle \tau : A : \dot{\mathcal{C}} \rangle + \langle \tau : B : \mathcal{C} \rangle = 0 \quad \text{for every } \tau \text{ in } SE \end{aligned}\right\} \tag{30}$$

which admits a unique solution $C(y,t)$. Coming back to $C(\lambda)$ we see that

$$C(\lambda) = \lambda C^{\star}(\lambda) + C^{B}$$

and

$$\hat{\sigma}(\lambda) = C(\lambda) : \hat{\Sigma} = C^{\star}(\lambda) : \lambda\hat{\Sigma} + C^{B} : \hat{\Sigma} \; .$$

Applying the inverse Laplace transform(+) yields

$$\sigma(t) = \int_0^t C(t-s) : \dot{\Sigma}(s)ds + C^{B} : \Sigma(t) \; .$$

This completes the localization procedure.

Homogenization

Noting that

$$\dot{\sigma}(t) = C^{A} : \dot{\Sigma}(t) + \int_0^t \dot{C}(t-s) : \dot{\Sigma}(s)ds$$

we obtain

$$\varepsilon(\dot{u}) = A : C^{A} : \dot{\Sigma} + \int_0^t [\, A : \dot{C}(t-s) + B : C(t-s)\,] : \dot{\Sigma}(s)ds$$

$$+ B : C^{B} : \Sigma(t)$$

where A , C^{A} , B , C^{B} and C depends on the microscopic variable y . Averaging gives

---

(+) Note that $\Sigma(t) = 0$ since $\sigma(t) = 0$.

$$\dot{E} = A^{hom} : \dot{\Sigma} + \int_0^t J(t-s) : \dot{\Sigma}(s)ds + B^{hom} : \Sigma(t) \tag{31}$$

where $J(\xi) = \langle A : \dot{C}(\xi) + B : C(\xi) \rangle$.

This completes the proof of proposition 3 .

CHAPTER 4

4. FAILURE OF DUCTILE HETEROGENEOUS MATERIALS

4.1 EXTREMAL YIELDING SURFACE

We assume that each constituent of the composite has an *extremal surface* which delimits the set $P(y)$ of all stress states that the material can physically admit

$$\sigma(y) \in P(y) \qquad y \in V \tag{32}$$

The behavior of the material is not further specified, the only useful information being the constraint (32). In most examples $P(y)$ is defined by means of a yield function $f(y,\sigma)$:

$$P(y) = \{\sigma \mid f(y,\sigma) \leqslant 0\}$$

Since the microscopic stress field is constrained its average the macroscopic stress has to be constrained too. More specifically let us assume that the yield locus P is defined by means of a (semi) norm $\| \ \|$

$$P(y) = \{\sigma \mid \|\sigma\| \leqslant \sigma_o(y)\} \quad .$$

Then

$$\|\Sigma\| \leqslant <\|\sigma\|> \leqslant <\sigma_o> \tag{33}$$

For instance if the norm under consideration is that of the equivalent stress

$$\|\sigma\| = \sigma_{eq} = (\frac{3}{2}\,\sigma^D_{ij}\,\sigma^D_{ij})^{1/2} \quad ,$$

then (33) amounts to

$$\Sigma_{eq} \leqslant <\sigma_o> \tag{34}$$

(33) provides a crude but simple upper bound for the macroscopic extremal yield locus, which remains to be defined in a more specific way.

For this purpose we note that, in order that a macroscopic stress Σ can be physically attained it must be possible to find a microscopic stress field σ fulfilling the following requirements :

i) $<\sigma> = \Sigma$

ii) $\operatorname{div} \sigma = 0$ and boundary conditions.

Note that i) ii) express that σ is in equilibrium with Σ .

iii) $\sigma(y) \in P(y)$ for every y in V .

It is therefore natural to consider the following set of macroscopic stresses :

$$P^{hom} = \{\Sigma \in \mathbb{R}^9_s \text{ such that there exists } \sigma \text{ satisfying}$$

$$<\sigma> = \Sigma \quad , \quad \sigma \in \varepsilon(\mathcal{V}_o)^{\perp} \quad , \quad \sigma(y) \in P(y) \quad \text{for every } y \text{ in } V\} \tag{35}$$.

Let us now assume that $P(y)$ exhibits further properties :

i) $P(y)$ is a closed convex set in $\mathbb{R}_s^9$. Then simple arguments show that P^{hom} is a closed convex set in $\mathbb{R}_s^9$. *Convexity is a stable property under homogenization.*

ii) For every y in V $P(y)$ contains a fixed ball of center 0 and of radius $k_o > 0$. Then P^{hom} is a non-empty set since it contains the ball of radius k_o and of center 0 .

Having shown that all physical macroscopic stress states Σ must lie within P^{hom} , the question arise whether all states Σ in P^{hom} are physical macroscopic stress states : the microscopic stress field associated with Σ should be related to a microscopic admissible strain field by the local constitutive law. If we do not further specify the constitutive law, the answer to the question is no : in the vocabulary of SALENCON [15] P^{hom} is the set of "potentially safe" Σ , and not of safe Σ . However if we consider elastic plastic constituents obeying the normality rule, it can be proved through rather technical functional analysis arguments that all stress states in the interior of P^{hom} can be attained. On the contrary, computing P^{hom} for elastic brittle constituents, as are most of the fibers in composite materials, could lead to a serious overestimate of the strength of the composite (see in this connection WEILL [16]) . Therefore the computation of P^{hom} will give a reliable prediction of the failure of a composite materials, only if the constituents are elastic plastic (or rigid plastic).

Throughout the following it will be assumed that $P(y)$ *is a closed convex set and that the constituents obey the normality rule.*

Rigid plastic constituents

Assume that the local constituents are rigid plastic and obey the normality rule. The inequality of maximal plastic work at the microscopic level, is valid for every y in V and reads as

$$\left.\begin{array}{l} \sigma(y) \in P(y) \\ \varepsilon(\dot{u}(y)) : \overline{\sigma} - \sigma(y) \leqslant 0 \quad \text{for every } \overline{\sigma} \text{ in } P(y) \end{array}\right\} \tag{36}$$

Let $\overline{\Sigma}$ be an element of P^{hom} to which corresponds $\overline{\sigma}(y)$ at the microscopic level by (35) . Then averaging (36) and applying the equality (7) of virtual work yields

$$\left.\begin{array}{l} \Sigma \in P^{hom} \\ \dot{E} : \overline{\Sigma} - \Sigma \leqslant 0 \quad \text{for every } \overline{\Sigma} \text{ in } P^{hom} \end{array}\right\} \tag{37}$$

If Σ is in the interior of P^{hom} we can take $\overline{\Sigma}$ in the form

$$\overline{\Sigma} = \Sigma + \Sigma^{\star}$$

where $\Sigma^{\star}$ is any vector in $\mathbb{R}^9_s$ with a sufficiently small norm, such that $\overline{\Sigma}$ lies in P^{hom} . Then (37) yields

$$\dot{E} : \Sigma^{\star} \leqslant 0$$

for every $\Sigma^{\star}$ with a sufficiently small norm. This last inequality applied to $\pm\Sigma^{\star}$ turns out to be an equality, and to be valid for every $\Sigma^{\star}$ (multiply it by any scalar value).

Thus $\dot{E} = 0$, and the composite is rigid if Σ is inside P^{hom} . The only possibility of straining occurs when Σ is on the boundary of P^{hom} .

Therefore the composite is rigid plastic, its domain of admissible stresses is exactly P^{hom} *, and it obeys the normality rule.*

4.2 DETERMINATION OF THE EXTREMAL SURFACE

It follows from its true definition (35) , that the determination of P^{hom} amounts to the resolution of a limit analysis problem on the r.v.e., where the loading parameters are the components of Σ . Classically this limit analysis problem can be solved either by the inside, through the construction of statically and plastically admissible fields, or by the outside through the evaluation of the plastic energy rate dissipated in strain fields leading to ruin.

Determination by the inside

A direction Σ° of macroscopic stresses is fixed and we consider the following onedimensional limit analysis problem

$$\lambda_o = \sup \{\lambda \text{ such that there exists } \sigma \text{ satisfying}$$

$$<\sigma> = \lambda\Sigma^\circ , \quad \sigma \in \varepsilon(V_o)^{\perp} , \quad \sigma(y) \in P(y) \text{ for every } y \text{ in } V\} \qquad (38)$$

$\lambda_o\Sigma^\circ$ is on the boundary of P^{hom} .

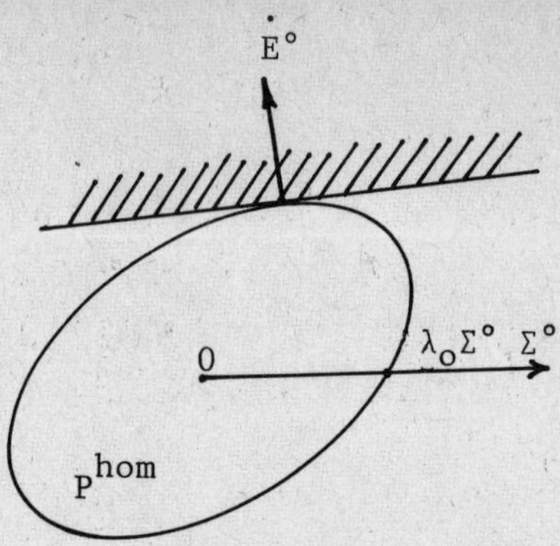

- Figure 8 -

Determination by the outside

Let us consider a macroscopic strain rate $\dot{E}$. Then P^{hom} is the intersection of the following half spaces

$$H(\dot{E}) = \{\overline{\Sigma} \mid \overline{\Sigma} : \dot{E} \leqslant \mathcal{D}(\dot{E})\} \tag{39}$$

where $\mathcal{D}$ denotes the energy-rate plastically dissipated at the macroscopic scale in the strain rate $\dot{E}$. $\mathcal{D}$ is computed by means of the averaging procedure :

$$\mathcal{D}(\dot{E}) = \langle d(y,\varepsilon(\dot{u}))\rangle = \underset{\overline{u} - Ey \in V_o}{\text{Inf}} \langle d(y,\varepsilon(\overline{u}))\rangle \tag{40}$$

where $d(y,e) = \sup\limits_{\overline{\sigma} \in P(y)} \overline{\sigma} : e$

Indeed we shall prove the following inclusion

$$P^{hom} \subset \bigcap_{\dot{E} \in \mathbb{R}^9_s} H(\dot{E}) \qquad (41)$$

letting the proof of the reverse inclusion to the reader. Let Σ be an element of P^{hom} and σ one possible microscopic stress field fulfilling the requirements of (35) . $\dot{E}$ being given, let $\overline{u}$ be any admissible displacement rate satisfying

$$\overline{u}^{\star} = \overline{u} - \dot{E}y \in V_o \qquad (42)$$

Then by the equality (7) of virtual work

$$\Sigma : \dot{E} = \langle \sigma : \varepsilon(\overline{u}) \rangle \leqslant \langle \sup_{\overline{\sigma} \in P(y)} \overline{\sigma} : \varepsilon(\overline{u}) \rangle = \langle d(y, \varepsilon(\overline{u})) \rangle$$

Taking the infimum over all admissible displacement rates $\overline{u}$ satisfying (42) yields

$$\Sigma : \dot{E} \leqslant \mathcal{D}(\dot{E}) \quad \text{for every} \quad \dot{E} \in \mathbb{R}^9_s$$

which proves that Σ belongs to $H(\dot{E})$ for every $\dot{E}$ in $\mathbb{R}^9_s$.

<u>Numerical determination of P^{hom}</u> .

In order to numerically determine P^{hom} we solve an auxilliary evolution problem for a fictitious elastic perfectly plastic material obeying the normality rule and admitting $P(y)$ as its local yield locus.

Two types of loadings can be considered. Either a direction of macroscopic stress Σ° is specified,and the evolution problem yields asymptotically a solution to (38) (MICHEL[12]). Or a direction of macroscopic strain rate is specified and the macroscopic stress of the evolution problem follows a path within P^{hom} which ends as t goes to $+\infty$ on the boundary of P^{hom}, at a point Σ_∞ which admits $\dot{E}^\circ$ as external normal to the extremal surface (SUQUET[17]).

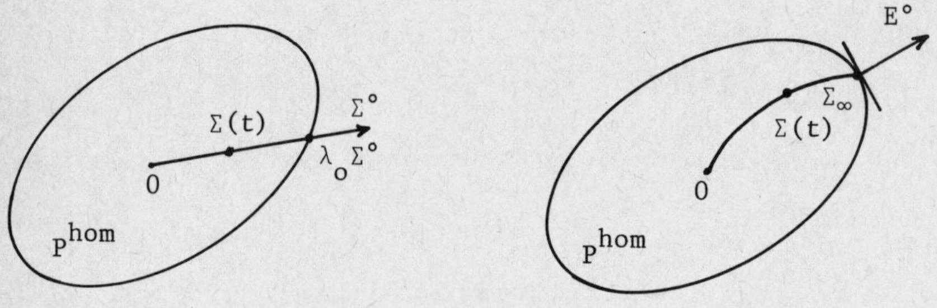

- Figure 9 -

For more details the reader is referred to DEBORDES & al where numerical computations show that the two above loadings give very similar results for P^{hom}.

Influence of the boundary conditions on ∂V.

The variety of boundary conditions which can be imposed on ∂V, leading to various possible choices of the space V_o, allows to define at least three different sets P^{hom} : $\hat{P}^{hom}$, P^{hom}_{per}, $\tilde{P}^{hom}$. In the definition of $\hat{P}^{hom}$ the stresses are supposed to be uniform on ∂V, in the definition of P^{hom}_{per} they are only supposed to be periodic, and no assumption on the stresses on ∂V are involved in the definition of $\tilde{P}^{hom}$. Since the other requirements contained in the definition (35) of P^{hom} are identical for the three sets, the following inclusions are easily stated

$$\hat{P}^{hom} \subset P^{hom}_{per} \subset \tilde{P}^{hom} \tag{43}$$

Using the embeddings $\tilde{\mathcal{V}} \subset \mathcal{V}_{per} \subset \hat{\mathcal{V}}$, we derive the following inequalities on the plastic dissipations which also result from (43) :

$$\hat{\mathcal{D}} \leqslant \mathcal{D}_{per} \leqslant \tilde{\mathcal{D}} \tag{44}$$

If periodic media are under consideration, (43) asserts that the assumption of uniform stresses on ∂V will give an underestimate of the strength, while the assumption of uniform strains on ∂V will overestimate the strength.

Comments :

P^{hom} has been introduced in the above form by the author [18], for periodic media. However previous works similarly based on limit analysis contained more or less explicitely the above definition (35) of P^{hom} : HILL [1], DRUCKER [19], SHU & al [20], M^c LAUGHLIN [21], LE NIZHERY [22] (+). More recently DE BUHAN [23,24] reached a similar result for multi layered media which amounts to (35) for periodic stratifications and illustrated his work by interesting analytical determinations of P^{hom} in connection

---

(+) further references might be found in HASHIN [4] section 7 .

with Soil Mechanics problems. In a different direction, GURSON [25] has proposed a yield criterion for porous materials, and his arguments almost amount to the computation of $\mathcal{D}(\dot{E})$. He did not used the entire space $\mathcal{V}_o$, but rather a Riesz approximation of it, splitting the field $\bar{u}$ which enters (40) on a basis of displacement rates derived from solutions of linear problems.

4.3 COMPARISON EXPERIMENTS/NUMERICAL COMPUTATIONS

We go back to the experiments by LITEWKA & al reported in the introduction and compare them with numerical results taken from MICHEL [12] and MARIGO & al [11] .

Rupture loads

In the tensile test reported on figure 1 the macroscopic stress tensor, when expressed in axis (1,2) , takes the form

$$\Sigma = \lambda \begin{pmatrix} \sin^2\alpha & \sin\alpha\ \cos\alpha & 0 \\ \sin\alpha\ \cos\alpha & \cos^2\alpha & 0 \\ 0 & 0 & 0 \end{pmatrix} = \lambda\Sigma^\circ(\alpha) \qquad (45)$$

We use the definition (38) of $\lambda_r(\alpha)$ where it appears as an upper bound :

$$\lambda_r(\alpha) = \sup \{\lambda \mid \lambda\Sigma^\circ(\alpha) \in P^{hom}\}$$

In order to solve (38) the computations are performed on the square perforation pattern, and the virgin material is idealized as an elastic-perfectly plastic one. Therefore the hardening part of the stress-strain curve is not correctly reproduced, but this lack of precision does not affect the value of the limit load. The elastic properties of the virgin material are specified in section 3, and we note on figure 1 that its ultimate equivalent stress is

$$\sigma_o = 159 \text{ MPa} .$$

It will be supposed to obey the Von Mises criterion

$$\sigma_{eq} \leqslant \sigma_o .$$

The specific numerical method used to solve (38) is described in details in [11]. Let us only comment briefly on the periodicity boundary conditions. In elastic problems on a r.v.e. exhibiting symmetries, we have reduced them to ordinary ones, mainly by means of the superposition principle. However, in the non-linear setting under consideration here, tensile stresses and shear stresses cannot be decoupled, and for the general stress $\Sigma^\circ(\alpha)$ we cannot get rid of the periodicity conditions. A survey of possible methods of resolution of problems involving periodic boundary conditions (penalty, elimination, Lagrangian...) is given in DEBORDES & al [10].

We have plotted on figure 10 the external stress strain curves, computed on the idealized material at various inclinations α. We can deduce from this figure the values of the ultimate loads $\lambda_r(\alpha)$.

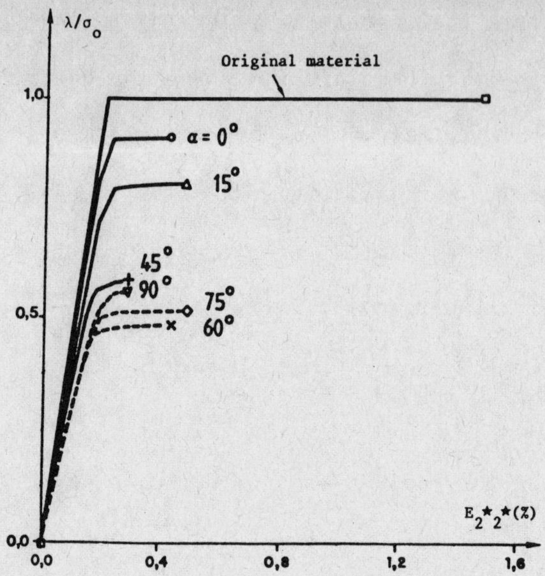

External stress / external strain (computed)

- Figure 10 -

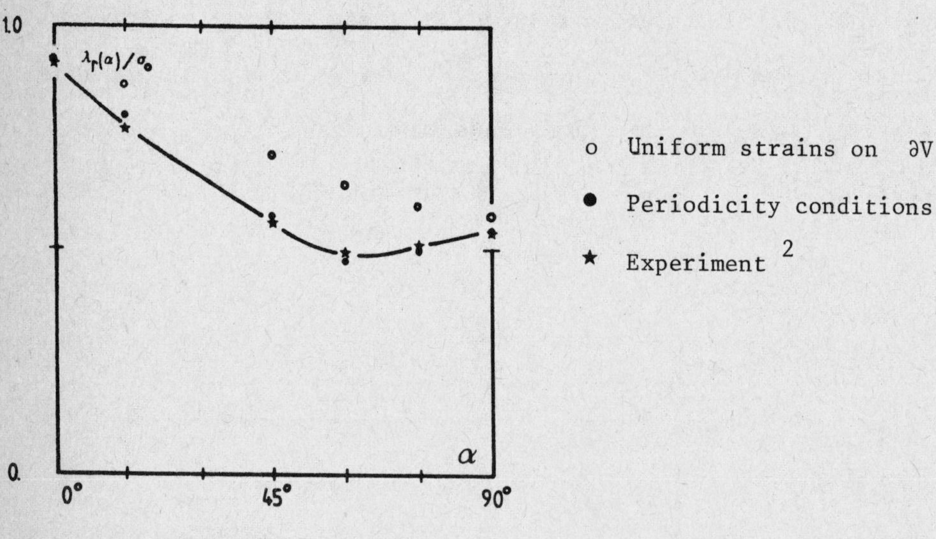

Failure loads / orientation

- Figure 11 -

Figure 11 reports the results of experiments, of the homogenization theory with periodicity conditions, and of the homogenization theory with uniform strain on ∂V . The agreement of the former theory with experiments is quite satisfactory, while the results of the latter (uniform strain) are overestimated in an obvious manner in agreement with previous considerations (43) .

Comments

This observation leads us to the following important comment.

In elasticity, the boundary conditions seemed to play a minor role since all local effects due to deviations in the boundary conditions were smoothed out. However, in plasticity local effects play an important role, and the deviations in boundary conditions are enlarged at the macroscopic scale. This fact has another interpretation. While it seems acceptable to model random distributions of elastic constituents by a periodic arrangement, such an idealization will be hazardous for elastic plastic constituents.

Rupture modes

Up to now the comparison experiments/computations was performed on macroscopic quantities (stiffnesses or rupture loads). This comparison can also been made at a more local level on the shape of the rupture modes. Figure 12 shows that the agreement is still good, but this holds true only for the periodic boundary conditions, since it is clear that when the plastic zone reaches the boundary of V the strain cannot be any more closed to a uniform strain on ∂V .

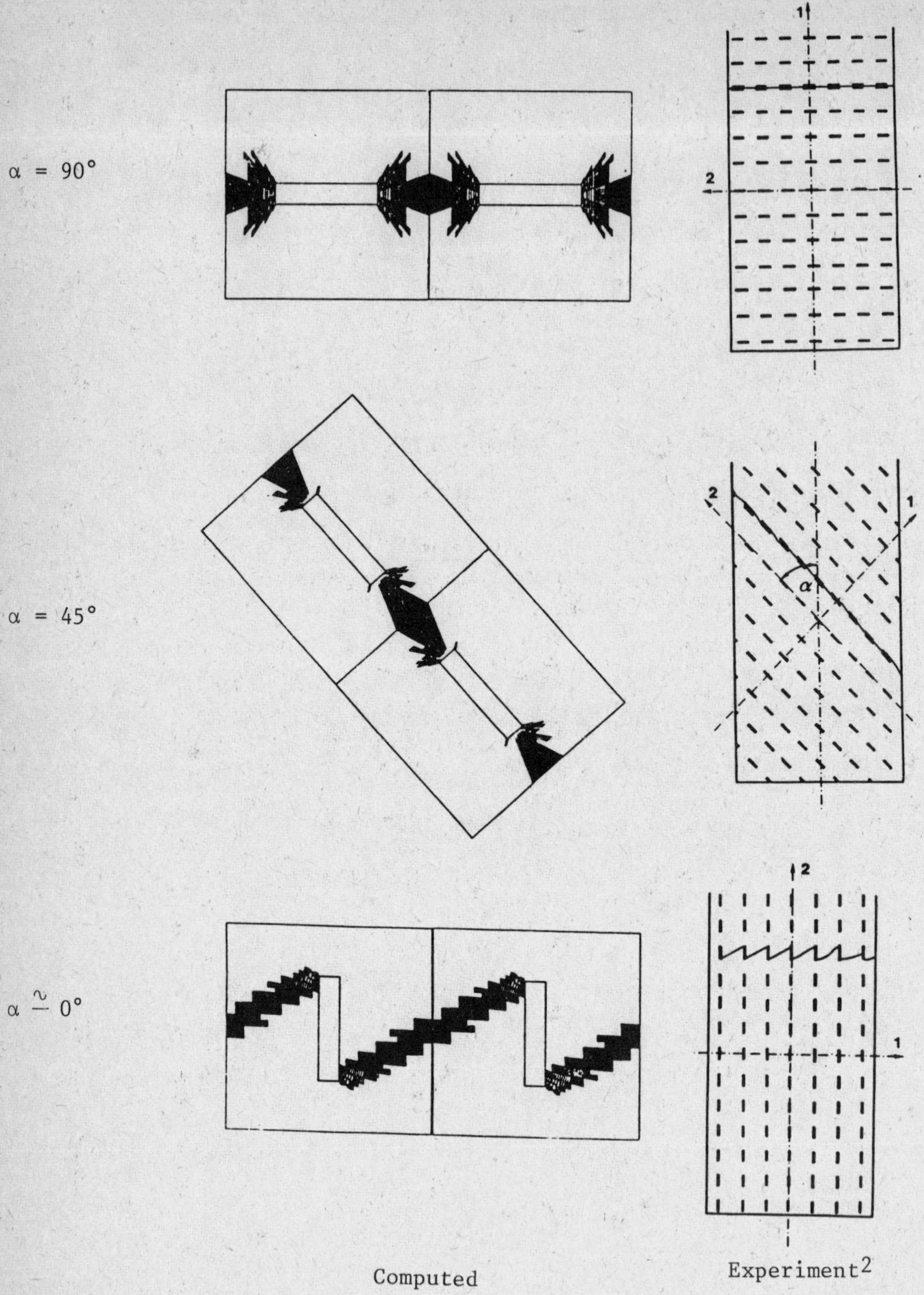

- Figure 12 : Failure Modes -

C H A P T E R 5

5. ELASTIC PERFECTLY PLASTIC CONSTITUENTS

We now turn to the more difficult problem of describing the overall behavior of a material made of the assembly of elastic perfectly plastic constituents. In the stress-strain curve of such a material a hardening part is strongly expected. This hardening effect, due to a micro-stored elastic energy, will be described qualitatively and approximate models will be proposed. Most of the developments presented here are also valid for viscoplastic constituents.

The micro constitutive law reads as

$$\varepsilon(u) = \varepsilon^e + \varepsilon^p \quad , \quad \varepsilon^e = A : \sigma \tag{46}$$

and $\sigma(y) \in P(y)$ for every y in V

$$\dot{\varepsilon}^p(y) : \bar{\sigma} - \sigma(y) \leqslant 0 \qquad \text{for every } \bar{\sigma} \text{ in } P(y)$$

5.1 MACROSCOPIC POTENTIALS

The major part of this paragraph follows the line of MANDEL's work[3] chap. 7 devoted to the macroscopic behavior of polycristalline aggregates.

Macroscopic plastic strain

We multiply (46) by the transposed tensor of elastic stress localization C^T, and we average on the r.v.e.

$$<C^T : \varepsilon> = <C^T : A : \sigma> + <C^T : \varepsilon^p> = <\sigma : A : C> + <C^T : \varepsilon^p>$$

A:C and σ are respectively an admissible strain field and an admissible stress field. By the equality of virtual power (7) we get

$$<C^T : \varepsilon> = <C^T> : <\varepsilon> = E \; ; \; <\sigma : A : C> = <\sigma> : <A : C> = A^{hom} : \Sigma$$

and

$$E = A^{hom} : \Sigma + <C^T : \varepsilon^p> \qquad (47)$$

We recognize in $A^{hom} : \Sigma$ the elastic part of the macroscopic strain

$$E^e = A^{hom} : \Sigma = <C^T : A : \sigma> = <C^T : \varepsilon^e> ,$$

and therefore the plastic part of the macroscopic strain is given by

$$E^p = <C^T : \varepsilon^p> \qquad (48)$$

It is worth noting that, generally speaking, neither the elastic part nor the plastic part of the macro strain is the average of its microscopic analogue (+).

(+) a noticeable exception occurs when ε^p is compatible (i.e. derives from a displacement field). It can be the case for large strains, when elastic strains are small.

Macro internal energy

The averaging procedure for additive quantities described in section 2 allows us to compute the macroscopic internal energy as the average of the microscopic internal energy. Since the processes under consideration in this section are isothermal the micro internal evergy reduces to the elastic energy :

$$\bar{\rho}\,\mathcal{E} = \langle \rho\ e \rangle = \frac{1}{2} \langle (\varepsilon(u)-\varepsilon^{p}) : a ;(\varepsilon(u) - \varepsilon^{p}) \rangle = \frac{1}{2}\langle \sigma : A : \sigma \rangle$$

We split the actual micro stress σ into two parts : the one which would occur if the material were perfectly elastic and a *self equilibrated residual stress tensor* (so called since it is the stress state under a null macro stress $\Sigma = 0$)

$$\sigma(y) = C(y) : \Sigma + \sigma^{r}(y) \qquad (49)$$

Using this decomposition we get

$$\bar{\rho}\,\mathcal{E} = \frac{1}{2} \langle \sigma : A : \sigma \rangle = \frac{1}{2}\Sigma : \langle C^{T} : A : C \rangle : \Sigma + \langle \sigma^{r} : A : C \rangle \Sigma + \frac{1}{2}\langle \sigma^{r} : A : \sigma^{r} \rangle$$

But $A:C$ is an admissible strain field, while σ^{r} is a self equilibrated field. Applying (7) we see that the cross term $\langle \sigma^{r} : A : C \rangle$ vanishes. We are let with

$$\bar{\rho}\,\mathcal{E} = \frac{1}{2}\,\Sigma : A^{hom} : \Sigma + \frac{1}{2} \langle \sigma^{r} : A : \sigma^{r} \rangle \qquad (50)$$

The first term in the above expression of $\mathcal{E}$ is the macroscopic elastic energy, and the second term is the *stored energy* : it is the elastic energy of the residual stresses and it is always positive except when the residual stresses vanish. It will be shown later on that this occurs only when the micro plastic strains are compatible, i.e. when they derive from a displacement field.

More about stored energy

In a recent work[26] CHRYSOCHOOS has reported microcalorimetric experiments performed on a AU4G in monotonic uniaxial tension. He has observed for this specific experiment that the stored energy reaches a threshold when the plastic strain increases. This means that the ratio stored energy/external work tends to 0 and not to 10 % as it is classically admitted by a somewhat hazardous interpretation of TAYLOR and QUINNEY's experiments. The limitation of the stored energy can receive an interpretation by means of the above arguments. Consider an assembly of elastic perfectly plastic constituents each of them obeying a Von Mises criterion (or any pressure insensitive criterion). It follows from the inequality (34) that the deviatoric part Σ^D of the macro stress tensor is limited. In a uniaxial tension test this deviatoric part can be expressed in terms of the only non vanishing component of Σ and this shows that Σ itself is bounded at any stage of the tension test. At the microscopic scale we know from the Von Mises criterion that σ^D is bounded. If we can prove that $\mathrm{Tr}\,\sigma$ is bounded at least in the space $L^2(V)$ then σ itself will be bounded in this space at any stage of the test. Therefore by (49) the residual stresses σ^r will be bounded in $L^2(V)$ and their elastic energy, i.e. the energy stored along the tensile test will be limited. In order to prove that $p = -\frac{\mathrm{Tr}\,\sigma}{3}$ is bounded in $L^2(V)$ we notice that the equilibrium equation yields

$$\frac{\partial p}{\partial x_i} = \frac{\partial \sigma^D_{ij}}{\partial x_j}$$

Since σ^D is bounded in $L^\infty(V)$, $\frac{\partial p}{\partial x_i}$ is bounded in $H^{-1}(V)$. A classical argument in the discussion of Navier-Stokes equations yields

$$|p - \langle p \rangle|_{L^2(V)} \leqslant C \left|\frac{\partial p}{\partial x_i}\right|_{H^{-1}(V)} \leqslant C'$$

But $\langle p \rangle = -\frac{\mathrm{Tr}\,\Sigma}{3}$ is bounded by the above arguments. Therefore we have shown that p is bounded in $L^2(V)$. This completes the proof of the following result, evidenced by CHRYSOCHOOS's experiments : *in a monotonic uniaxial tension test, the stored energy of an assembly of elastic perfectly plastic materials is limited*. It is quite interesting to note, following the lines of CHRYSOCHOOS, that the classical models of kinematical or isotropic hardening do not ensure this limitation of the stored energy, although the mechanical behavior (i.e. the stress strain curve) predicted can fit experiments in a satisfactory manner for simple loadings. If we remember the role played by the stored energy in shakedown or accommodation analysis it becomes obvious that the above property is not an academic one if it can be generalized to more complex loadings. It also evidences the role that thermal experiments should play in the determination of the mechanical behavior of an aggregate.

Plastic work

The averaging procedure for additive functions allows us again to compute the macroscopic dissipation as the average of the microscopic one :

$$\mathcal{D} = <d> = <\sigma : \dot{\varepsilon}^p> = <C\ \Sigma : \dot{\varepsilon}^p> + <\sigma^r : \dot{\varepsilon}^p>$$

$$= \Sigma : <C^T\ \dot{\varepsilon}^p> + <\sigma^r : \dot{\varepsilon}^p> = \Sigma : \dot{E}^p + <\sigma^r : \dot{\varepsilon}^p>$$

In order to carry on the computation of $\mathcal{D}$ we note that the field of residual stresses σ^r has the following properties, which result from its definition

$$\left. \begin{aligned} &\sigma^r \in SE \\ &A\sigma^r + \varepsilon^p = \varepsilon(u^r) \end{aligned} \right\} \qquad (51)$$

where $\varepsilon(u^r) = \varepsilon(u) - A : C : \Sigma$ is an admissible strain field. Therefore the macroscopic dissipation amounts now to :

$$\mathcal{D} = \Sigma : \dot{E}^p + <\sigma^r : \varepsilon(\dot{u}^r)> - <\sigma^r : A : \dot{\sigma}^r> \qquad (52)$$

Since σ^r is self equilibrated the second term in (52) vanishes

$$\mathcal{D} = \Sigma : \dot{E}^p - <\sigma^r : A : \dot{\sigma}^r> = <\sigma : \dot{\varepsilon}^p>$$

Therefore the macro plastic work-rate $\Sigma : \dot{E}^p$ does not reduce to the average of the micro one $<\sigma : \dot{\varepsilon}^p>$, and the difference between these quantities is the elastic energy-rate $<\sigma^r : A : \dot{\sigma}^r>$ due to the development of residual stresses. At the microscopic level the plastic work-rate is entirely dissipated, while at the macroscopic level it is partly dissipated (in the plastic micro mechanisms) and partly stored in the increase of the elastic energy of residual stresses :

$$\Sigma : \dot{E}^p = \langle \sigma : \dot{\varepsilon}^p \rangle + \langle \sigma^r : A : \dot{\sigma}^r \rangle .$$

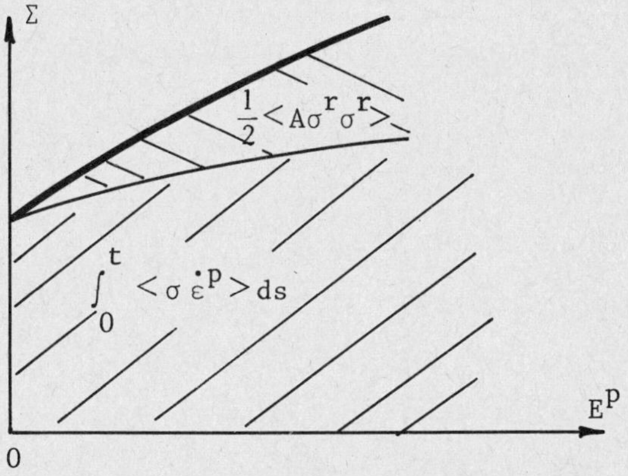

\- Figure 13 -

Stability in Drucker's sense

Since the constituents are assumed to be elastic perfectly plastic we know that at every microscopic point the following equality holds true

$$\dot{\sigma} : \dot{\varepsilon}^p = 0 .$$

The decomposition (49) of σ yields

$$\langle C\, \dot{\Sigma} : \dot{\varepsilon}^p \rangle + \langle \dot{\sigma}^r : \dot{\varepsilon}^p \rangle = \langle \dot{\sigma} : \dot{\varepsilon}^p \rangle = 0$$

But it follows from (51) and (48) that

$$<\dot{\sigma}^r : \dot{\varepsilon}^p> = -<\dot{\sigma}^r : A : \dot{\sigma}^r> \ , \ <C : \dot{\Sigma} : \dot{\varepsilon}^p> = \dot{\Sigma} : <C^T : \dot{\varepsilon}^p> = \dot{\Sigma} : \dot{E}^p$$

Therefore

$$\dot{\Sigma} : \dot{E}^p = <\dot{\sigma} : \dot{\varepsilon}^p> + <\dot{\sigma}^r : A : \dot{\sigma}^r> \geqslant 0 \tag{53}$$

This last inequality shows that the composite material is stable in Drucker's sense at the macroscopic level. It should be noted from (53) that, since $<\dot{\sigma}^r : A : \dot{\sigma}^r>$ is always non negative, we have

$$\dot{\Sigma} : \dot{E}^p \geqslant <\dot{\sigma} : \dot{\varepsilon}^p> \quad .$$

We could express this inequality by saying in a somewhat loosy manner that the change of scale stabilizes the material.

Macroscopic yield surface

We now assume that the composite material has been loaded up to microscopic stress state $\sigma(y)$ with residual stresses $\sigma^r(y)$. The macroscopic yield locus is the set of macroscopic stresses $\Sigma^\star$ which can be reached from the present state Σ by an elastic path, along which the residual stresses remain unchanged. The microscopic state $\sigma^\star$ satisfies

$$\sigma^\star(y) - \sigma(y) = C(y) : (\Sigma^\star - \Sigma) \tag{54}$$

and $\sigma^\star(y) = C(y) : \Sigma^\star + \sigma^r(y)$.

We notice that the condition

$$\sigma^{\star}(y) \in P(y) \quad \text{for every} \quad y \quad \text{in} \quad V$$

is equivalent to

$$\Sigma^{\star} \in C(y)^{-1} : (P(y) - \{\sigma^{r}(y)\}) \quad \text{for every} \quad y \quad \text{in} \quad V \quad \text{and therefore}$$

$$\Sigma^{\star} \in P^{hom}(\{\sigma^{r}\}) = \bigcap_{y \in V} C(y)^{-1} [P(y) - \{\sigma^{r}(y)\}] \tag{55}$$

The macro yield locus $P^{hom}(\{\sigma^{r}\})$ is a convex set (intersection of convex sets). Its determination at a given time t requires the knowledge of the whole set of residual stresses. Therefore it is not possible to entirely eliminate the microscopic level from the macroscopic behavior, as it is the case in the elastic setting. However we can analyse in a qualitative manner the way in which the macroscopic yield locus is obtained in the stress space :

★ The set $P(y) - \{\sigma^{r}(y)\}$ is translated from the original set $P(y)$, without change in shape, or size. This operation results in a kinematic hardening.

★ Multiplying the previous set by $C(y)^{-1}$ amounts to a rotation and an anisotropic expansion of this set. This operation does not reduce to isotropic hardening although it bears some resemblance with it.

★ The last operation is the intersection over all y in V. This is a complex operation including a change in shape, a change in size, and a change in the center of the convex set. If the intersection is to be taken over a finite set of points y (+) the boundary of the set $P^{hom}(\{\sigma^r\})$ will probably exhibit vertices. Such vertices will be smoothed off if the intersection is taken over an infinite set of points y (++). It should be noted that this smoothing effect is due to the non uniformity of the residual stresses and therefore take its origin mainly in the heterogeneous *elasticity* of the composite.

5.2 STRUCTURE OF THE MACROSCOPIC CONSTITUTIVE LAW

We now try to analyse in a qualitative way the structure of the macroscopic constitutive law. We claim that the state variables are :

★ the macro strain E

★ the *whole field* of micro plastic strains $\{\varepsilon^p(y), y \in V\}$. This means an infinite number of internal variables.

Indeed, once these variables are specified the actual micro stress state can be derived as follows :

a) Σ is deduced from E and $\{\varepsilon^p\}$ by (47)

b) σ^r can be computed as the solution of the elastic problem (51) where ε^p is considered as a known quantity (analogous to a thermal strain). The field σ^r is a linear functional of the field ε^p

(+) this is the case if yielding is likely to occur on a finite set of planes (monocristal) or if the residual stresses $\sigma^r(y)$, and the yield loci $P(y)$ are piecewise constant.

(++) this is the case if the r.v.e. is a polycristal with a large number of grains.

$$\sigma^r = - R : \varepsilon^p \quad \text{i.e.} \quad \sigma^r(y) = - \int_V R(y,y')\varepsilon^p(y')dy \tag{56}$$

The integro-differential operator R can be expressed easily in terms of the Green function of the elastic problem (51) , but we shall not need its exact expression.

Once the state variables are identified one has to compute the internal energy of the material. We already know by a previous computation that

$$\bar{\rho}\,\mathcal{E} = \frac{1}{2}\,\Sigma : A^{hom} : \Sigma + \frac{1}{2} < \sigma^r : A : \sigma^r >$$

and we now express $\mathcal{E}$ in terms of the state variables $(E,\{\varepsilon^p\})$

$$\bar{\rho}\,\mathcal{E} = \frac{1}{2}\,(E-E^p) : a^{hom} : (E-E^p) + \frac{1}{2} < \sigma^r : (\varepsilon(u^r) - \varepsilon^p) >$$

Taking into account (56) and the fact that σ^r is self equilibrated we get

$$\bar{\rho}\,\mathcal{E}(E,\{\varepsilon^p\}) = \frac{1}{2}\,(E-E^p) : a^{hom}(E-E^p) + \frac{1}{2} < R\varepsilon^p : \varepsilon^p > \tag{57}$$

The *state laws* relate the thermodynamical forces associated with the state variables, and the state variables themselves. The thermodynamical forces are defined as

$$\bar{\rho}\,\frac{\partial \mathcal{E}}{\partial E} \quad \text{and} \quad - \bar{\rho}\,\frac{\partial \mathcal{E}}{\partial \{\varepsilon^p\}} \quad ,$$

and their computation is immediate

$$\bar{\rho}\,\frac{\partial \mathcal{E}}{\partial E} = a^{hom}(E-E^p) = \Sigma$$

and for a virtual field of microscopic plastic strains $\delta\,\varepsilon^P$

$$<-\bar{\rho}\,\frac{\partial \mathcal{E}}{\partial\{\varepsilon^P\}} : \delta\,\varepsilon^P> \;=\; <a^{hom}(E-E^P)\frac{\partial E^P}{\partial\{\varepsilon^P\}} : \delta\,\varepsilon^P> - <R\,\varepsilon^P : \delta\,\varepsilon^P>$$

$$= <\Sigma : C^T : \delta\,\varepsilon^P> - <R\varepsilon^P : \delta\,\varepsilon^P> = <\delta\,\varepsilon^P : (C:\Sigma+\sigma^r)> = <\delta\,\varepsilon^P : \sigma>$$

The thermodynamical force associated with the state variable $\{\varepsilon^P\}$ is the micro stress field $\{\sigma\}$. Setting

$$\alpha = \{\varepsilon^P\}\ ,\quad A = \{\sigma\}\ ,\quad \mathcal{P} = \{\tau \in \varepsilon(\mathcal{V}_o)^{\perp}\ ,\tau(y) \in P(y) \ \text{ for every } y \text{ in } V\}$$

we can recognize the generalized standard form [27,28] of the macro constitutive law :

$$\left.\begin{array}{lll} \text{state laws} & \Sigma = \bar{\rho}\,\dfrac{\partial\mathcal{E}}{\partial E}\ , & A = -\,\bar{\rho}\,\dfrac{\partial\mathcal{E}}{\partial\alpha} \\ & A \in \mathcal{P} & \\ \text{complementary laws} & & \\ & (\dot{\alpha}\ ,\ A'-A) \leqslant 0 & \forall\, A' \in \mathcal{P} \end{array}\right\} \qquad (58)$$

However, this information on the structure of the macroscopic law is of little pratical importance since the constitutive law involves a infinite number of internal variables α . The next section will be devoted to the description of more useful, though approximate, models.

Remark. If the constituents are viscoelastic or viscoplastic a result similar to (58) holds true. Indeed the micro constitutive law now reads as

$$\varepsilon(\dot{u}) = \dot{\varepsilon}^{e} + \dot{\varepsilon}^{an} = A(y)\dot{\sigma} + \frac{\partial\varphi}{\partial\sigma}(y,\sigma) \qquad (59)$$

where φ is the potential which defines the anelastic part of the strain rate. The relations (47) and (48) still define the elastic and anelastic parts of the macroscopic strain. Moreover following RICE [29] it can be shown that the composite admits a *macroscopic potential* from which the anelastic part of the strain rate can be derived :

$$\dot{E}^{an} = \frac{\partial\Phi}{\partial\Sigma}(\Sigma,\sigma^{r}) \qquad (60)$$

where $\Phi(\Sigma,\sigma^{r}) = <\varphi(y,\sigma)> = <\varphi(y, C:\Sigma + \sigma^{r}>$.

The complete form of the macroscopic constitutive law is

$$\dot{E} = A^{hom}\,\dot{\Sigma} + \frac{\partial\Phi}{\partial\Sigma}\,(\Sigma,\sigma^{r}) \qquad (61)$$

where the residual stresses σ^{r} are found as the solution of the microscopic problem

$$\left.\begin{array}{l} \varepsilon(\dot{u}^{r}) = A\,\dot{\sigma}^{r} + \dfrac{\partial\Phi}{\partial\sigma^{r}}\,(\Sigma,\sigma^{r}) \\ \sigma^{r} \text{ self equilibrated, } \varepsilon(u^{r}) \text{ admissible strain field.} \end{array}\right\}$$

Once more the macroscopic and the microscopic levels are coupled by the presence of residual stresses.

5.3 APPROXIMATE MODELS

Once the complexity of the homogenized law is recognized we turn to approximate models in order to obtain more quantitative results. These approximate models are based on an a priori feeling of the microscopic distribution of plastic strains, or of residual stresses and more generally on the way in which yielding occurs at the microscopic level.

Piecewise constant plastic strains

In this first approximate model we replace the constraints

$$\sigma(y) \in P(y) \quad \textit{for every} \;\; y \;\; \textit{in} \;\; V \tag{62}$$

by the following weaker requirements

$$\Sigma^{(i)} \in P_i \qquad i = 1,\ldots,n \tag{63}$$

where V has been partitionned into $V_1,\ldots,V_n$, where Σ_i is the partial average of the microscopic stress on the ith phase V_i

$$\Sigma^{(i)} = \frac{1}{|V_i|}\int_{V_i} \sigma(y)\,dy = \langle\sigma\rangle_i \quad ,$$

and where P_i is the typical yield locus of V_i. We note that

$$\Sigma = c_i\,\Sigma^{(i)} \qquad \text{with} \qquad c_i = \frac{|V_i|}{|V|} \quad .$$

The set P reduces to

$$\mathcal{P} = \{\sigma \in \varepsilon(\mathcal{V})^{\perp} \; ; \; \Sigma^{(i)} = <\sigma>_i \in P_i \quad , \quad i = 1,\ldots,n\}$$

Then the normality law, expressed at the microscopic level and averaged over each phases V_i , reads as

$$\begin{aligned} &\sigma \in \mathcal{P} \\ &<\dot{\varepsilon}^p(y) : \bar{\sigma}(y) - \sigma(y)>_i \leqslant 0 \quad \text{for every } \bar{\sigma} \text{ in } \mathcal{P} \text{ , } i=1,\ldots,n \end{aligned} \tag{64}$$

Taking $\bar{\sigma} = \sigma + \sigma^{\star}$ where $\sigma^{\star}$ is chosen such that

$$<\sigma^{\star}>_i = 0 \quad , \quad \sigma^{\star} \in \varepsilon(\mathcal{V}_o)^{\perp} \tag{65}$$

We see that for every $i = 1,\ldots,n$, and for every $\sigma^{\star}$ satisfying (65)

$$<\dot{\varepsilon}^p : \sigma^{\star}>_i = 0 \tag{66}$$

The classical theory of Lagrange's multipliers shows that $\dot{\varepsilon}^p$ must be constant on each phase V_i

$$\dot{\varepsilon}^p(y) = \dot{E}_i^p \qquad \text{for all } y \text{ in } V_i \text{ .}$$

After a time integration we deduce that ε^p must be constant on V_i . The internal variables $(E,\{\varepsilon^p(y)\})$ reduce now to the finite set $(E,E_1^p,\ldots,E_n^p)$. For the sake of simplicity we shall assume that n equals 2 , i.e. that the plastic strain depends on only two independent variables E_f^p and E_m^p

$$\varepsilon^p(y) = E^p_m \, \theta_m(y) + E^p_f \, \theta_f(y) \quad , \tag{67}$$

where $\theta_m(y) = 1$ in the phase V_m and 0 in $V_f = V - V_m$ (similar definition for θ_f)(+). The microscopic constitutive law and the localization problem amount to find u and σ such that

$$\left.\begin{aligned} &\varepsilon(u) = E + \varepsilon(u^\star) = A : \sigma + E^p_m \, \theta_m + E^p_f \, \theta_f \\ &\text{div } \sigma = 0 \quad \text{and boundary conditions} \end{aligned}\right\} \tag{68}$$

This problem bears a strong resemblance with (11) and, since it is linear with respect to E , E^p_m , E^p_f its solution $u^\star$ can be split into

$$u^\star = E\,\chi + E^p_m \, \chi^p_m + E^p_f \, \chi^p_f \quad ,$$

where χ has been defined in section 3 , and where χ^p_m and χ^p_f are solutions of

$$\left.\begin{aligned} &\chi^p_m \in V_o \quad \text{and for every } v \text{ in } V_o \\ &\langle \varepsilon(v) : a : \varepsilon(\chi^p_m) \rangle = - \langle \varepsilon(v) : a : \theta_m \rangle \end{aligned}\right\} \tag{69}$$

(similar definition for χ^p_f) .

The microscopic strain reads as

(+) the use of the subscripts f and m indicates that we have in mind a matrix/fiber composite.

$$\varepsilon(u) = D : E + E_m^p \, \varepsilon(\chi_m^p) + E_f^p \, \varepsilon(\chi_f^p) \tag{70}$$

and this expression allows us to compute the macroscopic energy by the averaging procedure (2) . Computing the thermodynamical forces associated with E , E_m^p , E_f^p yields

$$\bar{\rho}\,\frac{\partial \mathcal{E}}{\partial E} = \langle \rho\,\frac{\partial e}{\partial E} \rangle = \langle (\varepsilon(\frac{\partial u}{\partial E}) - \frac{\partial \varepsilon^p}{\partial E}) : a : (\varepsilon(u) - \varepsilon^p) \rangle$$

$$= \langle D^T : a : (\varepsilon(u) - \varepsilon^p) \rangle = \langle D^T : \sigma \rangle = \Sigma \quad .$$

In the same way a straightforward computation shows that :

$$-\bar{\rho}\,\frac{\partial \mathcal{E}}{\partial E_m^p} = c_m \langle \sigma \rangle_m = c_m \, \Sigma^{(m)} \; , \; -\bar{\rho}\,\frac{\partial \mathcal{E}}{\partial E_f^p} = c_f \langle \sigma \rangle_f = c_f \, \Sigma^{(f)}$$

Moreover we derive from the averaged normality law (64) the following inequalities :

$$\dot{E}_m^p : \Sigma^{\star} - \Sigma^{(m)} = \langle \dot{\varepsilon}^p : \Sigma^{\star} - \sigma \rangle_m \leqslant 0 \quad \text{for every} \quad \Sigma^{\star} \in P_m$$

$$\dot{E}_f^p : \Sigma^{\star\star} - \Sigma^{(f)} = \langle \dot{\varepsilon}^p : \Sigma^{\star\star} - \sigma \rangle_f \leqslant 0 \quad \text{for every} \quad \Sigma^{\star\star} \in P_f$$

Now we can give the general form of the macroscopic law : after resolution of the problems (13) (69) we can compute the macroscopic internal energy $\bar{\rho}\,\mathcal{E}$ by the averaging process (2) . Then we set

$$\alpha = (E_m^p , E_f^p) \; , \; A = (c_m \, \Sigma^{(m)} \; , \; c_f \, \Sigma^{(f)}) \; , \; P = (c_m \, P_m \times c_f \, P_f)$$

the macroscopic law has the standard form :

$$\Sigma = \bar{\rho}\,\frac{\partial \mathcal{E}}{\partial E} \quad , \quad A = -\,\bar{\rho}\,\frac{\partial \mathcal{E}}{\partial \alpha}$$

$A \in P$ and for every $A^{\star}$ in P

$$(\dot{\alpha} : A^{\star} - A) \leqslant 0$$

Remark. The assumption that θ_m and θ_f are piecewise constant is not essential : therefore we can give to them more general values modelling a non uniform distribution of plastic strains at the microscopic level. The model is still valid provided that the equality (67) holds true.

Averaged criterion (MICHEL[12])

In the second approximate model we assume that the yield criterion is satisfied *in average*. Specifically it is assumed that the constraints (62) , which take the form

$$f(y,\sigma(y)) \leqslant 0 \quad \text{for every } y \text{ in } V$$

are replaced by one inequality :

$$\langle f(y,\sigma) \rangle \leqslant 0 \tag{71}$$

and the set P becomes

$$P = \{\sigma \in \varepsilon(V_o)^{\perp} \quad ; \quad \langle f(y,\sigma) \rangle \leqslant 0\} \quad .$$

The normality law for P reads as

$$\begin{aligned} &\sigma \in P \\ &\dot{\varepsilon}^P(y) = \dot{\lambda} \frac{\partial f}{\partial \sigma} (y,\sigma(y)) \end{aligned} \tag{72}$$

and there is only *one* plastic multiplier $\dot{\lambda}$ for the whole r.v.e. V since the yield condition is expressed by a single inequality (71). As pointed out by MICHEL [12] the problem (51) of computing the residual stresses takes a very simple form if we assume that the yielding is governed by the microscopic elastic energy

$$f(y,\sigma) = \frac{1}{2}\,\sigma : A : \sigma - k \, . \tag{73}$$

In this eventuality (72) gives

$$\dot{\varepsilon}^p = \dot{\lambda}\, A : \sigma$$

and the problem (51) of evaluating the residual stresses is equivalent to

$$\left.\begin{aligned} &\sigma^r \in SE \\ &A : \dot{\sigma}^r + \dot{\lambda}\, A : \sigma^r = \varepsilon(\dot{u}^r) - \dot{\lambda}\, A : C : \Sigma \end{aligned}\right\} \tag{74}$$

If we remember that $A : C : \Sigma$ is an admissible strain field we obtain that σ^r is solution of the following evolution problem

$$\left.\begin{aligned} &\sigma^r \in SE \\ &\langle \tau : A : \dot{\sigma}^r \rangle + \dot{\lambda} \langle \tau : A : \sigma^r \rangle = 0 \quad \text{for every } \tau \text{ in } SE \end{aligned}\right\} \tag{75}$$

the solution of which is

$$\sigma^r(t) = e^{-(\lambda(t)-\lambda_o)}\sigma^r(0) \ . \tag{76}$$

The remarkable fact in (76) is that we are able to compute the whole field of residual stresses as a function of a single parameter ξ :

$$\xi(t) = e^{-(\lambda(t)-\lambda_o)} \tag{77}$$

We note that the yield condition (71) becomes

$$\frac{1}{2}\langle (C : \Sigma + \sigma^r) : A : (C : \Sigma + \sigma^r)\rangle - \langle k\rangle \leqslant 0$$

$$\frac{1}{2}\,\Sigma : A^{hom} : \Sigma + \frac{1}{2}\,h\,\xi^2 - \langle k\rangle \leqslant 0 \tag{78}$$

where $h = \langle\sigma^r(0) : A : \sigma^r(0)\rangle$.

This last inequality (78) shows that the composite undergoes *isotropic hardening*, where the hardening parameter is ξ . It turns out that the state variables of this model are (E, E^p, ξ) , and the following lines will show that this choice of state variables lead to a generalized standard form of the macroscopic constitutive law. Indeed the general expression of the internal energy reduces here to

$$\bar{\rho}\ (E, E^p, \xi) = \frac{1}{2}\,(E-E^p) : a^{hom} : (E-E^p) + \frac{1}{2}\,h\,\xi^2$$

since $\langle\sigma^r : A : \sigma^r\rangle = h\,\xi^2$.

The thermodynamical forces associated with E , E^p and ξ are

$$\overline{\rho}\,\frac{\partial\mathcal{E}}{\partial E} = \Sigma \ , \quad -\,\overline{\rho}\,\frac{\partial\mathcal{E}}{\partial E^p} = \Sigma \ , \quad -\,\rho\,\frac{\partial\mathcal{E}}{\partial\xi} = -\,h\,\xi \ = A^{\xi} \quad . \tag{79}$$

We set

$$F(\Sigma,A^{\xi}) = \frac{1}{2}\,\Sigma : A^{hom} : \Sigma + \frac{1}{2}\,\frac{(A^{\xi})^2}{h} - \langle k\rangle \quad ,$$

and we note that according to (78)

$$F(\Sigma,A^{\xi}) \leqslant 0 \tag{80}$$

$$\begin{aligned}\dot{E}^p &= \langle C^T : \dot{\varepsilon}^p\rangle = \langle C^T : \dot{\lambda}\,A : \sigma\rangle = \\ &= \dot{\lambda}\,\langle C^T : A : C\rangle : \Sigma + \dot{\lambda}\,\langle C^T : A : \sigma^r\rangle\end{aligned}$$

But $\langle C^T : A : \sigma^r\rangle = \langle\sigma^r : A : C\rangle = 0$ by (7) .

We are let with

$$\dot{E}^p = \dot{\lambda}\,A^{hom} : \Sigma = \dot{\lambda}\,\frac{\partial F}{\partial\Sigma}\,(\Sigma,A_{\xi}) \tag{81}$$

On the other hand

$$\dot{\xi} = -\,\dot{\lambda}\,\xi \ = \dot{\lambda}\,\frac{\partial F}{\partial A_{\xi}}(\Sigma,A_{\xi}) \tag{82}$$

It can be checked that $\dot{\lambda}$ obeys the usual requirements of a plastic multiplier. The macroscopic constitutive law which consists in (79)-(82) is therefore a generalized standard law. We can give further interpretations of the parameters entering the model :

. the hardening modulus h is the elastic energy of the *initial residual stresses.*

. the size of the loading surface defined by $\mathcal{F}$ (or similarly by (78)) is

$$\langle k \rangle - \frac{1}{2} h \, \xi^2$$

But since $\dot{\lambda}$ is positive, λ increases and ξ decreases by (77) .

Therefore the size of the macroscopic loading surface increases and approaches k asymptotically :

. the plastic multiplier is proportional to the macroscopic dissipation. Indeed

$$D = \langle d \rangle = \langle \sigma : \dot{\varepsilon}^p \rangle = \langle \sigma : \dot{\lambda} : A : \sigma \rangle$$

$$= \dot{\lambda} \, \langle \sigma : A : \sigma \rangle$$

This last term either vanishes if $\langle \sigma : A : \sigma \rangle - \langle k \rangle < 0$ or is equal to $\dot{\lambda}\, k$ if $\langle \sigma : A : \sigma \rangle - \langle k \rangle = 0$. Thus

$$D = \dot{\lambda} \langle k \rangle$$

. The stored energy *decreases* along any loading path, since ξ decreases.

DVORAK & RAO's model for unidirectional fiber composites

A significant advance in the modelling of the plastic behavior of unidirectional fiber composite has been achieved by DVORAK & RAO[30] who proposed a model of kinematic hardening briefly described here below.

The matrix is supposed to be elastic perfectly plastic and to obey the normality rule, with yield function f . The fibers are elastic and aligned in the direction y_1 . The elementary volume is a composite cylinder of external radius b , in which a fiber of radius a is embedded in the matrix. The loadings under consideration are axisymmetric : traction or compression along y_1 , and equiaxial stresses in the plane (y_2, y_3) and this loading is represented by a two components vector $\underline{\Sigma}$

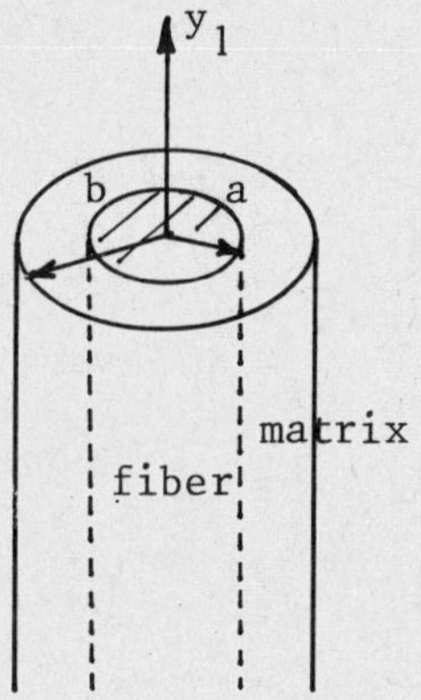

- Figure 14 -

$$\underline{\Sigma} = (\Sigma_{11} , \frac{1}{2} (\Sigma_{22} + \Sigma_{33}))$$

We shall denote by $\underline{C}$ the part of the localization tensor which yields the microscopic state in terms of $\underline{\Sigma}$ in the elastic range

$$\sigma(r) = \underline{C}(r)\underline{\Sigma}$$

where $r = (y_2^2 + y_3^2)^{1/2}$. In the elastic regime the maximal local stresses are located in the fiber and at the fiber-matrix interface. The first assumption of DVORAK & RAO's model is that this property still holds true in the elastic plastic range whatever is the field of (axisymmetric) initial residual stresses. More specifically numerical experiments performed by these authors show that it is reasonable to assume that *under any axisymmetric complex loading yielding occurs first at the fiber/matrix interface.* (83)

Assume that the composite is loaded from an elastic state in which the residual stresses are σ^r . Yielding at the macroscopic level occurs as soon as plasticity occurs at the microscopic level, i.e. by virtue of the above assumption as soon as the stresses at the interface reach the yield limit. Therefore the macroscopic yielding starts as soon as

$$f(\underline{C}(a)\underline{\Sigma} + \sigma^r(a)) = 0 \tag{84}$$

Let us set $\underline{X} = - \underline{C}(a)^{-1}\, \sigma^r(a)$. Then the condition (84) of macroscopic yielding reads as :

$$g(\underline{\Sigma} - \underline{X}) = 0 \ .$$

The above assumption, indeed rather weak, has allowed to derive the following remarkable result : the macroscopic yield surface undergoes *kinematic hardening*. Its center $\underline{X}$ moves in the space of axisymmetric loadings, while its shape (characterized by g) does not change. It is readily seen that this result is a general one, i.e. that if yielding at the microscopic level turns out to occur first in the same points, then the macroscopic yield surface undergoes kinematic hardening.

Hardening rule

The generality of the assumption (83) does not permit to compute the microscopic fields of plastic strains or of residual stresses as in the two previous examples. Therefore we cannot rely on thermodynamical arguments to derive the hardening rule and the flow rule. However this can be done by direct means. The loading condition expresses the orthogonality between $d\underline{\Sigma} - d\underline{X}$ and grad g .

$$0 = dg = \frac{\partial g}{\partial \underline{\Sigma}_1}(d\underline{\Sigma}_1 - d\underline{X}_1) + \frac{\partial g}{\partial \underline{\Sigma}_2}(d\underline{\Sigma}_2 - d\underline{X}_2) = 0$$

Therefore there exists a multiplier $d\mu$ such that

$$d\underline{X} = d\underline{\Sigma} - d\mu \begin{pmatrix} -\dfrac{\partial g}{\partial \underline{\Sigma}_2} \\ \\ \dfrac{\partial g}{\partial \underline{\Sigma}_1} \end{pmatrix} \tag{85}$$

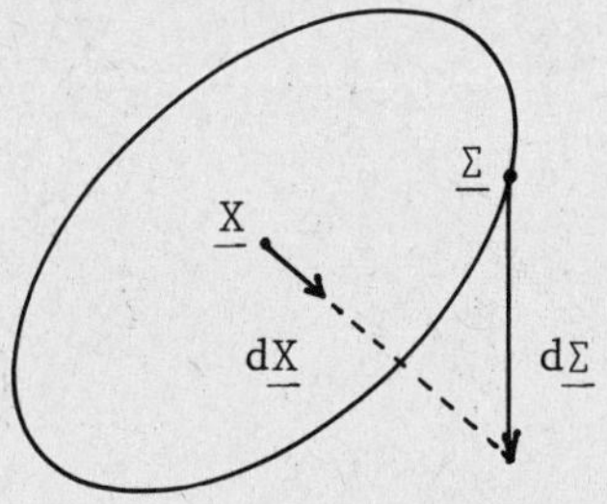

- Figure 15 -

In order to determine the multiplier $d\mu$ we assume that the increment $d\underline{X}$ of the internal stress $\underline{X}$, points from $\underline{X}$ in the direction of the new stress state $\underline{\Sigma} + d\underline{\Sigma}$ (see figure). The vectors $d\underline{X}$ and $\underline{\Sigma} - \underline{X} + d\underline{\Sigma}$ must be colinear. Neglecting the second order terms we obtain

$$d\mu = \frac{(\underline{\Sigma}_2 - \underline{X}_2)\, d\underline{\Sigma}_1 - (\Sigma_1 - \underline{X}_1)\, d\underline{\Sigma}_2}{(\underline{\Sigma}_1 - \underline{X}_1)\dfrac{\partial g}{\partial \underline{\Sigma}_1} + (\underline{\Sigma}_2 - \underline{X}_2)\dfrac{\partial g}{\partial \underline{\Sigma}_2}} \tag{86}$$

The hardening law is completely determined by (85) and (86) .

Flow rule

The plastic part of the macroscopic strain satisfies

$$\begin{pmatrix} d\underline{E}_1^p \\ \\ d\underline{E}_2^p \end{pmatrix} = d\lambda \begin{pmatrix} \frac{\partial g}{\partial \underline{\Sigma}_1} \\ \\ \frac{\partial g}{\partial \underline{\Sigma}_2} \end{pmatrix}$$

DVORAK & RAO determine the macroscopic plastic multiplier $d\lambda$ by assuming further properties of the localization of stresses during the loading process. Their assumptions, based on numerical calculations, yield the following expression for $d\lambda$:

$$d\lambda = \frac{1}{\frac{\partial g}{\partial \underline{\Sigma}_1}} \left(\frac{1}{c_f E_f} - \underline{A}_{12}^{hom}\right) d\underline{X}_1 + \left(\frac{-2\nu_f + (1-c_f)/c_f}{E_f}\right) d\underline{X}_2$$

where c_f denotes the fiber volume fraction and $\underline{A}_{12}^{hom}$ and $\underline{A}_{22}^{hom}$ the compliances of the composite relating $\underline{\Sigma}$ and $\underline{E}$.

CONCLUSIONS

The behavior of composites in the nonlinear and inelastic range is still a widely open subject where all further contributions would be greatly appreciated. We have tried to show that failure of ductile heterogeneous materials could be predicted in a satisfactory manner by means of a limit analysis study. Moreover we have proposed a few simplified models describing the macroscopic hardening of a composite. These models are based on crude approximation on the microscopic fields, or restrict the attention to specific loadings, and there is an important need of further work in this direction.

PROBLEMS

Part 2 2.1 Show that there exists infinitely many boundary conditions ensuring the validity of the equality of virtual work and leading to a well posed problem in (13). Consider for instance an arbitrary partition of V into $\partial_1 V$ and $\partial_2 V$ and impose uniform stresses on $\partial_1 V$ and uniform strains on $\partial_2 V$.

2.2 Consider a thin sheet periodically perforated in its thickness. Show that the natural boundary conditions on ∂V are

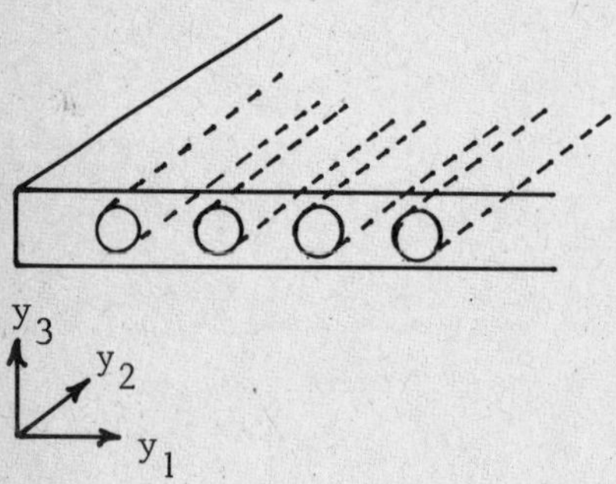

- Figure 16 -

$\sigma_{i3} = 0 \quad 1 \leqslant i \leqslant 3 \quad$ for $y_3 = \pm h/2$

$u_\alpha = E_{\alpha\beta} y_\beta + u^\star_\alpha \quad u^\star_\alpha$ periodic on ∂V

$\sigma_{\alpha\beta} n_\beta$ antiperiodic on ∂V

$(1 \leqslant \alpha, \beta \leqslant 2)$.

Prove that these boundary conditions ensure the validity of (7).

2.3 In the periodic setting can you relate (6) with the asymptotic expansion theory.

Part 3 3.1 Prove that $\varepsilon(V_o)$ is a Hilbert space, for the three possibilities $V_o = \hat{V}$, V_{per}, $\tilde{V}$, when endowed with the scalar product $<\varepsilon : \varepsilon'>$. In a first step prove that

$$(\exists\, c > 0)(\forall u \in H^1(V)^3)\ |u - <u>|_{L^2(V)^3} \leq c\ |\varepsilon(u)|_{L^2(V)^9_s} .$$

3.2 Prove that the solution $\varepsilon(u^\star)$ of (13) has the following variational property : $\varepsilon(u^\star)$ minimizes among all $\varepsilon(\vec{u}^\star)$ in $\varepsilon(V_o)$ the microscopic elastic energy

$$<(\varepsilon(\vec{u}^\star) + E) : a : (\varepsilon(\vec{u}^\star) + E)>$$

and prove that the minimum is $E : a^{hom} : E$. Using the inclusions $\tilde{V} \subset V_{per} \subset \hat{V}$ prove that for every E in $\mathbb{R}^9_s$

$$E : \hat{a}^{hom} : E \leq E : a^{hom}_{per} : E \leq E : \tilde{a}^{hom} : E .$$

3.3 Prove that the long memory entering (26) is fading

$$(\exists\, c > 0)(\exists\, k > 0) \qquad |J(t)| \leq c\ e^{-kt}$$

3.4 Consider a KELVIN-VOIGT's material

$$\sigma = a : \varepsilon(u) + b : \varepsilon(\dot{u})$$

Prove that the homogenized constitutive law is :

$$\Sigma(t) = a^{hom} : E(t) + \int_0^t K(t-s) : \dot{E}(s)ds + b^{hom} : \dot{E}(t) \quad \text{(cf. [14])}$$

where K is a kernel which you will specify.

Part 4   4.1   Consider a stratified two phase composite material, infinite and homogeneous in the directions (y_1, y_2) and periodically heterogeneous in the direction (y_3). Show that (cf. [24])

$$P^{hom} = \{\sigma \mid \Sigma = c_1 \Sigma^1 + c_2 \Sigma^2 \quad ; \quad \Sigma^1 \in P^1 \, , \Sigma^2 \in P^2 \, , \Sigma^1_{i3} = \Sigma^2_{i3}\}$$

where P^1 and P^2 are the yield locus of the phases 1 and 2.
Assume moreover that the two constituents are Tresca's materials

$$P^i = \{\sigma \mid \sup_{k,\ell} |\sigma_k - \sigma_\ell| \leq 2k^i\}$$

where σ_k denote the principal stresses of σ. Restricting your attention to 2 dim problems, show that

$$P^{hom} = \{\Sigma \mid \sup_{k,\ell} |\Sigma_k - \Sigma_\ell| \leq 2C(\alpha)\}$$

where α is the angle between Oy_1 and the direction of the major principal stress.

Part 5   5.1   In case of a MAXWELL's viscoelastic material relate the result (60)(61) with the homogenized constitutive law established in section 3. Compute the macroscopic dissipation with the help of (2), and show that it was not possible to derive it directly from (26).

REFERENCES

1. Hill, R., The essential structure of constitutive laws for metal composites and polycristals, *J. Mech. Phys. Solids*, 15, p. 79-95, 1967.

2. Litewka, A., Sawczuk, A., Stanislawska, J., Simulation of oriented continuous damage evolution, *J. Méca. Th. Appl.*, 3, p. 675-688, 1984.

3. Mandel, J., *Plasticité classique et viscoplasticité*, CISM Lecture Notes n° 97, Springer-Verlag, Wein , 1972.

4. Hashin, Z., Analysis of composite materials : a survey, *J. Appl. Mech.*, 50, p. 481-505, 1983.

5. Hill, R., A self consistent mechanics of composite materials, *J. Mech. Phys. Solids*, 13, p. 213-222, 1965.

6. Aboudi, J., Elastoplasticity theory for porous materials, *Mech. Materials*, 3, p. 81-94, 1984.

7. Hill, R., Elastic properties of reinforced solids : some theoretical principles, *J. Mech. Phys. Solids*, 11, p. 357-372, 1963.

8. Kröner, E., *Statistical Continuum Mechanics*, CISM Lecture Notes n° 92, Springer-Verlag, Wein , 1972.

9. Duvaut, G., Homogénéisation et matériaux composites. *Trends and Applications of Pure Mathematics to Mechanics*, Ed. Ciarlet, Roseau. Springer-Verlag, Wein , p. 35-62, 1984.

10. Debordes, O., Licht, C., Marigo, J.J., Mialon, P., Michel, J.C., Suquet, P., Analyse limite de structures fortement hétérogènes. Note Technique 85-3, Université de Montpellier, 1985.

11. Marigo, J.J., Mialon, P., Michel, J.C., Suquet, P., Plasticité et homogénéisation : un exemple de calcul des charges limites d'une structure hétérogène. Submitted.

12. Michel, J.C., Homogénéisation de matériaux élastoplastiques avec cavités. Thèse , Paris 6, 1984.

13. Sanchez Palencia, E., Sanchez Palencia, J., Sur certains problèmes physiques d'homogénéisation donnant lieu à des phénomènes de relaxation, *C.R. Acad. Sc. Paris*, (A), 286, 903-906, 1978.

14. Francfort, G., Leguillon, D., Suquet, P., Homogénéisation de milieu viscoélastiques linéaires de Kelvin Voigt, *C.R. Acad. Sc. Paris*, I, 296, 287-290, 1983.

15. Salençon, J., *Calcul à la rupture et analyse limite*, Presses ENPC, Paris, 1983.

16. Weill, F., Application de la théorie de l'homogénéisation à l'étude élastique et à la rupture de composites résine-fibres de verre, Thèse Doct. Ing. Nantes, 1984.

17. Suquet, P., Local and global aspects in the mathematical theory of plasticity, in *Plasticity Today*, Ed. Bianchi & Sawczuk, Elsevier Pub., 1984.

18. Suquet, P., Méthodes d'homogénéisation en Mécanique des solides, 15e Congrès du G.F.R. in *Comportements rhéologiques et structure des matériaux*, Ed. Huet & Zaoui, Presses ENPC, p. 87-128, 1981.

19. Drucker, D.C., On minimum weight design and strength of nonhomogeneous plastic bodies, in *Nonhomogeneity in Elasticity and Plasticity*, Ed. Olszak, Pergamon Press, 1959.

20. Shu, L.S., Rosen, B.W., Strength of fiber reinforced composites by limit analysis method., *J. Composite Mat.*, 1, p. 366-381, 1967.

21. Mc Laughlin, P.V., Limit behavior of fibrous materials, *Int. J. Solids Struct.*, 6, p. 1357-1376, 1970.

22. Le Nizhery, D., Calcul à la rupture des matériaux composites, Symposium Franco-Polonais, Cracovie 1977, in *Problèmes non linéaires de Mécanique*, Ed. Acad. Sc. Pologne, Varsovie, 1980, p. 359-370.

23. De Buhan, P., Homogénéisation en calcul à la rupture : le cas du matériau composite multicouche, *C.R. Acad. Sc. Paris* , II, 296, p. 933-936, 1983.

24. De Buhan, P., Salençon, J., Determination of a macroscopic yield criterion for a multilayered material, Colloque CNRS, *Critères de rupture des matériaux à structure interne orientée,* Ed. J.P. Boehler, To be published.

25. Gurson, A.L., Continuum theory of ductile rupture by void nucleation and growth : I. Yield criteria and flow rules for porous ductile media, *J. Eng. Mat. Tech.*, 99, 1977, p. 1-15.

26. Chrysochoos, A., Bilan énergétique en élastoplasticité grandes déformations, To be published in *J. Méca. Th. Appl.*

27. Halphen, B., Nguyen Q.S., Sur les matériaux standard généralisés, *J. Méca.*, 14, p. 39-63, 1975.

28. Germain, P., Nguyen, Q.S., Suquet, P., Continuum thermodynamics, *J. Appl. Mech.*, 105, p. 1010-1020, 1983.

29. Rice, J.R., On the structure of stress-strain relations for time dependent plastic deformations in metals, *J. Appl. Mech.*, 37, p. 728-737, 1970.

30. Dvorak, G., Rao, M., Axisymmetric plasticity theory of fibrous composites, *Int. J. Eng. Sc.*, 9, p. 971, 1971.

PART V

RANDOMLY INHOMOGENEOUS MEDIA

J.R. Willis
School of Mathematics
University of Bath
Bath BA2 7AY, England

The lectures that follow give a connected account of the overall behaviour of a composite, treated as a random medium. Unifying threads are the use of Green's functions to derive integral equations, and the use of variational principles. Insight into the type of behaviour that may be expected can be obtained from calculations based upon perturbation theory, either for weakly heterogeneous media or for media which are homogeneous except for the presence of a dilute distribution of inclusions. Both of these are described, but the main emphasis is on methods of approximation which may be used more generally. A statistical hierarchy of equations has to be closed: it is demonstrated that a common closure approximation (the QCA) follows naturally from an 'optimal' approach based upon an associated variational formulation. Most of the lectures are concerned with linear elastostatics but the initial formulation is general and extensions of the methods to dynamic problems and to nonlinear problems are outlined towards the end.

CHAPTER 1

1. - INTRODUCTION

It is a trivial statement that any sample of material is microscopically inhomogeneous, even if it appears homogeneous at some natural scale of observation. Inevitably, therefore, a description of any material in terms of continuum mechanics is an approximation, and any experimental determination of constitutive behaviour yields, in fact, a relationship between the "overall" properties measured in the experiment. Suppose, for example, the tensile stress-strain behaviour of a wire were under investigation. Either the wire would be loaded to some level and its extension measured, or else it would be extended by some amount and the load measured. Then (in simple terms), stress would be taken as load divided by area of cross-section and strain as extension divided by original length. This seems reasonable enough, except that the wire is not a perfectly homogeneous continuum and no precisely uniform stress and strain fields would exist within it. The stress and strain fields calculated from the experiment thus represent averages of the actual forces and displacements in the wire, and their use is justified by the fact that predictions based upon the constitutive property so obtained are, except under extreme conditions, borne out in practice. It is, perhaps, just worth mentioning that experiments such as this involve some preconception of the type of behaviour that is expected. If, for instance, stress σ were in fact related to strain e through

$$\sigma(z) = E\,e(z) + \alpha\,e'(z) \quad ,$$

z being measured along the length of the wire, the experiment described would determine E but would give no information at all on α : for the latter quantity, some macroscopically non-uniform field would need to be imposed, perhaps by propagating a tensile pulse down the wire.

Similar considerations apply to composite materials: there is a microscopic length scale, of the order of fibre thickness, or grain size or similar, relative to which a composite is explicitly heterogeneous, though it can still be modelled as a continuum; yet, for many practical purposes, it behaves as though it were homogeneous. The macroscopic 'overall' properties can be determined experimentally in much the same way as for the wire above, but it is also of interest to relate this 'overall' behaviour to more detailed behaviour at the level of the microstructure. The sections that follow give precise definitions for overall properties, which are consistent with the experimental notions mentioned above.

Throughout these lectures, a composite will be taken as a medium composed of n distinct constituents, or phases. Each phase will be taken as uniform, with known constitutive properties. Every point x of the composite is occupied by some phase (so that a set of cavities would be allowed for by including one phase with zero properties), and the phases are taken to be firmly bonded across interfaces. Behaviour of individual phases may be nonlinear in general, though the case of linear behaviour will receive most attention. Problems raised by imperfect bonding are of great interest but are not treated in these lectures, except perhaps through the possibility that imperfect bonds might be modelled in terms of an additional phase.

Overall properties via spatial averages

This initial discussion is carried out within the framework of finite deformations. There is no restriction to elasticity but a convenient reference source for the basic concepts of nonlinear continuum mechanics is Ogden.[1]

Suppose that material (not necessarily a composite) occupies a domain Ω, with boundary $\partial\Omega$, when it is subjected to no surface or body forces. One possible definition for its overall behaviour is obtained by imposing a displacement at each point of Ω, which is compatible with a uniform deformation within Ω. That is, a boundary point X (with Cartesian components X_α) is moved to a corresponding point x (with Cartesian components x_i), where

$$x = \bar{A}X \quad , \quad x_i = \bar{A}_{i\alpha} X_\alpha \quad , \quad X \in \partial\Omega \quad . \tag{1.1}$$

Here, $\bar{A} = (\bar{A}_{i\alpha})$ is a constant matrix and the summation convention applies to repeated suffixes. The actual deformation induced within Ω is not, in general, uniform but is defined by some function $x = \chi(X)$, with associated deformation gradient $A_{i\alpha} = \partial x_i/\partial X_\alpha$. However, application of Gauss' theorem gives

$$\int_\Omega A_{i\alpha}\, dX = \int_{\partial\Omega} x_i N_\alpha\, dA = \int_{\partial\Omega} \bar{A}_{i\beta} X_\beta N_\alpha\, dA = |\Omega|\, \bar{A}_{i\alpha} \quad , \tag{1.2}$$

where $N = (N_\alpha)$ is the outward normal to $\partial\Omega$ and $|\Omega|$ is the volume of Ω. Thus, the matrix $\bar{A}$ is exactly the mean value of the deformation gradient A over Ω, subject only to the requirement of perfect bonding so that the mapping $x = \chi(X)$ is continuous across any interface.

Now suppose that the boundary displacement (1.1) induces a nominal stress S with components $S_{\alpha i}$ ($S_{\alpha i}$ = i-component of force per unit initial area, initially in α-direction), throughout Ω. The nominal stress is not objective but it has the useful property that its mean value $\bar{S}$ over Ω can be inferred from surface measurements. In fact, again by Gauss' theorem,

$$\int_\Omega S_{\alpha i}\, dX = \int_{\partial\Omega} X_\alpha t_i\, dS \quad , \tag{1.3}$$

where $t_i = N_\alpha S_{\alpha i}$ is the i-component of nominal traction over $\partial\Omega$. Evidently, the tensor $\bar{S}$ depends only upon $\bar{A}$ (together with the shape and size of Ω and the properties of the material) and this relationship,

$$\bar{S} = \bar{S}(\bar{A}) \tag{1.4}$$

provides a definition of the overall response of the material occupying Ω. The components of $\bar{S}$ are not all independent: just as local nominal stress and deformation gradient satisfy a rotational balance condition, so do $\bar{S}$ and $\bar{A}$. The relationship

$$\bar{A}\bar{S} = \bar{S}^T \bar{A}^T \tag{1.5}$$

follows from the overall equilibrium requirement

$$\int_{\partial\Omega} \varepsilon_{ijk} x_j t_k\, dS = 0$$

together with (1.1), Gauss' theorem and the equilibrium equations

$$\frac{\partial S_{\alpha i}}{\partial X_\alpha} = 0 \quad , \quad x \in \Omega. \tag{1.6}$$

If an overall Kirchhoff stress tensor $\bar{\tau}$ is defined so that

$$\bar{\tau} = \bar{A}\bar{S} \quad , \tag{1.7}$$

it follows from (1.5) that $\bar{\tau}$ is symmetric. It is not the volume average of the local Kirchhoff stress tensor τ (defined as in (1.7) but without bars) but, if the mean Cauchy stress $\bar{\sigma}$ is defined so that

$$\bar{J}\bar{\sigma} = \bar{\tau} = \bar{A}\bar{S} \quad , \tag{1.8}$$

where

$$\bar{J} = \det(\bar{A}) \tag{1.9}$$

then $\bar{\sigma}$ is the mean value of the local Cauchy stress σ, taken over the domain D into which Ω is mapped under $x = \chi(X)$:

$$|D|\bar{\sigma} = \int_D \sigma \, dx \quad . \tag{1.10}$$

The proof of (1.10) is another exercise in the use of Gauss' theorem, coupled with Nanson's formula to relate elements of surface on ∂D and $\partial\Omega$.

The types of overall relations discussed above were first developed, in the context of nonlinear media, by Hill[2] and also by Ogden.[3] Corresponding relations between stress and strain rates can also be obtained; a suitable reference is Hill.[4] However, all such relations amount to re-statements of (1.4), which is all that is needed at present.

It is of interest to consider work-rates, as follows. During a quasi-static deformation, of the type described above but containing time t as a parameter, the rate of working of the surface tractions t_i is

$$\int_{\partial\Omega} N_\alpha S_{\alpha i} \dot{x}_i \, dS = \int_\Omega S_{\alpha i} \dot{A}_{i\alpha} \, dX \quad , \tag{1.11}$$

from Gauss' theorem and (1.6). But also, substituting (1.1) into the left side of (1.11) shows that $\dot{A}_{i\alpha}$ can be replaced by the mean value $\dot{\overline{A}}_{i\alpha}$. Evaluating the integral over Ω then gives

$$|\Omega|\overline{S}_{\alpha i}\,\dot{\overline{A}}_{i\alpha} = \int_{\Omega} S_{\alpha i}\,\dot{A}_{i\alpha}\,dX \quad , \tag{1.12}$$

showing that the volume average of the work-rate can be calculated directly from the mean stress and mean rate of deformation. This has an immediate implication if the material in fact is elastic, with energy density W(A) , for then

$$\overline{S}_{\alpha i}\,\dot{\overline{A}}_{i\alpha} = \frac{1}{|\Omega|}\int_{\Omega} \frac{\partial W}{\partial A_{i\alpha}}\,\dot{A}_{i\alpha}\,dX = \dot{\overline{W}}(\overline{A}) \tag{1.13}$$

where $\overline{W}(\overline{A})$ is the mean energy density

$$\overline{W}(\overline{A}) = \frac{1}{|\Omega|}\int_{\Omega} W(A)\,dX \quad . \tag{1.14}$$

These relations demonstrate, in this case, that the mean stress, defined by (1.3), is related to the mean energy density by

$$\overline{S}_{\alpha i} = \partial\overline{W}/\partial\overline{A}_{i\alpha} \quad . \tag{1.15}$$

Overall properties via ensemble averages

An alternative definition for overall behaviour is obtained by regarding any given composite specimen as one taken from an ensemble of media all of identical shape and size and subjected to the same boundary conditions. The fields S , A that are realised are now random (since they depend upon the realization of the medium) and it is interesting to seek their ensemble averages, $\langle S\rangle$ and $\langle A\rangle$. Little more can be said in general, but an 'overall' relation would express $\langle S\rangle$ in terms of $\langle A\rangle$.

Of course, for general boundary conditions, both <S> and <A> will depend upon position X and it might be expected that <S> is some functional of <A>. Whether such a relation always exists is not known, but a later lecture will demonstrate the validity of this notion, for linear elastic behaviour.

Relationships with homogenization

The theory of homogenization, initiated by Sanchez Palencia,[5] considers media with a microstructure of characteristic length εd, d being a macroscopic length associated with Ω. Then, for fixed ε, the deformation field produced by given boundary conditions can be called $x = \chi_\varepsilon(X)$ and the questions arise as to whether χ_ε has a limit (χ, say) as $\varepsilon \to 0$ and, if it does, whether χ is the actual deformation that the boundary conditions would induce in some 'homogenized' material. These questions have been answered, affirmatively, in the case of a periodic microstructure, for many types of constitutive response, but for random microstructures the answers so far are less complete. If the media are 'statistically uniform', it has been shown by Kozlov[6] and Papanicolaou and Varadhan[7] that, for linear constitutive response, $\langle\chi_\varepsilon\rangle$ tends to a limit as $\varepsilon \to 0$ and that this limit satisfies a set of linear equations corresponding to a uniform medium. Precise statements of these results are rather technical and readers are referred to the original articles for details. It is simply observed here that, if the result of homogenization is really to generate a uniform medium for any boundary conditions, then it does so in particular for the boundary condition (1.1) and this special problem must suffice for determining the 'homogenized' uniform medium. Also, with reference to the preceding section, the fields $\langle S \rangle$ and $\langle A \rangle$ will be uniform and, if $\langle S \rangle$ is a functional of $\langle A \rangle$ in general, it reduces to a function of $\langle A \rangle$ when $\langle A \rangle$ is constant over Ω, which will coincide with (1.4) because $\overline{A} = \langle \overline{A} \rangle = \langle A \rangle$ since $\langle A \rangle$ is uniform, and similarly for S.

CHAPTER 2

2. - LINEAR ELASTICITY

The preceding discussion was general but most of the subsequent lectures will concern the particular case of linear elasticity. This means that the deformation $x = \chi(X)$ will take the particular form

$$x = X + u \quad , \tag{2.1}$$

with the displacement u small, so that the matrix A is close to the identity I. Correspondingly, to lowest order, there is no distinction between S, τ or σ, nor any distinction between functions defined in terms of X over Ω (Lagrangian description) and functions defined in terms of x over D (Eulerian description). The linear elastic constitutive relation will be given in the form

$$\sigma = L\,e \quad , \quad \sigma_{ij} = L_{ijk\ell}\, e_{k\ell} \tag{2.2}$$

where the (infinitesimal) strain e has components

$$e_{ij} = \tfrac{1}{2}\left(\frac{\partial u_i}{\partial x_j} + \frac{\partial u_j}{\partial x_i}\right) \quad . \tag{2.3}$$

In the particular case of isotropic behaviour, the tensor of moduli L is expressible in terms of the bulk modulus K and the shear modulus μ:

$$L_{ijk\ell} = K\,\delta_{ij}\,\delta_{k\ell} + \mu\left(\delta_{ik}\,\delta_{j\ell} + \delta_{i\ell}\,\delta_{jk} - \frac{2}{3}\,\delta_{ij}\,\delta_{k\ell}\right) \quad . \tag{2.4}$$

The symmetry $L_{ijk\ell} = L_{k\ell ij}$ will always be assumed (it is realised automatically by (2.4)), which implies the existence of an energy density $W(e)$, given by

$$W(e) = \frac{1}{2}\sigma_{ij}e_{ij} = \frac{1}{2}L_{ijk\ell}e_{ij}e_{k\ell} \quad . \tag{2.5}$$

It should, perhaps, be recorded for completeness that the equilibrium equations take the form

$$\operatorname{div}\sigma + f = 0 \quad , \quad \frac{\partial\sigma_{ij}}{\partial x_j} + f_i = 0 \quad , \quad x \in D \quad , \tag{2.6}$$

which is equivalent to (1.6) but with the addition of body force $f = (f_i)$. Boundary conditions will be applied over ∂D, whose outward normal will be denoted $n = (n_i)$.

Mathematical specification of a composite

It has been mentioned already that a composite will be considered as a medium with n different constituents or phases firmly bonded across interfaces. A convenient way to describe its geometrical arrangement is to introduce indicator functions: $f_r(x)$ takes the value 1 if x lies in phase r and zero otherwise. If the composite is random, the functions f_r are random fields, so that they depend also upon a parameter α which belongs to a sample space A, over which a probability measure p is defined. This amounts to regarding the given composite as one particular member, with label α, of an ensemble. For any given α, the functions $f_r(x,\alpha)$ are defined explicitly, at least in principle. In practice, however, only mean properties such as

$$P_r(x) = \langle f_r(x) \rangle = \int_A f_r(x,\alpha)\,p(d\alpha) \tag{2.7}$$

will be known. The function $P_r(x)$ gives the probability of finding phase r at x; it is equal to the volume concentration c_r if the medium is statistically uniform. Another statistic, used explicitly in the sequel, is

$$P_{rs}(x,x') = \langle f_r(x) f_s(x') \rangle = \int_A f_r(x,\alpha) f_s(x',\alpha)\,p(d\alpha) \quad . \tag{2.8}$$

This represents the probability of finding phase r at x and phase s at x'; in the case of a statistically uniform medium (more exactly, statistically second-order stationary), P_{rs} depends on x,x' only in the combination $x - x'$ and can be estimated as the average of $f_r(x) f_s(x')$ over a large volume of one particular medium, with $x - x'$ fixed.

If the composite is linearly elastic, phase r will have (constant) tensor of moduli L_r and the (variable) tensor of moduli $L(x)$ for the composite can be expressed as

$$L(x) = \sum_{r=1}^{n} L_r f_r(x) \quad , \tag{2.9}$$

so that

$$\langle L(x) \rangle = \sum_{r=1}^{n} L_r P_r(x) \quad , \tag{2.10}$$

$$\langle L(x) \otimes L(x') \rangle = \sum_{r=1}^{n} \sum_{s=1}^{n} L_r \otimes L_s P_{rs}(x,x') \tag{2.11}$$

and so on.

Some composites consist of a matrix phase in which are embedded inclusions of known shape. This information is best incorporated by employing a slightly different description from that given above. In the case of a single population of identical inclusions, it is convenient to assign labels $A = 1, 2, \ldots$ to particular inclusions in a sample, and to define their positions by points x_A where their 'centres' (to be interpreted as some identifiable point in an inclusion) are located. It is then assumed that certain probability densities are known: P_A denotes the probability density for finding some inclusion centred at x_A, P_{AB} is the joint density for finding distinct inclusions at x_A and x_B, and so on. In the case of a statistically uniform medium, P_A is the number density of the inclusions while P_{AB} depends on $x_B - x_A$ only and can be estimated from observations over a large volume of one sample, varying x_A but keeping $x_B - x_A$ fixed.

Integral equation formulation

Suppose that the composite, with moduli $L(x)$, is subjected to body force f over D, together with some boundary condition over ∂D. For simplicity, it will be assumed that the displacement u is prescribed over ∂D, but this need not necessarily correspond to a uniform deformation, like (1.1), at this stage. The elastic moduli $L(x)$ vary in a complicated way with position x and there is some advantage in introducing a 'comparison' medium, with moduli L_o which, in practice, will be taken as uniform over D, though this is not a requirement for the formal development.

Let u, e, σ represent, respectively, the displacement, strain and stress in the actual composite, so that they satisfy (2.2), (2.3) and

(2.6), together with the boundary conditions. It is convenient to define a 'stress polarization' τ as

$$\tau = (L - L_o)e \quad , \tag{2.12}$$

so that

$$\sigma = L_o e + \tau \quad . \tag{2.13}$$

Then, since σ satisfies (2.6),

$$\mathrm{div}(L_o e) + (f + \mathrm{div}\,\tau) = 0 \quad , \quad x \in D \tag{2.14}$$

and a representation for u in terms of τ can be given in terms of the Green's function G_o for the comparison body. This satisfies the equations

$$\frac{\partial}{\partial x_j}(L_o)_{ijk\ell}\frac{\partial (G_o)_{kp}}{\partial x_\ell}(x,x') + \delta_{ip}\,\delta(x - x') = 0, \quad x \in D \tag{2.15}$$

$$(G_o)_{ip}(x,x') = 0 \quad , \quad x \in \partial D \quad , \tag{2.16}$$

so that

$$u_p(x') = u_o(x') + \int_D \frac{\partial \tau_{ij}}{\partial x_j}(G_o)_{ip}(x,x')\,dx \quad . \tag{2.17}$$

Here, u_o is the solution of (2.14), with the given boundary conditions, but with τ replaced by zero; thus u_o is the solution of the given problem for the comparison body. Integration by parts in (2.17), followed by differentiation with respect to x'_q, provides the representation

$$e_{pq}(x') = (e_o)_{pq}(x') - \int_D \tau_{ij}(x)\,(\Gamma_o)_{ijpq}(x,x')\,dx$$

or, symbolically,

$$e = e_o - \Gamma_o \tau \quad , \tag{2.18}$$

where

$$\left(\Gamma_o\right)_{ijpq}(x,x') = \left.\frac{\partial^2 \left(G_o\right)_{ip}(x,x')}{\partial x_j \partial x_q'}\right|_{(ij)(pq)} , \tag{2.19}$$

the suffix (ij) implying symmetrization on these indices. It should be noted that the well-known symmetry of G_o induces the symmetry

$$\left(\Gamma_o\right)_{ijpq}(x,x') = \left(\Gamma_o\right)_{pqij}(x',x) \quad . \tag{2.20}$$

The integral implied in (2.18) is singular and should be interpreted in the sense of generalized functions.

Integral equations may now be obtained by combining (2.12) and (2.18): eliminating e gives

$$\left(L - L_o\right)^{-1} \tau + \Gamma_o \tau = e_o \quad , \tag{2.21}$$

while eliminating τ gives

$$e = e_o - \Gamma_o \left[\left(L - L_o\right) e\right] \quad . \tag{2.22}$$

If either of these equations could be solved, the stress σ would also be known, through (2.2) or (2.13), and overall properties could be investigated directly.

Perturbation theory

In the particular case of a weakly inhomogeneous medium, L_o can be chosen so that $L - L_o$ is everywhere small and (2.22) can be solved by iteration. The first few terms are

$$e = \left[I - \Gamma_o\left(L - L_o\right) + \Gamma_o\left(L - L_o\right)\Gamma_o\left(L - L_o\right) \cdots\right] e_o \quad . \tag{2.23}$$

The snag with this form for the solution is that the operators Γ_o require integration over the whole of D so that, if D is large, difficulties with convergence of the series could arise unless $L - L_o$ were very small indeed. A related point is that G_o, and hence Γ_o, is unlikely to be known in practice for a finite body, and there is some incentive to consider the limit of infinite D in such a way as to allow Γ_o to be given its form for an infinite body, which is easy to find explicitly. Direct substitution of the infinite-body Γ_o into (2.23) definitely gives trouble, not just with convergence of the series but with convergence of the integrals in individual terms, and the best procedure is to approach (2.22) slightly differently.

Bearing in mind that L is a random function of position and that one possible 'overall' description would relate $\langle \sigma \rangle$ and $\langle e \rangle$, it is convenient to set

$$e = \langle e \rangle + e' \quad , \tag{2.24}$$

with similar definitions for other fields and, for simplicity, to take $L_o = \langle L \rangle$. Then, taking the expectation of (2.22) gives

$$\langle e \rangle = e_o - \Gamma_o \langle L' e' \rangle \tag{2.25}$$

and so, subtracting this equation from (2.22),

$$e' = -\Gamma_o L' \langle e \rangle - \Gamma_o (L' e' - \langle L' e' \rangle) \quad . \tag{2.26}$$

Formal iteration of (2.26) now gives

$$e' = -\Gamma_o L' \langle e \rangle + \Gamma_o (L' \Gamma_o L' - \langle L' \Gamma_o L' \rangle) \langle e \rangle + \ldots \quad . \tag{2.27}$$

Now from (2.2),

$$\langle \sigma \rangle = \langle L \rangle \langle e \rangle + \langle L' e' \rangle \tag{2.28}$$

and so, from (2.27) and (2.28),

$$\langle\sigma\rangle = \left\{\langle L\rangle - \langle L'\Gamma_o L'\rangle + \langle L'\Gamma_o L'\Gamma_o L'\rangle \ldots\right\}\langle e\rangle \quad . \tag{2.29}$$

The field <e> is still unknown. However, (2.29) gives

$$\langle\sigma\rangle = \hat{L}\langle e\rangle \ , \tag{2.30}$$

where $\hat{L}$ is the operator in brackets in (2.29), and this is the desired 'overall' relation. In fact, to find <e>, (2.30) is substituted back into the expectation value of (2.6), to yield

$$\mathrm{div}(\hat{L}\langle e\rangle) + f = 0 \quad , \quad x \in D \quad , \tag{2.31}$$

together with the boundary conditions applied to <u>.

It should be noted that $\hat{L}$ is a non-local operator. For example, the second term in the series in (2.29) gives, with (2.10) and (2.11),

$$\langle L'\Gamma_o L'\rangle\langle e\rangle(x') = \sum_{r=1}^{n}\sum_{s=1}^{n} L_r \int dx\ \Gamma_o(x,x')\Big[P_{rs}(x',x) - P_r(x')P_s(x)\Big] L_s\langle e\rangle(x) \quad . \tag{2.32}$$

The term $P_{rs} - P_r P_s$ can usually be expected to decay to zero when $|x - x'|$ is large relative to dimensions typical of the microstructure; equivalently, the decay rate defines a 'correlation length' ℓ for the microstructure, which is usually much smaller than a typical macroscopic dimension d of D. In this case, except in a 'boundary layer' close to ∂D, the operator Γ_o can be replaced by its infinite-body form without significant error, when $\varepsilon = \ell/d \ll 1$. It may be noted, too, that, if <e> varies smoothly with x relative to the length scale d, then it varies only slightly over any distance of order ℓ and <e>(x) in (2.32) can be replaced, without significant error, by <e>(x') when $\varepsilon = \ell/d \ll 1$. In this case, clearly, the non-local operator $\hat{L}$ reduces to a local tensor of moduli, given by evaluating the series (2.29) with <e> constant. This is the 'homogenization limit' referred to in Lecture 1. Clearly, in this limit, if boundary conditions corresponding to a uniform strain are imposed over ∂D, then <e> is uniform (except in a boundary layer) and equal to its mean value over D, and the overall stress-strain relation that this boundary value problem yields is precisely the local form of (2.29).

Perturbation theory of the type described here was pioneered by Keller.[8] The trick of subtracting off the mean value of an equation (in this context, (2.22)) is known as the 'method of smoothing'.

CHAPTER 3

3. - A MATRIX-INCLUSION COMPOSITE

Perturbation theory of a different type can be performed for a composite which consists of a matrix containing a dilute suspension of inclusions. For ease of exposition, only a single population of identical inclusions, distributed uniformly at number density n_1, will be considered. Thus, in the notation established in Lecture 2, P_A takes the constant value n_1 and P_{AB} depends on $x_B - x_A$ only, except when x_A (or x_B or both) is in a boundary layer close to ∂D. The inclusions will be taken to have tensor of moduli L_1 and the matrix to have moduli L_2. It is convenient also to let D_A be the region occupied by inclusion A and D_m be the region occupied by matrix material. Then, if τ^A, τ^m denote the restrictions of τ to D_A, D_m respectively, the integral equation (2.21) can be written in the form

$$\left(L_1 - L_o\right)^{-1}\tau^A + \Gamma_o\tau^A + \sum_{B \neq A} \Gamma_o\tau^B + \Gamma_o\tau^m = e_o \quad , \quad x \in D_A, \tag{3.1}$$

$$\left(L_2 - L_o\right)^{-1}\tau^m + \sum_B \Gamma_o\tau^B + \Gamma_o\tau^m = e_o \quad , \quad x \in D_m \quad . \tag{3.2}$$

The quantities of ultimate interest (at least for the moment) are $\langle\sigma\rangle$, $\langle e\rangle$ or, equivalently, $\langle\tau\rangle$, since $\langle\sigma\rangle = L_o\langle e\rangle + \langle\tau\rangle$ and $\langle e\rangle = e_o - \Gamma_o\langle\tau\rangle$. Now

$$\tau(x) = \sum_A \tau^A(x) + \tau^m(x) \qquad (3.3)$$

and so, taking expectations,

$$<\tau(x)> = \int dx_A\, \tau^A_A(x) P_A + \tau^m_m(x) P_m(x) \quad , \qquad (3.4)$$

where τ^A_A represents the expectation value of τ^A, conditional upon D_A being fixed and $\tau^m_m(x)$ is the expectation value of $\tau^m(x)$, conditional upon finding matrix at x. P_A has been defined already; $P_m(x)$ is the probability of finding matrix at x and is equal to the volume concentration c_2 of matrix, as the composite is statistically uniform. Corresponding to (3.4),

$$<e> = e_o - \int dx_A \left(\Gamma_o \tau^A_A\right) P_A - \Gamma_o \left(\tau^m_m P_m\right) \quad . \qquad (3.5)$$

An equation for τ^A_A can be found, hopefully, by averaging (3.1), conditionally upon D_A being fixed. Thus, also using (3.5) to eliminate e_o in favour of $<e>$, as in the method of smoothing,

$$\left(L_1 - L_o\right)^{-1} \tau^A_A + \Gamma_o \tau^A_A + \int dx_B\, \Gamma_o \left(\tau^B_{BA} P_{B|A} - \tau^B_B P_B\right)$$

$$+ \Gamma_o \left(\tau^m_{mA} P_{m|a} - \tau^m_m P_m\right) = <e> \;, \quad x \in D_A \quad . \qquad (3.6)$$

Here, τ^B_{BA} is the expectation value of τ^B, conditional upon inclusions centred at x_B and x_A and $\tau^m_{mA}(x')$ is the expectation value of $\tau^m(x')$, conditional upon finding matrix at x', with an inclusion occupying D_A. $P_{B|A}$, $P_{m|A}(x')$ are, respectively, the probability density for finding an inclusion centred at x_B and the probability of finding matrix at x', both conditional upon D_A being fixed.

Equation (3.6) contains, unfortunately, the new quantities τ^B_{BA}, τ^m_{mA} and attempts to find equations for these by further conditional averaging would simply generate a hierarchy of equations, for instance involving τ^{C}_{CBA}, τ^{m}_{mBA} and so on. In the case of a dilute suspension, however, progress can be made. First, choosing $L_o = L_2$ implies $\tau^m = 0$ precisely, so that all quantities containing the superscript m can be set to zero. Furthermore, the integral with respect to x_B in (3.6) is of order $n_1 \ell^3$, where n_1 is the number density and ℓ is a 'correlation length' defining the rate at which $P_{B|A}$ approaches P_B as $(x_B - x_A)$ increases. Then, if n_1 is small, to lowest order, (3.6) gives

$$\left(L_1 - L_2\right)^{-1} \tau^{A}_{A} + \Gamma_2 \tau^{A}_{A} = \langle e \rangle \ , \qquad x \in D_A \quad , \tag{3.7}$$

while (3.4) yields

$$\langle \tau \rangle = n_1 \int dx_A \, \tau^{A}_{A} = c_1 \overline{\tau^{A}_{A}} \quad , \tag{3.8}$$

where c_1 is the volume concentration of inclusions, equal to n_1 times the volume of D_A, and $\overline{\tau^{A}_{A}}$ is the mean value of τ^{A}_{A} over D_A. The operator Γ_2 in (3.7) is Γ_o, evaluated when $L_o = L_2$. Equation (3.8) is exact, while the solution of (3.7) provides an estimate for τ^{A}_{A}, which is correct to zeroth order in c_1. Equation (3.7) defines, in fact, the polarization which would be produced in a single inclusion occupying D_A, in an infinite matrix in which the strain field would be $\langle e \rangle$ exactly, if the inclusion were not present. This basic inclusion problem is of fundamental significance and is discussed next.

The single inclusion problem

It is worthwhile at this point to make the integral equation (3.7) more explicit. First, for an isotropic matrix, so that L_2 is characterized by bulk modulus K_2 and shear modulus μ_2, Green's function G_2 takes the form

$$\left(G_2\right)_{ij}(x) = \frac{1}{4\pi\mu_2}\left\{\frac{\delta_{ij}}{|x|} - \frac{3K_2+\mu_2}{2\left(3K_2+4\mu_2\right)}\,|x|_{,ij}\right\} . \tag{3.9}$$

Correspondingly, the kernel of Γ_2 has components

$$\left(\Gamma_2\right)_{ijk\ell} = \frac{-1}{4\pi\mu_2}\left\{\frac{1}{4}\left[\delta_{ik}\left(|x|^{-1}\right)_{,j\ell} + \delta_{jk}\left(|x|^{-1}\right)_{,i\ell} + \delta_{i\ell}\left(|x|^{-1}\right)_{,jk} + \delta_{j\ell}\left(|x|^{-1}\right)_{,ik}\right] - \frac{3K_2+\mu_2}{2\left(3K_2+4\mu_2\right)}\,|x|_{,ijk\ell}\right\} . \tag{3.10}$$

Then, for any polarization τ, in terms of the potentials

$$\phi_{k\ell} = \int \frac{\tau_{k\ell}(x')\,dx'}{|x-x'|} , \quad \psi_{k\ell} = \int \tau_{k\ell}(x')\,|x-x'|\,dx' , \tag{3.11}$$

$\Gamma_2\tau$ has components

$$\left(\Gamma_2\tau\right)_{ij} = \frac{1}{4\pi\mu_2}\left\{\frac{1}{2}\left[\phi_{i\ell,j\ell} + \phi_{j\ell,i\ell}\right] - \frac{3K_2+\mu_2}{2\left(3K_2+4\mu_2\right)}\,\psi_{k\ell,ijk\ell}\right\} , \tag{3.12}$$

so that the integral equation (3.7) can be expressed in terms of harmonic and biharmonic potentials associated with τ_A^A over D_A. The celebrated solution of Eshelby[9] for an ellipsoidal inclusion perturbing a uniform field in an isotropic matrix was obtained precisely from this representation, exploiting known properties of the potentials (3.11) in the case of constant τ over an ellipsoidal region D_A.

An alternative and more general approach to (3.7) can be made, starting from the plane-wave decomposition[10]

$$\delta(x) = \frac{-1}{8\pi^2}\int_{|\xi|=1} \delta''(\xi.x)\,dS \tag{3.13}$$

of the three-dimensional Dirac delta. Equation (3.13) is easily established by noting that

$$\int_{|\xi|=1} \delta(\xi.x)\,dS = \int_0^{2\pi} d\phi \int_0^{\pi} \delta\big(|x|\cos\theta\big)\sin\theta\,d\theta = \frac{2\pi}{|x|} \quad , \tag{3.14}$$

having used an obvious parameterization for the unit vector ξ. Taking the Laplacian of (3.14) then gives (3.13). Green's function G_o satisfies the equations

$$\big(L_o\big)_{ijk\ell}\big(G_o\big)_{kp,j\ell} + \delta_{ip}\,\delta(x) = 0 \tag{3.15}$$

and this motivates defining G_o^{ξ} so that

$$\big(L_o\big)_{ijk\ell}\big(G_o^{\xi}\big)_{kp,j\ell} + \delta_{ip}\,\delta''(\xi.x) = 0 \quad . \tag{3.16}$$

Equation (3.16) has a solution G_o^{ξ} which depends on x only in the combination $(\xi.x)$. This satisfies

$$\big(L_o\big)_{ijk\ell}\xi_j\xi_\ell\big(G_o^{\xi}\big)''_{kp} + \delta_{ip}\,\delta''(\xi.x) = 0 \quad , \tag{3.17}$$

giving immediately

$$G_o^{\xi} = -\Big[L_o(\xi)\Big]^{-1}\delta(\xi.x) \quad , \tag{3.18}$$

where the matrix $L_o(\xi)$ has components $\big(L_o\big)_{ijk\ell}\xi_j\xi_\ell$. The desired G_o can now be obtained by performing the superposition implied by (3.13):

$$G_o(x) = \frac{1}{8\pi^2}\int_{|\xi|=1}\Big[L_o(\xi)\Big]^{-1}\delta(\xi.x)\,dS \quad . \tag{3.19}$$

Correspondingly, the kernel of Γ_o has components

$$\big(\Gamma_o\big)_{ijk\ell} = \frac{1}{8\pi^2}\int_{|\xi|=1}\xi_{(i}\Big[L_o(\xi)\Big]^{-1}_{j)(k}\xi_{\ell)}\,\delta''(\xi.x)\,dS \quad . \tag{3.20}$$

It follows that $\Gamma_o\tau$ can be given in the form

$$\big(\Gamma_o\tau\big)_{ij} = \frac{1}{8\pi^2}\int_{|\xi|=1}\xi_{(i}\Big[L_o(\xi)\Big]^{-1}_{j)k}\,\xi_\ell\,\hat{\tau}''_{k\ell}(\xi.x,\xi)\,dS \quad , \tag{3.21}$$

where

$$\hat{\tau}_{k\ell}(p,\xi) = \int \tau_{k\ell}(x)\,\delta(p - \xi.x)\,dx \qquad (3.22)$$

is the Radon transform[11] of $\tau_{k\ell}(x)$ and the primes denote differentiations with respect to p which, in (3.21), takes the value ($\xi.x$). The interesting feature of this result is that, if D_A is an ellipsoid and τ_A^A is constant over D_A, then the second derivative of its Radon transform, as required in (3.21), is constant for $x \in D_A$. This is easy to illustrate in the case when D_A is the sphere $|x| \le a$. Then,

$$\int_{D_A} \delta(p - \xi.x)\,dx = \pi(a^2 - p^2)H(a^2 - p^2) \quad ,$$

since the value of the integral is just the area the disc formed by the intersection of the plane $\xi.x = p$ with D_A. Differentiating twice with respect to p gives

$$\int_{D_A} \delta''(p - \xi.x)\,dx = -2\pi H(a^2 - p^2) + 2\pi a\,\delta(a - p) + 2\pi a\,\delta(a + p) \quad ,$$

which takes the constant value -2π when $p = \xi.x$ and $|x| < a$.

The result just noted permits the solution of (3.7) when $\langle e\rangle$ is uniform over D_A and D_A is an ellipsoid. τ_A^A is constant over D_A and satisfies the algebraic system

$$\left[(L_1 - L_2)^{-1} + P\right]\tau_A^A = \langle e\rangle \quad , \qquad (3.23)$$

where the constant tensor P has components

$$P_{ijk\ell} = \left[4\pi \det(A)\right]^{-1} \int_{|\xi|=1} \xi_{(i}\left[L_o(\xi)\right]^{-1}_{j)(k}\xi_{\ell)} \left[\xi^T\left(A^T A\right)^{-1}\xi\right]^{-3/2} dS \;.(3.24)$$

In (3.24), the ellipsoid D_A is taken as $D_A = \left\{x:\; x^T A^T A x < 1\right\}$.

Substitution back through (3.8) and using $\langle\sigma\rangle = L_2\langle e\rangle + \langle\tau\rangle$ gives

$$\langle\sigma\rangle = \hat{L}\langle e\rangle \quad , \tag{3.25}$$

where

$$\hat{L} = L_2 + c_1\left[\left(L_1 - L_2\right)^{-1} + P\right]^{-1} \quad . \tag{3.26}$$

This result is correct to first order in c_1. The overall operator $\hat{L}$ given by (3.26) is local because it was obtained under the assumption that $\langle e\rangle$ was constant over D_A; the same result would not be obtained if significant variation of $\langle e\rangle$ over D_A were assumed.

Higher order perturbation theory

Equation (3.1) (with $L_o = L_2$ for simplicity) could also be averaged conditionally upon inclusions occupying D_A and D_B. The result of this, using also (3.5), is

$$\left(L_1 - L_2\right)^{-1}\tau^A_{AB} + \Gamma_2\tau^A_{AB} + \Gamma_2\tau^B_{BA}$$
$$+ \int dx_C\, \Gamma_2\left[\tau^C_{CAB}\, P_{C|AB} - \tau^C_C P_C\right] = \langle e\rangle \quad , \qquad x \in D_A \quad . \tag{3.27}$$

A similar equation holds when $x \in D_B$. To lowest order, the term involving the integral with respect to x_C can be neglected. Then, (3.27) and its counterpart for $x \in D_B$ form a closed system for τ^A_{AB} and τ^B_{BA}, which is correct to zeroth order in c_1. Substituting this lowest order approximation for τ^B_{BA} into (3.6) (with $L_o = L_2$) then provides an equation from which τ^A_A can be estimated correct to first order in c_1. The corresponding estimate for $\hat{L}$ is then correct to order $c_1{}^2$. The details are tedious and are not recorded. It is noted, however, that the term of order c_1, as in (3.26), involves only the single (one point) statistic c_1 whereas the term of order $c_1{}^2$ involves an integral containing the two-particle statistic $P_{B|A}$. An expansion of $\hat{L}$ up to some higher power of c_1 could in principle be obtained by an extension of the method outlined but it would rapidly become impractical, both through the length of the calculations and because conditional densities such as $P_{C|AB...}$ would be needed which would not be available.

CHAPTER 4

4. - CLOSURE ASSUMPTIONS

For the matrix-inclusion composite discussed in Lecture 3, suppose that the inclusions are distributed at number density n_1 and corresponding volume concentration c_1 but that c_1 is no longer small. The perturbation theory which led to (3.26) is no longer helpful and some allowance has to be made for the terms that were neglected in (3.6), for example. As a first attempt, consider (3.6) when $L_o = L_2$ so that $\tau^m = 0$. One method for closing this equation is to make the assumption, or approximation, that

$$\tau^B_{BA} = \tau^B_B \quad . \tag{4.1}$$

This is not entirely unreasonable: it would be true exactly if the inclusions were arranged on a lattice and for this reason (4.1) is called a quasicrystalline approximation (QCA). It is also true, asymptotically, when x_B is far from x_A. At present, however, the only honest justification for adopting (4.1) is that it reduces (3.6) (with $L_o = L_2$) to the closed equation

$$\left(L_1 - L_2\right)^{-1}\tau^A_A + \Gamma_2\tau^A_A + \int dx_B\, \Gamma_2\, \tau^B_B\left(P_{B|A} - P_B\right) = \langle e\rangle \quad , \quad x \in D_A \quad , \tag{4.2}$$

which can be solved for τ^A_A if $P_{B|A}$ is known. Formally, suppose the solution of (4.2) yields

$$\bar{\tau}_A^A(x) = \frac{1}{|D_A|}\int dx_A \, \tau_A^A(x) P_A = T_1 \langle e \rangle \quad . \tag{4.3}$$

The relation $\langle\sigma\rangle = L_o \langle e\rangle + \langle\tau\rangle$ then yields

$$\hat{L} = L_2 + c_1 T_1 \tag{4.4}$$

as an estimate for $\hat{L}$. Note that this $\hat{L}$ is in general non-local since T_1 will be an operator if $\langle e\rangle$ varies over D. It will reduce, however, to local form so long as $\langle e\rangle$ varies slowly relative to the correlation length ℓ associated with $\left(P_{B|A} - P_B\right)$.

Equation (4.2) is of no direct help if $P_{B|\bar{A}}$ is not known. All that can be guaranteed about $P_{B|A}$ is that it is zero whenever $x_B - x_A$ is such that D_A and D_B would overlap, and it is usually the case that $P_{B|A} \sim P_B$ when $|x_B - x_A| \gg \ell$. The simplest function with both of these properties is

$$P_{B|A} = P_B = n_1 \quad \text{if} \quad D_A \cap D_B = \phi$$
$$= 0 \quad \text{otherwise.} \tag{4.5}$$

It is open to criticism (particularly if applied to wave propagation) but still provides an explicit form for (4.2), in the case that only n_1 $\left(\text{or } c_1\right)$ is given.

An approximation similar to (4.1) can also be devised for a general n-phase composite. If the restriction of τ to phase r is called τ^r, the general equation (2.21) becomes

$$\left(L_r - L_o\right)^{-1}\tau^r + \sum_s \Gamma_o \tau^s = e_o \quad , \quad x \in \text{phase } r \tag{4.6}$$

and (2.18) gives

$$e = e_o - \sum_r \Gamma_o \tau^r \quad . \tag{4.7}$$

If (4.7) is averaged, there results

$$\langle e\rangle = e_o - \sum_r \Gamma_o \left(\tau_r^r P_r\right) \quad , \tag{4.8}$$

where τ_r^r is the expectation value of $\tau^r(x)$, conditional upon finding phase r at x and $P_r(x)$ is the probability of finding phase r at x; thus, in terms of the indicator function f_r ,

$$\tau_r^r P_r = \langle \tau^r f_r \rangle \quad . \tag{4.9}$$

Now average (4.6) conditionally on finding phase r at x (equivalently, multiply by $f_r(x)$ and average). This gives

$$\left(L_r - L_o\right)^{-1} \tau_r^r + \sum_s \Gamma_o \left(\tau_{sr}^s P_{s|r} - \tau_s^s P_s\right) = \langle e\rangle \quad , \tag{4.10}$$

having also used (4.8) as in the method of smoothing. In (4.10), the conditional probability $P_{s|r}(x',x)$ is defined so that

$$P_{sr}(x',x) = P_{s|r}(x',x) P_r(x) \quad . \tag{4.11}$$

Equation (4.10) can be closed by making the approximation, analogous to the QCA, that

$$\tau_{sr}^s = \tau_s^s \quad . \tag{4.12}$$

If the resulting solution for τ_r^r is written

$$\tau_r^r = T_r \langle e\rangle \quad , \tag{4.13}$$

it follows that $\hat{L}$ is estimated as

$$\hat{L} = L_o + \sum P_r T_r \quad . \tag{4.14}$$

This estimate still depends on L_o which, at the moment, can be chosen in any way.

The self-consistent QCA

Continuing for the moment with (4.14), there is no formal reason why (if necessary) L_o could not be a non-local operator and it might be postulated that the 'best' approximation to $\hat{L}$ would be obtained if L_o were actually equal to $\hat{L}$, since then at least the mean polarization vanishes:

$$\langle\tau\rangle = 0 \quad . \tag{4.15}$$

Whether or not this is believed, equation (4.15) defines a so-called self-consistent approximation; the easiest way to obtain the solution is probably to iterate (4.14).

Now revert to the matrix-inclusion composite, described by (3.1) and (3.2). Taking conditional expectations has already given (3.6) and this can be reduced when $L_o \neq L_2$, by making the two assumptions

$$\tau^B_{BA} = \tau^B_B , \quad \tau^m_{mA} = \tau^m_m \quad . \tag{4.16}$$

The second assumption is clearly false even for a periodic structure, when x is close to D_A, but these two relations serve to close (3.6), upon use of the additional exact relation

$$\langle e\rangle = \left(L_1 - L_o\right)^{-1} \int dx_A \, \tau^A_A P_A + \left(L_2 - L_o\right)^{-1} \tau^m_m P_m \tag{4.17}$$

which follows from averaging $e = \left(L - L_o\right)^{-1} \tau$.

An estimate for $\hat{L}$ results from solving for τ^A_A , which depends on the choice of L_o. The 'self-consistent' choice of L_o is the one for which (4.15) is satisfied.

A version of this procedure can be followed even if $P_{B|A}$ is not known, for the observation that there is either matrix or inclusion at x implies

$$P_m(x) + \int dx_B P_B(x)\, f_B(x) = 1$$

and

$$P_{m|A}(x) + \int dx_B\, P_{B|A}(x)\, f_B(x) = 1 \quad ,$$

where $f_B(x) = 1$ if an inclusion centred at x_B would contain x, and $f_B(x) = 0$ otherwise. Thus, $P_{m|A}$ would also be fixed if $P_{B|A}$ were chosen to have the form (4.5).

An alternative self-consistent QCA

It is, actually, possible to proceed without making the second of assumptions (4.16) for a matrix-inclusion composite. This is because (3.6) is not the only information that follows from averaging (3.1) and (3.2) with D_A fixed. There is also an equation which follows when $x \notin D_A$:

$$\left(L_1 - L_o\right)^{-1} \int dx_B\, \tau^B_{BA} P_{B|A} + \left(L_2 - L_o\right)^{-1} \tau^m_{mA} P_{m|A} + \Gamma_o \tau^A_A$$
$$+ \int dx_B\, \Gamma_o \left[\tau^B_{BA} P_{B|A} - \tau^B_B P_B\right] + \Gamma_o\left(\tau^m_{mA} P_{m|A} - \tau^m_m P_m\right)$$
$$= \langle e\rangle \quad , \quad x \notin D_A \quad . \tag{4.18}$$

Remarkably, equations (4.17) and (4.18), both of which are exact, can be solved exactly for $\tau^m_{mA} P_{m|A}$ and $\tau^m_m P_m$, in terms of τ^B_{BA} and τ^A_A, so that only the single QCA (4.1) is unavoidable, even when a general L_o is employed. Still more remarkable is the fact that when this system is solved with the single approximation (4.1), the estimate that results for $\hat{L}$ is always the 'ordinary' QCA estimate (4.4), independently of the choice of L_o. The final requirement of self-consistency, that $\langle\tau\rangle = 0$, thus adds nothing though, if invoked, it does give $L_o = \hat{L}$ as it should. The details of these calculations were given by Willis.[12]

Special case: isotropic two-point functions

This lecture is concluded with a short discussion of what happens when the geometry of the composite is isotropic and statistically uniform, though the moduli L_r remain general. First, for the general n-phase composite, suppose that $P_{s|r}(x',x)$ is isotropic, and so a function of $|x'-x|$ only, while $P_r = c_r$, volume concentration of phase r. If attention is restricted to local behaviour, <e> can be taken uniform and τ_r^r is then independent of x. With the closure approximation (4.12), the term involving Γ_o, written explicitly as an integral, becomes

$$\Gamma_o\left(P_{s|r} - P_s\right)\tau_s^s = \int dx'\,\Gamma_o(x-x')g(|x-x'|) \quad , \tag{4.19}$$

where $g = \left(P_{s|r} - P_s\right)\tau_s^s$ has the dependence shown because P_s and τ_s^s are independent of x'.

Now recall one of the results of Lecture 3, that

$$\Gamma_o \tau = P\tau \quad , \quad x \in D_A$$

for any $\tau(x)$ which is constant over a sphere D_A and zero elsewhere. A consequence is that

$$\int_{|x'|<a} dx'\,\Gamma_o(x') = P \quad , \tag{4.20}$$

independently of a and it follows that

$$\int_{|x'|=r} dS\,\Gamma_o(x') = 0 \tag{4.21}$$

for all $r > 0$. The integration in (4.19) therefore needs only to extend over a sphere of arbitrarily small radius, centred at x, and (4.20) then gives

$$\int dx' \, \Gamma(x-x') g(|x-x'|) = P\, g(0) \quad , \tag{4.22}$$

provided only that g is continuous. But, when $x' = x$, $P_{s|r}(x',x) = \delta_{rs}$, since phase r at x is guaranteed. Therefore,

$$\Gamma_o \left(P_{s|r} - P_s \right) \tau_s^s = P \left(\delta_{rs} - c_s \right) \tau_s^s$$

and (4.10) reduces to

$$\left[\left(L_r - L_o \right)^{-1} + P \right] \tau_r^r - P\langle\tau\rangle = \langle e\rangle \quad , \tag{4.23}$$

where $\langle\tau\rangle = \Sigma\, c_r \tau_r^r$. Equations (4.23) can be solved explicitly to give

$$\hat{L} = \left\{ \Sigma\, c_r \left[I + \left(L_r - L_o \right) P \right]^{-1} \right\}^{-1} \Sigma\, c_s \left[I + \left(L_s - L_o \right) P \right]^{-1} L_s \quad . \tag{4.24}$$

The most interesting property of (4.24) is that it contains only the volume concentrations c_r. The information on the two-point statistics has disappeared, though the isotropy of $P_{s|r}$ was used in an essential way in deriving (4.24).

A similar reduction occurs for a matrix-inclusion composite, if $\langle e\rangle$ is taken uniform and $P_{B|A}$ is a function of $|x_B - x_A|$ only and the inclusions are identical spheres. It can be demonstrated that τ_A^A is constant over D_A, τ_m^m is constant over D_m and that the precise form of $P_{B|A}$ is unimportant. Thus, in this particular case, the form (4.5) leads to the same result as any other isotropic $P_{B|A}$. Furthermore, since τ_A^A and τ_m^m are constant, the solution for this particular composite exactly fits the pattern already established for an n-phase composite, so that $\hat{L}$ can be given in the form (4.24), with $n = 2$. Matrix-inclusion composites have been discussed explicitly by Willis.[13]

The self-consistent prescription $L_o = \hat{L}$ can be given in a variety of forms, including

$$\Sigma c_s \left[I + \left(L_s - L_o\right)P\right]^{-1}\left(L_s - L_o\right) = 0 \quad , \tag{4.25}$$

$$\Sigma c_s \left[I + \left(L_s - L_o\right)P\right]^{-1} = I \quad , \tag{4.26}$$

$$\Sigma c_s \left[I + \left(L_s - L_o\right)P\right]^{-1} L_s = L_o \quad . \tag{4.27}$$

These can also be obtained directly from estimating the strain in one spherical inclusion by embedding it directly in a matrix with 'overall' properties $\hat{L}$. Details of this approach are likely to be pursued by Professor Zaoui. The embedding approach has also been reviewed by Willis.[14, 15] Similar results can also be obtained for ellipsoidal inclusions, if the geometry of the entire composite can be reduced to that of an isotropic distribution of spheres by an affine transformation.

CHAPTER 5

5. - VARIATIONAL PRINCIPLES

The minimum energy principle of linear elastostatics states that the functional

$$\mathcal{F}(u^*) = \int_D \left\{ \tfrac{1}{2} e^* Le^* - \sigma_o e^* \right\} dx \tag{5.1}$$

is minimized, amongst displacement fields u* that satisfy any given displacement conditions on ∂D, by the equilibrium displacement field u. In (5.1), e* denotes the strain corresponding to u*, $e^* Le^* = e^*_{ij} L_{ijk\ell} e^*_{k\ell}$, $\sigma_o e^* = (\sigma_o)_{ij} e^*_{ij}$ and σ_o is any stress field that satisfies the equilibrium equations (2.6) and whatever traction boundary conditions may be prescribed. The form (5.1) admits any combination of traction and displacement boundary conditions but not 'compliant' conditions, such as $\sigma_{ij} n_j = K_{ij} u_i$. The principle may be verified by direct calculation:

$$\mathcal{F}(u^*) - \mathcal{F}(u) = \int_D \left\{ \tfrac{1}{2}(e^* - e) L (e^* - e) + (e^* - e) Le - \sigma_o (e^* - e) \right\} dx, \tag{5.2}$$

having used the symmetry of L. Now $Le = \sigma$ and

$$\int_D \left(\sigma - \sigma_o\right)(e^* - e)\, dx = \int_D \frac{\partial}{\partial x_j} \left(\sigma - \sigma_o\right)_{ij} \left(u^*_i - u_i\right) dx \quad , \tag{5.3}$$

since σ and σ_o both satisfy (2.6) so that $\operatorname{div}\left(\sigma - \sigma_o\right) = 0$. The integral in (5.3) may now be transformed by Gauss' theorem; it is zero because either $u_i^* = u_i$ or $\sigma_{ij} n_j = (\sigma_o)_{ij} n_j$ at each point of ∂D. Hence,

$$F(u^*) - F(u) = \int_D \tfrac{1}{2}(e^* - e)L(e^* - e)\,dx \geq 0 \quad , \tag{5.4}$$

as claimed.

The complementary principle may be given in a similar form: the functional

$$G(\sigma^*) = \int_D \left(\tfrac{1}{2}\sigma^* L^{-1} \sigma^* - e_o\, \sigma^* \right) dx \tag{5.5}$$

is minimized, amongst stress fields σ^* that satisfy the equilibrium equations (2.6) and whatever traction boundary conditions are given, by the actual stress field σ. In (5.5), e_o is the strain field corresponding to any displacement u_o which takes any prescribed boundary values. The proof corresponds very closely to that for the minimum energy principle and is not given.

It has been demonstrated that the displacement boundary value problem is characterized by the integral equation (2.21):

$$\left(L - L_o\right)^{-1} \tau + \Gamma_o \tau = e_o \quad . \tag{2.21}$$

This formal equation in fact characterizes the same set of boundary value problems as does (5.1) if the Green's function G_o is chosen appropriately. The operator Γ_o is self-adjoint and so, immediately, (2.21) shows that the functional

$$H(\tau^*) = \int \left(\tfrac{1}{2}\tau^* \left(L - L_o\right)^{-1} \tau^* + \tfrac{1}{2}\tau^* \,\Gamma_o \tau^* - e_o \tau^* \right) dx \tag{5.6}$$

is stationary when $\tau^* = \tau$, the solution of (2.21).

The functional H can be related to F, as follows. Any chosen τ^*

generates an admissible displacement field u* (c.f. (2.17)) and corresponding strain field

$$e^* = e_o - \Gamma_o \tau^* \tag{5.7}$$

and stress field

$$\sigma^* = L_o e^* + \tau^* \quad . \tag{5.8}$$

The integral equation (2.21) will not be satisfied, but it is useful to define the 'error'

$$\varepsilon = e^* - \left(L - L_o\right)^{-1} \tau^* \quad . \tag{5.9}$$

Now by elementary algebra,

$$\tfrac{1}{2} e^* L\, e^* = \tfrac{1}{2} e^* L_o\, e^* + \tfrac{1}{2}\varepsilon \left(L - L_o\right) \varepsilon - \tfrac{1}{2}\tau^* \left(L - L_o\right)^{-1} \tau^* + \tau^* e^* \quad . \tag{5.10}$$

Therefore, expanding e* using (5.7)

$$F(u^*) = \int_D \Big\{ \tfrac{1}{2} e_o L_o e_o - \sigma_o e_o + \tau^* \left(\tfrac{1}{2}\Gamma_o L_o \Gamma_o - \Gamma_o\right) \tau^*$$

$$- \tfrac{1}{2}\tau^* \left(L - L_o\right)^{-1} \tau^* + \tau^* e_o + \left(\sigma_o - L_o e_o\right) \Gamma_o \tau^* + \tfrac{1}{2}\varepsilon \left(L - L_o\right) \varepsilon \Big\}\, dx \quad . \tag{5.11}$$

It may be noted that any displacement field u and associated stress field $\sigma = L_o e$ in the comparison body, which satisfy homogeneous boundary conditions, can be expressed in the form

$$e = -\Gamma_o \tau \quad ,$$

provided only that τ is such that

$$\operatorname{div} \tau = -\operatorname{div}\left(L_o e\right) \quad .$$

In particular, therefore,

$$e = \Gamma_o L_o e \quad .$$

Applying this result to Γ_o gives

$$\Gamma_o = \Gamma_o L_o \Gamma_o \quad . \tag{5.12}$$

Then, with the natural choice $\sigma_o = L_o e_o$, (5.11) yields

$$F(u^*) = -\int_D \tfrac{1}{2} e_o L_o e_o \, dx - H(\tau^*) + \int_D \tfrac{1}{2}\varepsilon\left(L - L_o\right)\varepsilon \, dx \tag{5.13}$$

and it follows immediately that

$$F(u) \le F(u^*) \le -\int_D \tfrac{1}{2} e_o L_o e_o \, dx - H(\tau^*) \quad , \tag{5.14}$$

so long as the quadratic form $\varepsilon\left(L - L_o\right)\varepsilon$ is negative (semi-) definite at each point of D. The inequalities in (5.14) become equalities when $\tau^* = \tau$, the solution of (2.21).

Now suppose that $\varepsilon\left(L - L_o\right)\varepsilon$ is positive (semi-) definite. A comparison with the complementary functional G can be carried out, similar to that performed with F, but the details are tricky and are left as a fairly difficult exercise for the reader (or, alternatively, see Hill[16] or Willis[15]). It is easy to prove, however, that the operator Γ_o is positive semi-definite, as follows. For any τ^*, let

$$e' = -\Gamma_o \tau^* \ , \qquad \sigma' = L_o e' + \tau^* \quad .$$

Then, $\operatorname{div}\sigma' = 0$ and σ' and the corresponding displacement u' satisfy homogeneous boundary conditions on ∂D. Now

$$\int_D \tau^* \Gamma_o \tau^* \, dx = \int_D \left(L_o e' - \sigma'\right) e' \, dx = \int_D e' L_o e' \, dx \ge 0 \quad ,$$

the integral of $\sigma' e'$ contributing nothing since it can be transformed to an integral over ∂D by Gauss' theorem, and σ', u' satisfy homogeneous boundary conditions.

It follows that $\left(L - L_o\right)^{-1} + \Gamma_o$ is positive (semi-) definite and so

$\mathcal{H}$ is minimized by the solution τ of (2.21). Hence, in this case, since the stationary value is still related to $\mathcal{F}(u)$ through the equalities in (5.14), this time,

$$-\mathcal{H}(\tau^*) - \int_D \tfrac{1}{2} e_o L_o e_o \, dx \leq -\mathcal{H}(\tau) - \int_D \tfrac{1}{2} e_o L_o e_o \, dx = \mathcal{F}(u) \quad . \tag{5.15}$$

The functional $\mathcal{H}$ may thus be used, with suitable choices for L_o, to bound the energy functional $\mathcal{F}(u)$ from above or below. The variational principle for $\mathcal{H}$ was established by Hashin and Shtrikman.[17]

Application to composites

The variational principles will be used in Lecture 6 to obtain bounds on the overall behaviour of a composite. The definition based on ensemble averaging, $\langle\sigma\rangle = \hat{L}\langle e\rangle$, will be used, so that non-local behaviour can be considered, the corresponding local overall moduli following when $\langle e\rangle$ is uniform. It is desirable that $\langle e\rangle$ should be reasonably general and, since $\langle e\rangle$ cannot be fixed throughout D by surface constraints alone, a body force f is imposed. Thus, the discussion will be based upon the boundary value problem

$$\operatorname{div} \sigma + f = 0 \quad , \quad x \in D \quad , \tag{5.16}$$

together with the boundary condition

$$u = 0 \quad , \quad x \in \partial D \quad . \tag{5.17}$$

This is imposed for the purely technical reason that it facilitates the use of the variational principles. It is restrictive in that the influence of the boundary is suppressed, but 'boundary layers' are in any case beyond the scope of these lectures.

For the problem defined by (5.16) and (5.17), together with the stress-strain relation (2.2), the energy stored in D can be expressed in the form

$$\int_D \tfrac{1}{2}\sigma_{ij} e_{ij}\, dx = \int_D \tfrac{1}{2} f_i u_i\, dx \quad , \tag{5.18}$$

by Gauss' theorem. Now take the ensemble average of (5.18):

$$\int_D \tfrac{1}{2}<\sigma_{ij} e_{ij}> dx = \int_D \tfrac{1}{2} f_i <u_i> dx \quad . \tag{5.19}$$

But ensemble averaging (5.16) shows that $f = -\operatorname{div} <\sigma>$ and so, by further application of Gauss' theorem,

$$\int_D \tfrac{1}{2}<\sigma_{ij} e_{ij}> dx = \int_D \tfrac{1}{2}<\sigma_{ij}><e_{ij}> dx$$

or, with the notation $<\sigma> = \hat{L}<e>$,

$$\int_D \tfrac{1}{2}<\sigma e> dx = \int_D \tfrac{1}{2}<\sigma><e> dx = \int_D \tfrac{1}{2}<e>\hat{L}<e> dx \quad . \tag{5.20}$$

Equation (5.20) demonstrates how bounds for the energy yield directly bounds on the quadratic form associated with $\hat{L}$. The similarity of (5.20) to equation (1.12), which involved volume averages suitable for investigation of local behaviour, may be noted.

CHAPTER 6

6. - SOME ELEMENTARY BOUNDS

As a first example, let u_o be the displacement induced in a comparison body with moduli L_o by the body force f. Thus, if e_o is the corresponding strain and $\sigma_o = L_o e_o$, σ_o satisfies (5.16) and u_o satisfies (5.17). The minimum energy principle can be applied, adopting this particular σ_o and taking $e^* = e_o$, to give

$$F(u) \le \int_D \left(\tfrac{1}{2} e_o L e_o - \sigma_o e_o\right) dx$$

or, since inequalities are not disturbed by averaging,

$$\langle F(u)\rangle \le \int_D \left(\tfrac{1}{2} e_o \langle L\rangle e_o - \sigma_o e_o\right) dx \quad . \tag{6.1}$$

Progress is difficult unless L_o is chosen equal to $\langle L\rangle$. Then, with its left side also given explicitly, (6.1) yields

$$-\tfrac{1}{2}\int_D \langle e\rangle \hat{L} \langle e\rangle dx \le -\tfrac{1}{2}\int_D e_o \langle L\rangle e_o \, dx \quad . \tag{6.2}$$

Now let G_o be the Green's function corresponding to L_o and let $\hat{G}$ correspond to $\hat{L}$. Then, (6.2) implies

$$\int_D f G_o f \, dx \le \int_D f \hat{G} f \, dx \tag{6.3}$$

or, symbolically,

$$G_o \le \hat{G} \quad , \tag{6.4}$$

this order relation implying (6.3) for any body force f. Continuing symbolically, (6.4) implies

$$\hat{G}^{-1} \le G_o^{-1} \tag{6.5}$$

or, explicitly,

$$\int_D u\,\hat{G}^{-1} u\,dx \le \int_D u\,G_o^{-1} u\,dx \quad , \tag{6.6}$$

for any displacement field u in the range of $\hat{G}$ and G_o; that is, for any displacement field which vanishes on ∂D. Note that (6.6) does not require u to be realized by any particular body force over D, in any particular material: it is implied by (6.5), which follows from the algebraic identity

$$\left(G_o^{-1} - \hat{G}^{-1}\right)^{-1} = G_o + G_o\left(\hat{G} - G_o\right)^{-1} G_o \quad .$$

Notice now that $u = G_o f$ implies $f = -\mathrm{div}\left(L_o e\right)$, so $G_o^{-1} u = -\mathrm{div}\left(L_o e\right)$; similarly, $\hat{G}^{-1} u = -\mathrm{div}(\hat{L}\,e)$. Therefore, by substituting into (6.6) and using Gauss' theorem,

$$\int_D e\,\hat{L}\,e\,dx \le \int_D e\langle L\rangle e\,dx \quad , \tag{6.7}$$

where e is defined from any displacement field u which vanishes over ∂D. The quadratic form on the right side of (6.7) involves just the Voigt average $\langle L\rangle$. The result that it bounds the local form for $\hat{L}$ (obtained by taking e uniform) is well-known, but (6.7) is not. A corresponding lower bound for $\hat{L}$ could be obtained similarly, starting from the complementary energy principle, the local 'lower bound' operator being the Reuss average, $\langle L^{-1}\rangle^{-1}$.

These bounds are generally rather crude but they have the interesting property that they involve the statistics of the composite only through the one-point probabilities P_r or, if the composite is statistically uniform, through the phase concentrations c_r. They must, therefore, apply for all composites with these properties, regardless of more detailed structure.

Hashin-Shtrikman bounds

Bounds which are sensitive to the two-point probabilities may be obtained by substituting a trial polarization of the form

$$\tau^*(x) = \sum_{r=1}^{n} f_r(x)\,\tau_r(x) \tag{6.8}$$

into the Hashin-Shtrikman functional (5.6), the n functions τ_r being non-random. The ensemble average of the energy can be bounded in terms of the ensemble average of H which, with (6.8), takes the form

$$\langle H(\tau^*)\rangle = \int_D dx \left\{ \tfrac{1}{2}\sum_{r=1}^{n} P_r(x)\,\tau_r(x)\left(L_r - L_o\right)^{-1}\tau_r(x) \right.$$
$$\left. + \tfrac{1}{2}\sum_{r=1}^{n}\sum_{s=1}^{n} \tau_r(x)\int_D dx'\,\Gamma_o(x,x')\,\tau_s(x')\,P_{rs}(x,x') - e_o \sum_{r=1}^{n} P_r(x)\,\tau_r(x) \right\} \tag{6.9}$$

directly from the definitions (2,7), (2.8) of $P_r(x)$, $P_{rs}(x,x')$. The expression (6.9) has stationary value

$$-\tfrac{1}{2}\int_D e_o \langle\tau^*\rangle dx = -\tfrac{1}{2}\int_D e_o \sum_{r=1}^{n} P_r\,\tau_r\,dx,$$

in which τ_r satisfy the equations

$$P_r\left(L_r - L_o\right)^{-1}\tau_r + \sum_{s=1}^{n}\Gamma_o\left(P_{rs}\tau_s\right) = P_r e_o \quad , \quad r = 1,2\ \dots\ n \quad . \tag{6.10}$$

The associated estimate <e*> for the mean strain is given by

$$\langle e^*\rangle = e_o - \sum_{r=1}^{n}\left(\Gamma_o P_r \tau_r\right) \tag{6.11}$$

and so, combining this with (6.10),

$$P_r\left(L_r - L_o\right)^{-1}\tau_r + \sum_{s=1}^{n}\Gamma_o\left(P_{rs} - P_r P_s\right)\tau_s = \langle e^*\rangle \quad . \tag{6.12}$$

Suppose these equations have solution

$$\tau_r = T_r\langle e^*\rangle \quad , \quad r = 1,2\ \dots\ n \quad , \tag{6.13}$$

for some set of operators T_r . Then, by substituting back into (6.11) and rearranging,

$$\langle e^*\rangle = \left[I + \Gamma_o\langle T\rangle\right]^{-1} e_o \quad , \tag{6.14}$$

so

$$\langle \tau^*\rangle = \langle T\rangle\left[I + \Gamma_o\langle T\rangle\right]^{-1} e_o \quad . \tag{6.15}$$

Here, I represents the identity operator.

Nothing so far has been said about L_o. Suppose first that L_o is such that $L_o - L_r$ is positive (semi-) definite for each r. Then, from the ensemble average of (5.14),

$$\langle F(u)\rangle \le -\int_D \tfrac{1}{2} e_o\left[L_o + \langle T\rangle\left[I + \Gamma_o\langle T\rangle\right]^{-1}\right] e_o\, dx \quad . \tag{6.16}$$

Now, in terms of the Green's function G_o , $u_o = G_o f$ and, correspondingly,

$$e_o = E_o f \quad , \tag{6.17}$$

where E_o is the strain associated with G_o. The inequality (6.16) may therefore be written

$$-\tfrac{1}{2}\int_D f\hat{G}\; f\,dx \le -\tfrac{1}{2}\int_D f\tilde{G} f\,dx \quad , \tag{6.18}$$

where

$$\tilde{G} = G_o - E_o^{\dagger}\langle T\rangle\left[I + \Gamma_o\langle T\rangle\right]^{-1}E_o \quad , \tag{6.19}$$

the operator $E_o^{\dagger}$ being the adjoint of E_o.

In symbolic notation, (6.18) can be written

$$\tilde{G} \le \hat{G}$$

and so, by reasoning like that which gave (6.7),

$$\int_D e\hat{L}e\,dx \le \int_D e\tilde{L}e\,dx \quad , \tag{6.20}$$

where $\tilde{L}$ is the effective nonlocal operator that corresponds to $\tilde{G}$. It can be shown, in fact, that

$$\tilde{L} = L_o + \langle T\rangle \quad , \tag{6.21}$$

which is consistent with the overall relation $\langle\sigma^*\rangle = L_o\langle e^*\rangle + \langle\tau^*\rangle$. To verify (6.21), differentiate (6.19) to obtain

$$\tilde{E} = E_o - \Gamma_o\langle T\rangle\left[I + \Gamma_o\langle T\rangle\right]^{-1}E_o \quad , \tag{6.22}$$

where $\tilde{E}$ is the strain corresponding to $\tilde{G}$. Equation (6.22) can be rearranged to give

$$\tilde{E} = E_o - \Gamma_o\langle T\rangle\tilde{E} \tag{6.23}$$

so, in view of the relations

$$\operatorname{div}\left(L_o\Gamma_o t\right) = \operatorname{div} t \tag{6.24}$$

for any t, and

$$\operatorname{div}\left(L_o E_o\right) + I = 0 \quad , \tag{6.25}$$

it follows that

$$\operatorname{div}(\tilde{L}\tilde{E}) + I = 0 \quad , \tag{6.26}$$

where $\tilde{L}$ is given by (6.21).

A lower bound for $\hat{L}$, of precisely the same form, follows by taking L_o so that $L_r - L_o$ is positive (semi-) definite for each r.

Several remarks can be made about the expression (6.21) for $\tilde{L}$. First, it provides bounds which depend upon the two-point probabilities P_{rs}. Next, it should be noted that the closure approximation (4.12) led precisely to the estimate $\tilde{L}$ as an approximation to $\hat{L}$; equation (4.10) for τ_r^r is precisely the same as (6.12) and (6.21) is the same as (4.14). Finally, $\langle H \rangle$ is stationary when (6.10) are satisfied, even when $L_r - L_o$ is not definite in the same sense for each r. Furthermore, the stationary principle still applies if L_o is a non-local operator. The self-consistent condition

$$\langle T \rangle = 0 \quad ,$$

which was previously discussed in relation to the QCA, thus also has a variational interpretation.

A similar development can be given for a matrix-inclusion composite, starting from the trial polarization

$$\tau^*(x) = \sum_A f_A(x)\, \tau^A(x) + \left(1 - \sum_A f_A(x)\right) \tau^m(x) \quad ,$$

where τ^A is non-random and depends upon x and x_A , and τ^m is non-random and depends only on x. Again, bounds result (when L_o is chosen appropriately), that can also be obtained as direct approximations by making the extended QCA (4.16).

Bounds of higher order

If more statistical information happened to be given, then this could be incorporated to obtain bounds better than (6.21). A detailed development is not given, both through lack of time and because higher-order statistics are very seldom known. However, if the strain field $e^* = e_o - \Gamma_o \tau^*$, with τ^* given by (6.8), is substituted into the classical principles, bounds which allow for third-order statistics ($P_{rst}(x,x',x'')$ in obvious notation) are obtained. A more extended outline is given by Willis.[15]

It seems to be a general feature that the classical principles employ information up to some odd order, while the Hashin-Shtrikman functional provides bounds of even order. The matter is further discussed by Kröner,[18] whose starting point was to substitute a formal perturbation series for e into the classical principles, and by Willis,[14] who also substituted into $\langle H \rangle$ a perturbation series for τ.

CHAPTER 7

7. - DYNAMIC PROBLEMS

Although a full discussion of dynamic problems is beyond the scope of these lectures, the brief outline which follows will demonstrate how all of the ideas already introduced are also applicable in the dynamic situation; the structure is the same but the details are more complicated.

First, note that the equations of motion for a medium occupying a region D can be given in the form

$$\mathrm{div}\,\sigma - \dot{p} + f = 0 \quad , \tag{7.1}$$

where p denotes the momentum density, related to velocity $\dot{u}$ through

$$p = \rho\,\dot{u} \quad , \tag{7.2}$$

ρ being the mass density. Equation (7.1) generalizes (2.6). Although not usually considered in this way, (7.2) is a constitutive relation, of status similar to $\sigma = L\,e$. Of course the system (7.1), with constitutive relations, has to be augmented by both boundary and initial conditions.

If a comparison medium with moduli L_o and density ρ_o is introduced, it is natural to define, in addition to the stress polarization τ, a momentum polarization

$$\pi = (\rho - \rho_o)\dot{u} \quad , \tag{7.3}$$

so that

$$p = \rho_o \dot{u} + \pi \quad . \tag{7.4}$$

Substitution back into (7.1) gives

$$\operatorname{div}(L_o e) - \rho_o \ddot{u} + f + (\operatorname{div} \tau - \dot{\pi}) = 0 \quad , \tag{7.5}$$

which generalizes (2.14). Then, if $G_o(x,x',t)$ denotes the dynamic Green's function for the comparison medium,

$$u(x,t) = u_o(x,t) + \int_D G_o * (\operatorname{div} \tau - \dot{p})\, dx' \quad , \tag{7.6}$$

where u_o is the solution of (7.5), with the boundary and initial conditions, when $\tau = \pi = 0$, and $*$ denotes the operation of time-convolution. Then, integrating by parts and differentiating with respect to x or t to find e or $\dot{u}$, the integral equations

$$\begin{bmatrix} (L - L_o)^{-1} \tau \\ (\rho - \rho_o)^{-1} \pi \end{bmatrix} + \begin{bmatrix} S_x & S_t \\ M_x & M_t \end{bmatrix} \begin{bmatrix} \tau \\ \pi \end{bmatrix} = \begin{bmatrix} e_o \\ \dot{u}_o \end{bmatrix} \tag{7.7}$$

are obtained upon eliminating $e, \dot{u}$ in favour of τ, π and, equivalently

$$\begin{bmatrix} e \\ \dot{u} \end{bmatrix} + \begin{bmatrix} S_x & S_t \\ M_x & M_t \end{bmatrix} \begin{bmatrix} (L - L_o) e \\ (\rho - \rho_o) \dot{u} \end{bmatrix} = \begin{bmatrix} e_o \\ \dot{u}_o \end{bmatrix} \tag{7.8}$$

follow by eliminating τ, π in favour of $e, \dot{u}$. These equations generalize (2.21), (2.22) respectively. The operators S_x , S_t , M_x , M_t involve various second derivatives of G_o. They involve integration with respect to x' over D and convolution with respect to time.

Perturbation theory and closure approximations can be applied to

(7.7) and (7.8), in exactly the same way as outlined in Lectures 2, 3 and 4. It was noted by Willis[14] that equations (7.7) have an associated variational structure: the symmetries of the operators imply that the expression

$$H(\tau,\pi) = \int_D \left\{ \tfrac{1}{2}[\tau \;\; \pi] * \begin{bmatrix} (L - L_o)^{-1}\tau \\ (\rho - \rho_o)^{-1}\pi \end{bmatrix} + \tfrac{1}{2}[\tau \;\; \pi] * \begin{bmatrix} S_x & S_t \\ M_x & M_t \end{bmatrix} \begin{bmatrix} \tau \\ \pi \end{bmatrix} - [\tau \;\; \pi] * \begin{bmatrix} e_o \\ \dot{u}_o \end{bmatrix} \right\} dx \qquad (7.9)$$

is stationary when (7.7) are satisfied. This $H(\tau,\pi)$ generalizes (5.6) but, since convolutions appear, its value is a (generalized) function of time, rather than a scalar. Willis[19] has further shown that $H(\tau,\pi)$ can be related to some stationary principles for dynamic problems due to Gurtin,[20] in much the same way as $H(\tau)$ was related to $F(u)$ and $G(\sigma)$ in Lecture 5. Furthermore, if the convolutions are removed by Laplace transforming and the transform variable is taken to be real, maximum and minimum principles result which are direct extensions of those given in Lecture 5.

Analysis similar to that given in Lecture 6 can be performed, which demonstrates that the dynamic QCA also follows naturally from the variational structure. A third-order approximation for dynamic problems, based upon the variational principles of Gurtin,[20] was outlined by Willis.[21]

This section is concluded by noting a recent observation of

Weaver.[22] Application of Gauss' theorem shows that, at the stationary point, if homogeneous initial and boundary conditions are assumed,

$$H(\tau,\pi) = \tfrac{1}{2}\int_D f * \left(u - u_o\right) dx \quad . \tag{7.10}$$

If, in particular, f is taken to have components $f_i = \delta_{ip}\,\delta\left(x - x_o\right)\delta(t)$, so that $u_o(x,t) = G_o\left(x,x_o,t\right)$ and $u(x,t) = G\left(x,x_o,t\right)$, the Green's function for the composite, then (7.10) gives

$$H(\tau,\pi) = \tfrac{1}{2}\left[G\left(x_o,x_o,t\right) - G_o\left(x_o,x_o,t\right)\right] \tag{7.11}$$

at the stationary point. Thus, any approximate (τ,π), substituted into H, provides a particularly accurate estimate of the 'admittance' $G\left(x_o,x_o,t\right)$.

Nonlinear problems

Very little work has been performed for nonlinear problems and (apart from a single exception mentioned below) none to date makes any allowance for statistics other than volume concentrations. Clearly, perturbation theory can be performed, either for an almost uniform medium or for a matrix containing a dilute suspension of inclusions but the only methods at present available for arbitrary concentrations are variants of the self-consistent procedure described by Professor Zaoui.

The Hashin-Shtrikman variational principle discussed in Lecture 5 has, however, recently been generalised to a class of nonlinear problems by Willis[23] and developed further by Talbot and Willis.[24] Applications to nonlinear elasticity have not yet been pursued, but results for a nonlinear dielectric composite have been obtained by Willis.[25] Since the study of nonlinear behaviour constitutes one of the present frontiers of research in composites, this set of lectures is concluded with a brief account of the variational structure for nonlinear problems.

The structure is best displayed by adopting an abstract approach which can be specialized later. Let F be a functional from a Banach space B to $\mathbb{R}$ and suppose that it is required to minimize F over some subset K of B; just in case a minimizer does not exist in K, the problem is generalized slightly to

$$\mathcal{P}: \ \mathrm{Inf}\{F(e) : e \in K\} \quad . \tag{7.12}$$

It is assumed that this infimum is finite. Now introduce 'comparison' functionals, $\underline{F}_o$ and $\overline{F}_o$ and define

$$\underline{f} = F - \underline{F}_o \quad , \quad \overline{f} = F - \overline{F}_o \quad . \tag{7.13}$$

Further, define

$$\underline{f}^*(\tau) = \mathrm{Sup}\ \{(\tau,e) - \underline{f}(e) : e \in B\} \quad , \tag{7.14}$$

and

$$\overline{f}_*(\tau) = \mathrm{Inf}\ \{(\tau,e) - \overline{f}(e) : e \in B\} \quad , \tag{7.15}$$

for any $\tau \in B^*$, the space dual to B, (τ,e) denoting the bilinear pairing between elements of B^* and elements of B.

It follows immediately from these definitions that, for any $\tau \in B^*$ and $e \in B$,

$$(\tau,e) + \underline{F}_o(e) - \underline{f}^*(\tau) \le F(e) \le (\tau,e) + \overline{F}_o(e) - \overline{f}_*(\tau) \quad . \tag{7.16}$$

Therefore, by taking infima over $e \in K$,

$$\underset{e\in K}{\mathrm{Inf}} \left\{(\tau,e) + \underline{F}_o(e) - \underline{f}^*(\tau)\right\} \le \underset{e\in K}{\mathrm{Inf}}\ F(e)$$

$$\le \underset{e\in K}{\mathrm{Inf}} \left\{(\tau,e) + \overline{F}_o(e) - \overline{f}_*(\tau)\right\} \quad . \tag{7.17}$$

These inequalities contain, as special cases, the Hashin-Shtrikman inequalities (5.14), (5.15). Furthermore, their derivation shows clearly their close relationship with the 'classical' variational problem $\mathcal{P}$. To reproduce (5.14), (5.15), identify F(e) with the functional $\mathcal{F}(u)$ given by (5.1). The space B is then identified with $(L^2(D))^6$, while K is the set of strain fields which are derived from displacements $u \in (H^1(D))^3$ which also satisfy any given displacement boundary conditions. The comparison functionals $\underline{F}_o$, $\overline{F}_o$ are chosen like F, except that L is replaced by L_o; for $\underline{F}_o$, $L - L_o$ is positive (semi-)definite while for $\overline{F}_o$, $L_o - L$ is positive (semi-)definite. The functionals $\underline{f}^*(\tau)$, $\overline{f}_*(\tau)$, defined by (7.14), (7.15) respectively, then take the form

$$\underline{f}^*(\tau) \quad (\text{or } \overline{f}_*(\tau)) = \int_D \tfrac{1}{2}\tau\left(L - L_o\right)^{-1} dx \quad , \tag{7.18}$$

with L_o chosen appropriately.

The infimum in the lower bound in (7.17), for example, now takes the explicit form

$$\operatorname*{Inf}_{e\in K} \int_D \left(\tfrac{1}{2}e\, L_o\, e + \tau e - \sigma_o e\right) dx \tag{7.19}$$

and this is attained when

$$\operatorname{div}\left(L_o\, e\right) + \operatorname{div}\tau + f = 0 \quad , \quad x \in D, \tag{7.20}$$

since $\operatorname{div}\sigma_o + f = 0$. Additionally, e must be derived from a displacement u and

$$\text{either } u_i = \left(u_o\right)_i \quad \text{or} \quad \left(L_o\right)_{ijk\ell}e_{k\ell}n_j = \left(\sigma_o\right)_{ij}n_j \tag{7.21}$$

for each i, at each point of ∂D. Equations (7.20), (7.21) follow, of course, only if the fields are sufficiently smooth; more generally, they only have to be satisfied in a weak sense. In either event, the solution of (7.20), (7.21) is expressible in the form

$$e = e_o - \Gamma_o \tau \qquad (7.22)$$

and substitution back into the lower bound in (7.17) reproduces (5.15) (with τ^* written as τ). The upper bound similarly reproduces (5.14).

A simple nonlinear generalization is obtained by considering a 'physically nonlinear' material, with energy function W(e). Then,

$$F(e) = \int_D \left[W(e) - \sigma_o e \right] dx \qquad (7.23)$$

and comparison functionals are defined with energy functions $\underline{W}_o$ or $\overline{W}_o$ replacing W. In particular (with the bars suppressed) W_o could be chosen as the quadratic,

$$W_o(e) = \tfrac{1}{2} e L_o e \quad , \qquad (7.24)$$

corresponding to linear behaviour, even if the actual medium is non-linear. The bounds (7.17) follow, so long as the functionals $\overline{f}_*, \underline{f}_*$ turn out to be finite (if one of these is infinite, the corresponding bound is also infinite and so true, but of no use). The easiest way of ensuring this is to choose sets of moduli L_o so that $F - \underline{F}_o$, $\overline{F}_o - F$ are convex. If the form of F makes one or both of these impossible with quadratic F_o's, a class of stationary principles can still be defined,[23] with an f* taken as the Legendre dual of an $F - F_o$, but these are likely not to be maximum or minimum principles.

Results have been obtained for an analogous problem for nonlinear dielectric behaviour by Willis.[25] The possible application of (5.17) to nonlinear elastic behaviour remains to be investigated. In particular, complications associated with loss of convexity during finite deformations have not yet been addressed, though it should be noted that the inequalities (5.17) were derived under minimal assumptions, without reference to convexity.

REFERENCES

1. Ogden, R.W., *Non-Linear Elastic Deformations*, Ellis Horwood, Chichester, 1984.

2. Hill, R., On constitutive macro-variables for heterogeneous solids at finite strain, *Proc. R. Soc.*, A326, 131, 1972.

3. Ogden, R. W., On the overall moduli of non-linear elastic composite materials, *J. Mech. Phys. Solids*, 22, 541, 1974.

4. Hill, R., On macroscopic effects of heterogeneity in elastoplastic media at finite strain, *Math. Proc. Camb. Phil. Soc.*, 95, 481, 1984.

5. Sanchez-Palencia, E., Comportements local et macroscopique d'un type de milieux physiques hétérogènes, *Int. J. Engng. Sci.*, 12, 331, 1974.

6. Kozlov, S.M., Averaging of random structures (in Russian), *Doklady Akad. Nauk. SSSR*, 241, 1016, 1978, (English translation: *Soviet Math. Dokl.*, 19, 950, 1978).

7. Papanicolaou, G.C. and Varadhan, S.R.S., Boundary value problems with rapidly oscillating coefficients, *Colloquia Mathematica Societatis János Bolyai 27. Random Fields*, North-Holland, Amsterdam, 835, 1982.

8. Keller, J.B., Stochastic equations and wave propagation in random media, *Proceedings of Symposia in Applied Mathematics,* Vol. XVI, *Stochastic Processes in Mathematical Physics and Engineering,* American Mathematical Society, Providence, R.I., 145, 1964.

9. Eshelby, J.D., The determination of the elastic field of an ellipsoidal inclusion and related problems, *Proc. R. Soc.*, A241, 376, 1957.

10. Gel'fand, I.M. and Shilov, G.E., *Generalized Functions,* Vol. 1, *Properties and Operations,* Academic Press, New York, 1964.

11. Gel'fand, I.M. Graev, M.I. and Vilenkin , N.Ya., *Generalized Functions,* Vol. 5, *Integral Geometry and Representation Theory,* Academic Press, New York, 1966.

12. Willis, J.R., Some remarks on the application of the QCA to the determination of the overall elastic response of a matrix/inclusion composite, *J. Math. Phys.,* 25, 2116, 1984.

13. Willis, J.R., Relationships between derivations of the overall properties of composites by perturbation expansions and variational principles, *Variational Methods in Mechanics of Solids,* edited by S. Nemat-Nasser, Pergamon, New York, 59, 1980.

14. Willis, J.R., Variational and related methods for the overall properties of composites, *Advances in Applied Mechanics,* Vol 21, edited by C.S. Yih, Academic Press, New York, 1, 1981.

15. Willis, J.R., Elasticity theory of composites, *Mechanics of Solids, the Rodney Hill 60th Anniversary Volume,* edited by H.G. Hopkins and M.J. Sewell, Pergamon, Oxford, 653, 1982.

16. Hill, R., New derivations of some elastic extremum principles *Progress in Applied Mechanics, The Prager Anniversary Volume,* Macmillan, New York, 99, 1963.

17. Hashin, Z. and Shtrikman, S., On some variational principles in anisotropic and non-homogeneous elasticity, *J. Mech. Phys. Solids,* 10, 335, 1962.

18. Kröner, E., Bounds for effective elastic moduli of disordered materials, *J. Mech. Phys. Solids,* 25, 137, 1977.

19. Willis, J.R., Variational principles for dynamic problems for inhomogeneous elastic media, *Wave Motion,* 3, 1, 1981.

20. Gurtin, M.E., Variational principles in linear elastodynamics, *Arch. Ration. Mech. Anal.*, 16, 34, 1964.

21. Willis, J.R., Variational principles for waves in random composites, *Continuum Models of Discrete Systems 4,* edited by O. Brulin and R.K.T. Hseih, North-Holland, Amsterdam, 471, 1981.

22. Weaver, R.L., A variational principle for waves in discrete random media, *Wave Motion,* 7, 105, 1985.

23. Willis, J.R., The overall elastic response of composite materials, *J. Appl. Mech.*, 50, 1202, 1983.

24. Talbot, D.R.S. and Willis, J.R., Variational principles for inhomogeneous non-linear media, *IMA J. Appl. Math.*, 34, 1985 (to appear).

25. Willis, J.R., Variational estimates for the overall response of an inhomogeneous non-linear dielectric, *Proceedings, IMA Workshop on Homogenization and Effective Properties of Composite Materials,* edited by D. Kinderlehrer, (to appear).

PART VI

APPROXIMATE STATISTICAL MODELLING AND APPLICATIONS

André Zaoui
Laboratoire P.M.T.M. - CNRS
Université de Paris XIII
Villetaneuse, France

APPROXIMATE STATISTICAL MODELLING AND APPLICATIONS

André ZAOUI
Laboratoire P.M.T.M. - CNRS
Université de Paris XIII - Villetaneuse (France)

CHAPTER 1
INTRODUCTION

This chapter is concerned with randomly inhomogeneous media, too, but, at variance with the foregoing one, several simplifications and approximations are introducted into the statistical treatment, in order to pay special attention to the elastic-plastic behaviour of metals and to deal with some applications in the field of metal forming. As a matter of fact, the theory of plastic microinhomogeneous media is far from being as firmly based and developed as the linear elastic one, because of the specific difficulties of plasticity. On the one hand, a systematic variational approach, yielding more and more narrow bounds for the overall behaviour, is still missing in the general case ; on the other hand, even if it was available, it would surely not be easy to use since the instantenous moduli are continuously changing during the plastic flow and so is the statistical description of the microstructure itself. Moreover, when the crystallographic nature of plastic glide in metals has to be taken into account, as it is really needed for many problems of metal forming, one has to deal with an additional degree of complexity which, at the time being, is hardly compatible with any rigourous and general statistical treatment.

Consequently, most of the modelling attempts in this field are restricted to approaches which are insensitive to the actual space statistical distribution of the constituent phases, whether they start from definite quite ordered situations in order to get extreme reference estimates and they introduce further more or less empirical fits or modifications so as to soften the initial assumptions, or they deal with implicity highly disordered microstructures and they progressively insert some elements of order. Both of these approaches, which belong to the Taylor and the self-consistent frames respectively, will be successively reported in the following (sections 3 and 4 resp.), after a short presentation of the mechanical and physical foundations of crystalline plasticity (section 2).

What is aimed at in this chapter is to point out the specific difficulties which arise when one deals with plastic inhomogeneous media, to set out the present "state of the art" in the field and to show how, in spite of the insatisfactory character of many approximations and simplifications, several aspects of metal forming which are of the first practical importance can be fairly well described and understood thanks to such crude models. This will be illustrated with the case of the prediction of crystallographic as well as morphological deformation textures of elastic-plastic polycrystals and of the associated induced plastic anisotropy. Further applications will be concerned with the plasticity of two-phase metals which, in addition to their intrinsic technological interest, present the theoretical advantage of a simpler physical constitution which allows some investigations of space phase distribution effects : this will be suggested in the last section (section 5) thanks to generalized, "multi-site" self-consistent schemes.

CHAPTER 2

BASIC PHYSICAL DATA OF POLYCRYSTAL PLASTICITY

Single phase polycrystals can be considered as composite materials from a mechanical point of view as soon as the elementary mechanical properties which are involved are anisotropic ones in each constituent crystal. Due to the misorientations between the crystal lattices, these mechanical proporties measured in a fixed exterior frame vary from one crystal to the other, as in a composite inhomogeneous material. Of course, they can also be considered as composites at a smaller scale if subgrain boundaries, dislocation cells and so on are taken into account. But, in the following, the granular nature only of the polycrystal will be considered as responsible for the inhomogeneity and each grain will be dealt with as a volume of uniform lattice orientation, despite the fact that it could slightly wary within it, whereas sharp lattice misorientations occur from one grain to its neghbours. Since plastic glide is mainly determined by crystallographic conditions, which lead to the activation of definite sets of slip systems, the intragranular plastic behaviour is essentially anisotropic, what confers to the plastically flowing polycrystal the character of a composite material. But, generally speaking, the phases of such a composite are not defined only by their lattice orientation : grains with the same orientation may also differ in shape and size which should be characterized by adequate parameters for any exhaustive statistical description of the granular structure. As far as the lattice orientation only is considered, the statistical description may be restricted to the crystalline orientation distribution function (C.O.D.F., or the so-called "texture function"), $f(\Omega)$ say, where Ω stands for the angular parameters of any lattice orientation - for instance the three Euler angles - and $f(\Omega)\, d\Omega$ denotes the volume fraction of grains with the lattice orientation Ω in the range $d\Omega$. Of course, such a texture function gives no indication at all on any correlation between orientation and position and can only be used in space distribution independent models.

2.1. The plasticity of single crystals. The first step of any mechanical analysis of the plasticity of polycrystals is concerned with the plastic behaviour of single crystals. Considerable advances have been performed in the field of the physical understanding of the elementary mechanisms of plastic deformation in metallic single crystals, but there is still a wide empty gap between the dislocations and point defects level and the scale adapted to a macroscopic mechanical description of the overall behaviour. If low temperature plasticity of pure single crystals only is considered, the dislocation glide play the prominent part in the plastic deformation ; it results in macroscopic planar glide distributed over the whole crystal and acting on definite slip systems ($\vec{n}^g$, $\vec{m}^g$), where g identifies the concerned slip system and $\vec{n}$ and $\vec{m}$ are unit vectors of the slip plane and slip direction respectively. If γ^g denotes the corresponding shear strain, the resulting plastic strain tensor $\underset{\sim}{\varepsilon}^p$ is given by :

$$\underset{\sim}{\varepsilon}^p = \sum_g \gamma^g (\vec{n}^g \times \vec{m}^g)^s = \sum_g \gamma^g \underset{\sim}{R}^g \qquad \text{(VI.1)}$$

when the small strain formalism is used (the superscript s refers to the symmetrization operation). Similarily, the plastic rotation $\underset{\sim}{\omega}^p$ involves the antisymmetrical part of the same tensorial product.

The identity of the slip systems which may contribute to the plastic deformation depends on the material ; it is determined by many physical parameters which control the dislocation multiplication, mobility and movement : the crystal lattice symmetry, the friction stress opposed by the lattice to the dislocation movement, the stacking fault energy and the splitting of dislocations into partial ones etc... Within a given temperature range, this results in a definite set of easy glide systems, which may be of a pure crystallographic nature (such as the twelve {111}<110> slip systems in FCC metals) or have a mixed crystallographic and mechanical definition (such as the "pencil glide" mode in some BCC metals where the slip direction is <111> and the slip plane is determined by a maximum shear stress condition). In any case, the set of slip systems which will be able to be active must be known as a primary information, deriving from an adequate preliminary experimental investigation or theoretical analysis.

The yield criterion may be correlated with the condition that the force per unit length of a mobile dislocation be critical ; at the more macroscopic scale of the slip systems, it is expressed by the Schmid and Boas law, according to which an easy glide system (g) may be active as soon as the resolved shear stress τ^g on it reaches a critical value, namely :

$$\tau^g = \vec{m}^g \cdot \underset{\sim}{\sigma} \cdot \vec{n}^g = \tau_c^g \qquad \text{(VI.2)}$$

where $\underset{\sim}{\sigma}$ is the local stress tensor and τ_c^g the critical resolved shear stress (CRSS) on the system (g). Note that even for a well annealed and homogenized initial state, τ_c^g may be different on different slip systems families (e.g. the {110}<111> and {112}<111> families in pure α-Fe crystals) and may also differ according to the slip sense, for a given direction (as for {112}<111> in iron crystals). Similar conditions of a critical shear stress are also valid for twinning initiation.

A more debated question is the hardening law, i.e. the evolution law of the τ_c^g's. For the solid state physicist, the natural hardening parameters are the dislocation densities. The initial Taylor's answer was concerned with an isotropic hardening, in the form :

$$\tau_c^g = \alpha\, \mu\, b \sqrt{\rho} \ , \qquad \forall g \qquad \text{(VI.3)}$$

where μ is the elastic shear modulus, b the Burgers parameter, α a dimensionless coefficient and ρ the total dislocation density.[1] Subsequent more detailed investigations have led to anisotropic relations, such as :

$$\tau_c^g = \mu \ b \sum_h (a^{gh} \ \rho^h)^{1/2} \qquad \text{(VI.4)}$$

where the plastic activity of any system (h) modifies the yield stress on any other system (g), through an interaction matrix a^{gh} which includes the whole physics of dislocations anisotropic interactions - long range as well as short range ones -. As it may be concluded from latent hardening investigations, the anisotropic character of the a^{gh} matrix is tightly correlated with the geometrical possibility for intersecting dislocations to build junctions or not and with the glide ability of such junctions, according as they are sessile or glissile, in connection with the stacking fault energy intensity.[2] Nevertheless, in the present state, such theories which certainly suit best as long as hardening only is concerned fail to lead to a whole description of the plastic flow, due to the difficulty of connecting plastic strain and dislocation densities. As a matter of fact, the famous old Orowan's formula :

$$\dot{\varepsilon} = \rho_m \ b \ v \qquad \text{(VI.5)}$$

which connects the strain rate $\dot{\varepsilon}$, the dislocation velocity v and the mobile dislocation density ρ_m is far from giving the key of this difficult problem : in addition to its scalar nature, it does not solve the thorny question of how to determine the mobile dislocation density and velocity. Despite several more recent attempts to clear up such questions, including more rigorous tensorial definitions of the dislocation densities which are quite useful for the calculation of their associated internal stress fields but still inadequate to an analysis of the plastic flow, it cannot be concluded that the dislocation theory is ready for use for a mechanical description of the plastic behaviour of single crystals.

Consequently, more phenomenological approaches have been developed in order to supply this failing, according to which the slip amounts γ^g themselves play the role of hardening parameters : since their connection with the plastic strain is quite straightforward (eqn. VI.1), the question to be solved is merely that of the relationship between the CRSS's and the slip amounts. The natural answer lies in an incremental hardening rule of the following form :

$$\dot{\tau}_c^g = \sum_h h^{gh} \dot{\gamma}^h \qquad \text{(VI.6)}$$

where the hardening matrix h^{gh} allows to describe the influence of the slip activity on any system (h) on the hardening behaviour of any other system (g).[3,4] Informations on this hardening matrix may be deduced from various experiments on single crystals (see e.g. Fig. 1), but many difficulties have to be overcome, both from experimental and theoretical points of view : the experimental identification of slip directions is uneasy ; the h^{gh} matrix is strongly dependent on the strain and stress path and the mutual hardening of two systems differs according as both of them are active at the same time or not ; the plastic activity of some secondary systems may have a large influence on the hardening behaviour whereas their contribution to the plastic flow is almost negligible etc...[2,5] Thus, even if such an approach is a crude approximation of the actual plastic behaviour, dealing with convenient, but physically inadequate hardening parameters (the γ^g's), it has still to be simplified when polycrystals are considered since many other pheno-

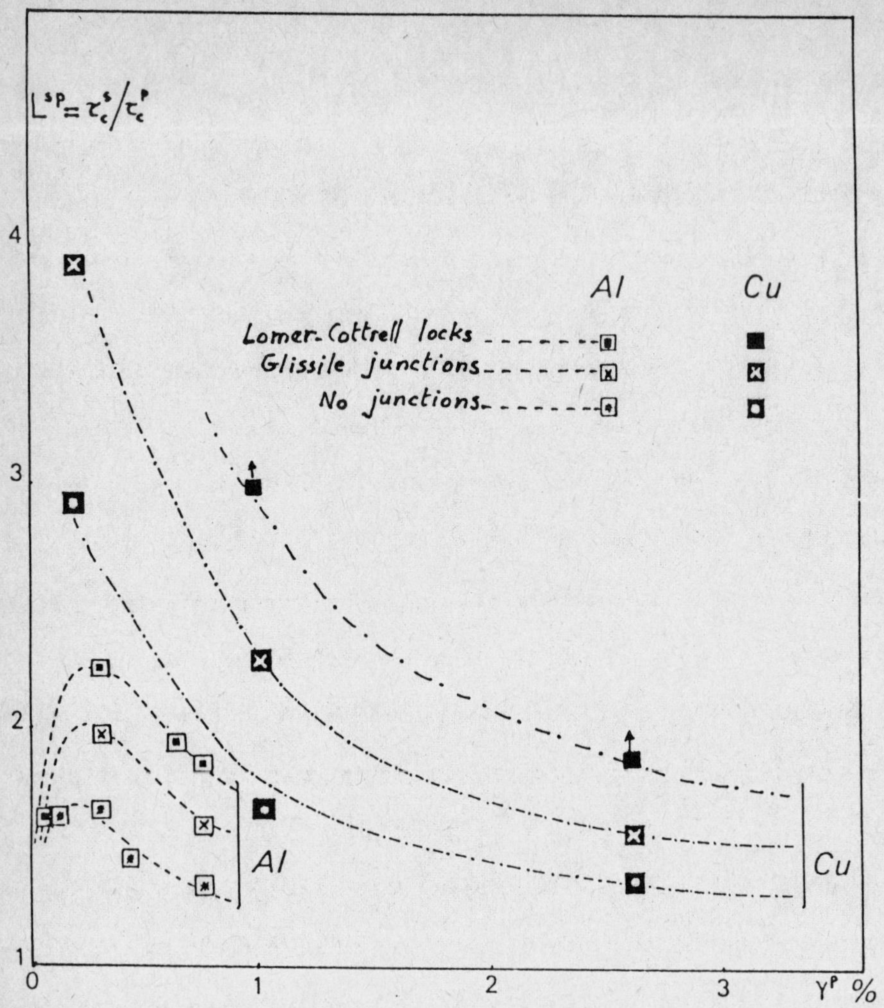

Fig. 1 : Latent hardening ratios vs the primary shear strain (Al and Cu crystals).

mena and parameters have to be taken into account in addition to the intragranular ones.

2.2. The influence of grain boundaries. The composite nature of polycrystals is expressed by the existence of grain boundaries which separate differently oriented adjoining grains. Whereas the microstructure of such boundaries may be a very complicated fabric of lattice defects, which are resolved as usual dislocations in simple cases only, their mechanical influence may be analysed with less details, but according to two distinct and complementary points of view.

On the one hand, they act as joining areas between grains which are compelled to deform as a whole, in spite of their specific individual characteristics, so that no decohesion occurs between them : this misorientation effect is then responsible for internal stresses deriving from the incompatible character of the plastic strain field and the compatibility requirement for the total strain field, associated with the perfect fit condition at the grain boundaries.[6] This condition may be softened if grain boundary sliding can occur but, in any case, it is generally associated with long range mechanical interactions between grains which force local stresses and strains to differ from the applied (or average) ones.[7] If the grain lattice orientations and the grain boundaries geometry is known, this kind of problem may find a solution according to the classical residual stress methods. Note that, in this case, a specific question must be solved at the same time, as far as lattive orientations are concerned. As a matter of fact, while a free plastic glide causes no lattice rotation in an isolated single crystal, the continuity conditions across the grain boundaries involve lattice rotations now within each grain, as an indirect consequence of plastic glide in polycrystals : that is the very origin of the development of deformation crystallographic textures in these materials.

On the other hand, grain boundaries act as obstacles to the free development of plastic glide from one grain to the other. This is responsible for specific phenomena which do not reduce to the foregoing misorientation effects : plastic glide within the grains are then disturbed at a lower scale so that slip inhomogeneities develop near the grain boundaries. Depending on the latent hardening characteristics of the involved material, intragranular interfaces may develop, seperating areas where distinct slip systems are activated within the same grain.[8,9] The resulting slip pattern within the grains may be rather complicated, as a combined result of a number of parameters and mechanisms, including the particular influence of grain junctions as well as of the specific orientation mapping of the neighbouring grains.[10]

As a conclusion of the hereabove reported complexity, which is only concerned with the most prominant facts which can be pointed out from a brief and schematic description of crystalline (rate-independent) plasticity, it can be suspected that homogenization techniques in this field cannot be so elaborate as they are, indeed, in other cases and, moreover, that their development needs adequate additional simplifications of the reported phenomena and properties.

2.3. Simplified description of polycrystal plasticity. The foregoing discussion has stressed the particular importance of the intragranular slip inhomogeneity, a feature which is the more significant, the more difficult the intersection of dislocation families (and then the lower the stacking fault energy). Nevertheless, an integration of such a characteristic of intragranular plastic deformation mode looks out of range in the present state of polycrystal modelling since, at variance with the elastic case, the instantaneous elastic plastic moduli which depend on the plastic strain could no longer be considered as uniform within the grains. So, we have to accept an omission of this important feature, in order to work out reasonably tractable treatments. These shortcomings can be attenuated if we somehow modify the intragranular plastic behaviour with respect to the one of isolated single crystals in shuch a way that an averaged, homogenized "grain behaviour" be considered which implicitely includes the slip inhomogeneities effects ; for instance, the diagonal components of the h^{gh} hardening matrix (VI.6), which should characterize the self-hardening moduli within a consistent local description, have to be raised up to cross-hardening values in such an homogenized grain description since single glide is quite seldom in polycrystal grains and double slip, at least, occurs very soon in them. Similarily, even if it may appear conceptually illogical, grain size dependent intragranular yield stresses are often considered whereas such a dependence should result from the whole homogenization process of the polycrystal behaviour if non uniform intragranular plastic strain fields and moduli could be dealt with through a more rigorous analysis.

Finally, an even if any physical investigation leads to the obvious necessity of considering the hardening moduli h^{gh} as strain path and active slip systems set dependent, most of the present treatments are restricted to constant hardening matrices - let us quote them as H^{gh} matrices in the following - and the discussions are focused on their symmetry, positiveness and anisotropy. For instance, latent hardening characteristics of FCC crystals may be accounted for by two kinds of components - say H_1 and H_2, with $A = H_2/H_1$, higher than unity -, the higher ones being assigned to strongly interacting slip system pairs (such as those whose intersection can result in dislocation junctions).

Now, even if such drastic simplifications are assumed, many difficult questions, which are unusual when elasticity only is considered, have to be answered in the case of elastic-plastic polycrystals, namely, at each step of any strain or stress path, and within each grain family : which are the critical systems ? which will be the active ones ? with what shear strain rate ? resulting in what plastic rotation, and then what lattice rotation ? which is the new grain shape ? which is the new orientation distribution function ? an so on ... We are indebted to G.I. Taylor for having approached the first, almost fifty years ago, such questions and, despite the very simplified character of his treatment, he is still inspiring most of the present applications in the field of metal forming analysis, what justifies the place that falls to him by right in the following section.

C H A P T E R 3

TAYLOR-TYPE MODELLING OF POLYCRYSTAL PLASTICITY

3.1. The original Taylor's analysis. If the initial Sachs' first contribution to the prediction of yield stress of polycrystals, according to an assumption of a simultaneous reach of the critical value by all the resolved shear stresses on the primary systems, is omitted, it may be said that Taylor's analysis was the first significant "homogenization" approach of polycrystal plasticity.[11,12] Extending Voigt's uniform strain assumption for elastic inhomogeneous media, Taylor's treatment is based upon a uniform plastic strain assumption throughout the polycrystal, within a rigid-perfectly plastic framework, namely :

$$\underset{\sim}{\varepsilon}^p(\Omega) = \langle \underset{\sim}{\varepsilon}^p \rangle_\Omega = \underset{\sim}{E}^p \qquad \text{(VI.7)}$$

where $\underset{\sim}{\varepsilon}^p$ and $\underset{\sim}{E}^p$ are the local and average plastic strain tensors respectively. If $\underset{\sim}{E}^p$ is assigned and results from crystallographic plastic glide in each grain family defined by its lattice orientation parameters Ω, (VI.7) consists in five conditions for each Ω (due to the plastic incompressibility property) which can only be satisfied in the general case by the activation of five independent slip systems. But there may be a lot of possible combinations of five slip systems in highly symmetric crystal lattices (e.g. $C_{12}^5 = 792$ sets in FCC crystals, with the {111}<110> easy glide systems), even if independent systems only are considered. Since the deformation mode (VI.7), which is a compatible one, corresponds to a kinematically admissible trial field and then overestimates the associated plastic work (the "upper bound" theorem is valid since the Schmid criterion, which is assumed here, makes the considered polycrystal a standard material), the best Taylor's approximation is given, for each Ω, by the systems set which minimizes this plastic work.

For instance, in the case of a tensile test of a polycrystal, with a uniforme CRSS τ_o, the estimated yield stress Σ_o is given by :

$$V\, \Sigma_o\, E^p = \tau_o \int_V \sum_g |\gamma_\Omega^g|\, dv \qquad \text{(VI.8)}$$

where $\sum_g |\gamma_\Omega^g|$ is minimum in the grain family with the orientation Ω.

Using the definition of the texture function $f(\Omega)$, this gives :

$$\Sigma_o = \tau_o \int \frac{\sum_g |\gamma_\Omega^g|}{E^p} f(\Omega)\, d\Omega = \tau_o \langle M \rangle_\Omega \qquad (VI.9)$$

where the "Taylor factor" $M(\Omega)$ is used. The classical value $\langle M \rangle = 3.06$ stands for the case of an isotropic FCC polycrystal. In any case, such a calculation yields an overestimate of the actual value.

Since the original Taylor's analysis, many developments have been performed according to the same scheme. Let us only quote Bishop and Hill's analysis, focusing on the stress aspects of Taylor's treatment, Chin and Mammel's results on the extremal nature of the Taylor model etc..., and even much more recent studies on specific weak points of this model.[13,14] This is particularly the case of ambiguous situations which exist concerning the identity of the active slip systems, after the minimization procedure : for certain lattice symmetry classes (such as the FCC one), several slip systems sets yield the same minimum value of the plastic work, so that the plastic rotation field is not unique. Since the compatibility conditions imply a uniform total rotation field when the strain field is uniform, the lattice rotations, which, a constant apart, just oppose the plastic ones, are no more unique. This is a source of difficulty for the prediction of crystallographic textures which, according to the authors, is overcome thanks to various methods : direct averaging of all the admissible solutions ; introduction of twinning modes and of physical arguments in order to favour this or that systems set ; second order and stability analyses ; considera-

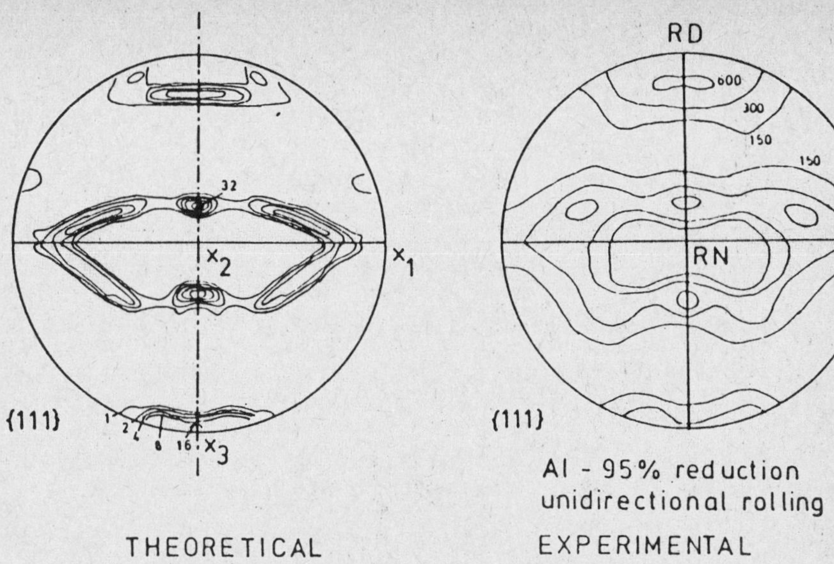

Fig. 2 : Texture prediction using the Taylor model.[15]

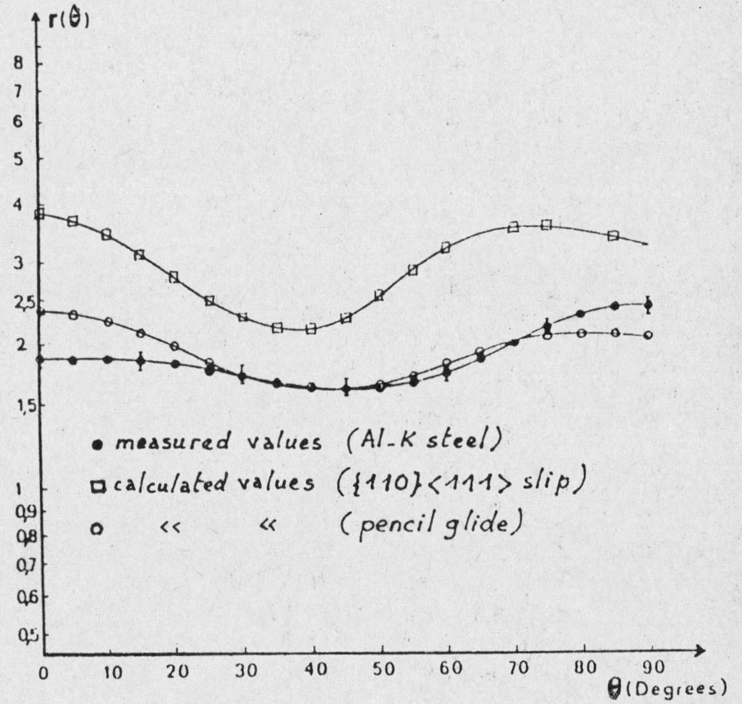

Fig. 3 : Lankford coefficient prediction using the Taylor model.[16]

tion of rate-dependent behaviour etc... Of course, the resulting texture prediction is strongly dependent on the way to deal with this ambiguity problem, what, after all, endows the Taylor model with some flexibility and physical adaptability which is missing in its foundations ...

Somehow and other, many simulations of texture development according to the Taylor model have been performed for various cases of metal forming and, generally speaking, it may be admitted that most of the main characteristics of the experimental texture patterns, regardless of a quite definite quantitative agreement, have been thus recognized. Powerful computing programs and numerical treatments, especially using the development of the texture function into spherical harmonics series, are now operating currently, including the description of plastic anisotropy resulting from a given initial texture. For this purpose, specific minimization procedures are needed, since the macroscopic average plastic strain $\underset{\sim}{E}^{P}$ is not known beforehand. For instance, the characterization of the orthotropic plastic anisotropy of a rolled thin sheet by secondary tensile tests at various angles θ with the rolling direction involves the well known Lankford coefficient $r(\theta)$: for each θ value, the hereabove described minimization procedure must first be performed for any value of an auxiliary parameter involving the unknown Lankford coefficient $r(\theta)$; after that, a second minimization operation is needed in order to yield finally the sought coefficient. Fig. 2 and 3 report illustrative examples of texture and Lankford coefficient predictions using the Taylor model.[15,16]

For most of these studies which aim at comparing experimental data and simulation results, it is concluded that the general trends are well accounted for, all the better as the plastic flow is more intense, but that the theoretical predictions are more marked and stronger than the observed features, as well as not dependent enough on the considered material, within a given crystal symmetry class. These conclusions, added to the common admission that real grains hardly slip according to five homogeneously activated systems as claimed from the basic assumption of the Taylor model, are generally imputed to the extremal character of this model. So, considerable subsequent research work has been devoted to some softening of the basic rules of the original Taylor model, either by inclusion of elasticity into the material behaviour or by relaxing the strain uniformity initial requirement.

3.2. Modified Taylor modelling. The first extension of Taylor's original analysis has been performed by Lin who considered an elastic-plastic behaviour instead of a rigid-plastic one.[17] The Lin-Taylor model is thus the exact Voigt analogue for the case of elastic plastic materials. It may be admitted that it has not been extensively used and applied, but its main interest lies in the correlation which may be easily derived from its basic assumption between internal stresses and plastic strain deviation. In the case of isotropic uniform elasticity, we get :

$$\left\{\begin{array}{l} \underset{\sim}{\varepsilon}^{e} + \underset{\sim}{\varepsilon}^{p} = \underset{\sim}{E}^{e} + \underset{\sim}{E}^{p} \\ \underset{\sim}{\varepsilon}^{e} = \frac{1}{2\mu}\left(\underset{\sim}{\sigma} - \frac{\nu}{1+\nu}\,\mathrm{Tr}\,\underset{\sim}{\sigma}\,\underset{\sim}{I}\right) \\ \underset{\sim}{E}^{e} = \frac{1}{2\mu}\left(\underset{\sim}{\Sigma} - \frac{\nu}{1+\nu}\,\mathrm{Tr}\,\underset{\sim}{\Sigma}\,\underset{\sim}{I}\right) \end{array}\right. \qquad \text{(VI.10)}$$

where $\underset{\sim}{I}$ is a unit second order tensor and ν the Poisson ratio. The plastic incompressibility results in traceless internal stresses (i.e. $Tr(\underset{\sim}{\sigma} - \underset{\sim}{\Sigma}) = 0$), so that (VI.10) yields what will be called the Lin "interaction law", namely :

$$\underset{\sim}{\sigma} = \underset{\sim}{\Sigma} + 2\mu(\underset{\sim}{E}^p - \underset{\sim}{\varepsilon}^p) \qquad \text{(VI.11)}$$

what means that very high internal stresses would develop following any non zero plastic strain deviation. From a practical point of view, considering usual very low values of the ratio of the flow stress to the elastic shear modulus, local plastic strains cannot really differ from one grain to the other, so that the original Taylor model is not noticeably modified by Lin's extension.

A more significant modification has been recently suggested by Honeff and Mecking, Kocks and Chandra and others.[18,19] It is concerned with the introduction of some grain shape sensitivity at very large strain into the Taylor model. Consider for instance a very heavily deformed rolled sheet whose grains are, consequently, quite flattened and look, according to the case, lath-shaped or pancake-shaped. If (x_1, x_2, x_3) refer to the rolling, transverse and normal direction respectively, the compatibility requirements may be restricted to the continuity of ε^p_{11}, ε^p_{22} and ε^p_{12} components (ε^p_{23} also in the lath model) only, so that the remaining components are free and optimized in order to get the lowest value of the overall plastic work. Similarily, a modified Bishop-Hill formulation may be developed with some vanishing corresponding stress components.[20] Such a procedure, which may be made progressive along the plastic flow by a gradual transition from a "full constraints" to a"relaxed constraints" description, makes the original Taylor model less rigid and more adaptable to specified situations ; apparently, it gives predictions which fit better most of the usual experimental data, especially concerning texture development of heavily deformed polycrystals.

Finally, let us briefly quote a more recent original attempt by Asaro and Needleman, which combines and develops in a more rigorous way all these modifications of the initial Taylor model.[21] It consists in a finite strain treatment of a rate-dependent polycrystal - i.e. with an elastic-viscoplastic behaviour -, where the overall deformation gradient is calculated as the (numerical) solution of a mixed homogeneous boundary value problem and then assigned to be undergone by any grain of the polycrystal. Such a model, whose extensive application is now in progress, offers the advantage of a rigorous finite strain formulation, of well-defined boundary conditions as well as of an elegant short-circuit of the ambiguity difficulties mentioned hereabove concerning the active slip systems determination, thanks to the consideration of rate-dependence. Moreover, it allows a more direct and rigorous investigation of the localization modes (e.g. localized necking in biaxially stretched sheets).

3.3. Conclusion and discussion. Compared with other homogeneization techniques which have been developed in various fields of solid and fluid mechanics, the Taylor approach could certainly appear rather crude and primitive and its wide and persistent popularity among metal forming people somewhat undue. This success is probably due to the fact that, thanks to the well developed numerical methods of linear programming, this model is relatively easy to be implemented,as well to its kinematic character, which makes it more intuitive. As a matter of fact, the dual model which would be based upon an assumption of stress uniformity - let us say the "static model" -, originally intimated by Sachs, but explicitely formulated by Batdorf and Budiansky twenty years later, has been much less appreciated and used.[22]

Its main advantage lies in its extremal character, which makes it the upper bound analogue of Voigt's approximation in the case of linear elasticity. For this reason, it plays the important role of giving reference estimates and it may be deplored that its predictions are not systematically compared with those deriving from the static model, in order to yield complementary bounds for the actual behaviour. But, at the same time, this extremal character is responsible for its main drawbacks too, when a comparison with experimental data is aimed at. Of course, a good quantitative agreement may only be expected for very specifically ordered polycrystal structures for which the uniform strain assumption would be valid. In this respect, Taylor's approach cannot be blamed for yielding too sharp textures, too stiff moduli, too high yield stresses, and so on, since all these characteristics derive from its foundation itself. It may even be wondered whether the various "relaxed Taylor theories", which frequently offer a better fit with experimental data, have not, in a way, lost the main advantage of the s.s. Taylor model, namely its extremal nature, as well as a theoretical justification of their minimization procedure ... In any case, if they surely take in better account one geometrical property of the actual polycrystals (even if it is in a very abrupt fashion), by considering grain shape effects, they do not really depart from the general common nature of all these Taylor-type models, i.e. a fundamental insensitiveness to the actual space statistical distribution of the constituent phases. This crucial point will also deserve discussing in what follows, concerning the self-consistent modelling.

CHAPTER 4
SELF-CONSISTENT MODELLING OF POLYCRYSTAL PLASTICITY

Whereas the Taylor-type models were essentially dealing with very ordered situations, the self-consistent approach is mainly concerned with disorder. Before reporting applications of this approach to the field of elastoplasticity, it looks whorthwhile recalling briefly its definition in the case of linear elasticity, for which its statistical meaning has been clearly settled and rigorously demonstrated.

4.1. Self-consistent scheme and perfect disorder. The self-consistent scheme can be defined formally from the systematic statistical theory in the case of linear elasticity as a rigorous model for "perfectly disordered" materials, in the sense defined by Kröner.[23] Let us take as the starting point the so-called Lippmann-Schwinger equation which writes :

$$\underset{\sim}{\varepsilon} + \underset{\sim}{\Gamma}\, \underset{\sim}{\delta c}\, \underset{\sim}{\varepsilon} = \underset{\sim}{\varepsilon}^{o} \qquad \text{(VI.12)}$$

Here, $\underset{\sim}{\varepsilon}$ is the strain field of the considered disordered material, $\underset{\sim}{\varepsilon}^{o}$ is the strain field which would be the solution of the same boundary value problem for a reference medium with the elastic moduli $\underset{\sim}{C}^{o}$ (which can be considered uniform and local, for sake of simplicity), $\underset{\sim}{\delta c} = \underset{\sim}{c} - \underset{\sim}{C}^o$ with $\underset{\sim}{c}$ the elastic tensor field of the real inhomogeneous material and $\underset{\sim}{\Gamma}$ is the modified Green operator for the reference medium. Note that (VI.12) is an operator equation where products (such as $\underset{\sim}{\Gamma}\ \underset{\sim}{\delta c}\ \underset{\sim}{\varepsilon}$) indicate operator products. If $\underset{\sim}{I}$ is the unit operator, which has, in space representation, the form : $\frac{1}{2}\ (\delta_{ik}\ \delta_{jl} + \delta_{il}\ \delta_{jk})\ \delta(\vec{r}, \vec{r}')$ where δ_{ij} is the Kronecker symbol and $\delta(\vec{r}, \vec{r}')$ the Dirac delta function, the integral equation (VI.12) may be solved formally in the form :

$$\underset{\sim}{\varepsilon} = (\underset{\sim}{I} + \underset{\sim}{\Gamma}\ \underset{\sim}{\delta c})^{-1}\ \underset{\sim}{\varepsilon}^o \tag{VI.13}$$

Taking the ensemble average of (VI.13), one gets, since $\underset{\sim}{\varepsilon}^o$ is not random :

$$\langle\underset{\sim}{\varepsilon}\rangle = \langle(\underset{\sim}{I} + \underset{\sim}{\Gamma}\underset{\sim}{\delta c})^{-1}\rangle\ \underset{\sim}{\varepsilon}^o \tag{VI.14}$$

or, by formal inversion :

$$\underset{\sim}{\varepsilon}^o = \langle(\underset{\sim}{I} + \underset{\sim}{\Gamma}\ \underset{\sim}{\delta c})^{-1}\rangle^{-1}\ \langle\underset{\sim}{\varepsilon}\rangle \tag{VI.15}$$

When (VI.15) is inserted into (VI.13), we find :

$$\underset{\sim}{\varepsilon} = (\underset{\sim}{I} + \underset{\sim}{\Gamma}\ \underset{\sim}{\delta c})^{-1}\ \langle(\underset{\sim}{I} + \underset{\sim}{\Gamma}\underset{\sim}{\delta c})^{-1}\rangle^{-1}\ \langle\underset{\sim}{\varepsilon}\rangle \tag{VI.16}$$

From the definition of the effective moduli $\underset{\sim}{C}^{eff}$:

$$\langle\underset{\sim}{\sigma}\rangle = \langle\underset{\sim}{c}\ \underset{\sim}{\varepsilon}\rangle = \underset{\sim}{C}^{eff}\langle\underset{\sim}{\varepsilon}\rangle \tag{VI.17}$$

we finally get the classical result :

$$\underset{\sim}{C}^{eff} = \langle\underset{\sim}{c}(\underset{\sim}{I} + \underset{\sim}{\Gamma}\ \underset{\sim}{\delta c})^{-1}\rangle\langle(\underset{\sim}{I} + \underset{\sim}{\Gamma}\ \underset{\sim}{\delta c})^{-1}\rangle^{-1} \tag{VI.18}$$

which may also be put in the form :

$$\langle(\underset{\sim}{c} - \underset{\sim}{C}^{eff})(\underset{\sim}{I} + \underset{\sim}{\Gamma}\ \underset{\sim}{\delta c})^{-1}\rangle = 0 \tag{VI.19}$$

If, now, we decide to take the reference medium as the effective one (i.e. the medium with the moduli $\underset{\sim}{C}^{eff}$ and the modified Green operator $\underset{\sim}{\Gamma}^{eff}$ associated with $\underset{\sim}{C}^{eff}$, while $\delta\underset{\sim}{c}$ is now $(\underset{\sim}{c} - \underset{\sim}{C}^{eff})$), the equation for the effective moduli $\underset{\sim}{C}^{eff}$ is :

$$< \delta\underset{\sim}{c}(\underset{\sim}{I} + \underset{\sim}{\Gamma}^{eff}\ \delta\underset{\sim}{c})^{-1} > = 0 \qquad \text{(VI.20)}$$

An equivalent equation could be obtained for the effective compliances $\underset{\sim}{S}^{eff} = \underset{\sim}{C}^{eff-1}$ by use of the second modified green operator $\underset{\sim}{\Delta}$, defined by :

$$\underset{\sim}{\Delta} = \underset{\sim}{c} - \underset{\sim}{c}\,\underset{\sim}{\Gamma}\underset{\sim}{c} \quad , \quad \underset{\sim}{\Gamma} = \underset{\sim}{s} - \underset{\sim}{s}\,\underset{\sim}{\Delta}\underset{\sim}{s} \qquad \text{(VI.21)}$$

with $\underset{\sim}{s} = \underset{\sim}{c}^{-1}$, namely :

$$<\delta\underset{\sim}{s}(\underset{\sim}{I} + \underset{\sim}{\Delta}^{eff}\ \delta\underset{\sim}{s})^{-1}> = 0 \qquad \text{(VI.22)}$$

Let us now introduce, after Kröner, the fundamental decomposition of $\underset{\sim}{\Gamma}$ into two parts, namely a local part $\underset{\sim}{E}$ of the form $\underset{\sim}{\mathcal{E}}\,\delta(\vec{r}, \vec{r}')$, where $\underset{\sim}{\mathcal{E}}$ is a usual (constant) fourth-order tensor, and a long range part $\underset{\sim}{F}$ (of the form $\underset{\sim}{\mathcal{F}}/|\vec{r} - \vec{r}'|^3$) which connects points of finite separation. According to Kröner, a perfectly disordered elastic material satisfies the following property :

$$<\underset{\sim}{c}'(\underset{\sim}{F}\ \underset{\sim}{c}')^n> = 0 \qquad \forall\, n, \text{ positive integer} \qquad \text{(VI.23)}$$

where $\underset{\sim}{c}' = \underset{\sim}{c} - <\underset{\sim}{c}>$ is the deviation of the random elastic moduli. Eqn (VI.23) may also be written in the form :

$$<\underset{\sim}{c}(\underset{\sim}{F}\underset{\sim}{c})^n> = <\underset{\sim}{c}>(\underset{\sim}{F}<\underset{\sim}{c}>)^n \qquad \forall\, n, \text{ positive integer} \qquad \text{(VI.24)}$$

If now (VI.24) is introduced into (VI.20), it can be proved, after some lengthy calculations, that the equation for the effective moduli of perfectly disordered materials, $\underset{\sim}{C}^{PD}$ say, reduce to :

$$<\underset{\sim}{\delta c}(\underset{\sim}{I} + \underset{\sim}{E}^{PD}\ \underset{\sim}{\delta c})^{-1}> = 0 \qquad \text{(VI.25)}$$

where the local part $\underset{\sim}{E}^{PD}$ of the Green operator only is to be considered ($\underset{\sim}{\delta c}$ and $\underset{\sim}{E}^{PD}$ are now referring to the effective reference medium with moduli $\underset{\sim}{C}^{PD}$). A similar procedure may be adopted from the decomposition of the $\underset{\sim}{\Delta}$ operator defined in (VI.21), what results in an equation for the compliances $\underset{\sim}{S}^{PD}$ which coïncide with $\underset{\sim}{C}^{PD-1}$.

As it will be stated in the next section, these elastic perfectly disordered materials are materials for which the self-consistent scheme applies rigorously. In other words, the self-consistent scheme, which will be introduced now according to a more classical definition as an approximate, rather empirical model, has been proved to yield an exact estimate of the overall elastic behaviour of perfectly disordered materials, so that its statistical meaning has been well elucidated in the linear case. But, whereas an extension of the systematic theory to plasticity, which would allow to deal with perfectly disordered elastic-plastic materials, still seems out of reach, extensions of the self-consistent scheme, as defined from inclusion patterns, to non linear behaviour look more tractable. That is the reason why such developments will be reported now within this more familial framework, in view of applications to polycrystal plasticity. Nevertheless, any attempt at somehow leaving the self-consistent logic itself, in order to integrate some space distribution sensitiveness into more general models, will need to come back to the statistical approach, as it will be tried in the last section.

4.2. The classical definition of the self-consistent scheme. The self-consistent scheme, as originally defined by Hershey and Kröner for the case of elastic polycrystals, is known to approximate the mechanical interactions between any element of an aggregate and all the others by the interactions between this element and an homogeneous matrix, the behaviour of which is the very overall behaviour which is to be determined thanks to this homogenization procedure.[24,25] In the case of macrohomogeneity which will be only treated from now on, the matrix may be considered infinite with uniform boundary conditions. We are thus left with a classical inclusion/matrix problem, the solution of which may, in principle, be obtained as a function of the characteristic parameters of the unknown effective behaviour (i.e. : the behaviour of the "homogeneous equivalent medium"). Finally, these parameters can be identified when the inclusion/matrix problem has been solved for each element by turns (with the same homogeneous matrix at each time) by performing an averaging process which allows to identify the average stress and strain (or their rates) of the elements with the macroscopic ones. When the inclusions may be considered with an ellipsoïdal shape, their resulting stress and strain field is uniform, as shown by Eshelby.[26] This leads to very simplified treatments, which are still more manageable if the overall behaviour is isotropic (i.e. in the case of macroisotropy) since explicit analytical results are then available.

This point may be briefly verified at once, in the case of a spherical inhomogeneity with isotropic elastic moduli (μ^*, k^*) - where k^* is the bulk modulus) embedded within an infinite matrix, with isotropic elastic moduli (μ, k), subjected to uniform strain $\underset{\sim}{\varepsilon}^o$ at infinity.[27] The classical Eshelby's solution of this problem can be derived thanks to the auxiliary fictitious stress-free strain $\underset{\sim}{\varepsilon}^T$ of an "equivalent homogeneous inclusion" with the moduli (μ, k) ; $\underset{\sim}{\varepsilon}^T$ is found to be given by :

$$e^T = A\ e^o \qquad\qquad e^T_{ij} = B\ e^o_{ij} \tag{VI.26}$$

where e and e_{ij} are the scalar and deviatoric parts of $\underset{\sim}{\varepsilon}$ respectively, and :

$$A = -\frac{\delta k}{k+\alpha\delta k} \qquad\qquad B = -\frac{\delta\mu}{\mu+\beta\delta\mu} \tag{VI.27}$$

with :

$$\alpha = \frac{3k}{3k+4\mu} \qquad\qquad \beta = \frac{6(k+2\mu)}{5(3k+4\mu)} \tag{VI.28}$$

From the solution of the spherical inclusion problem, namely :

$$e^c = \alpha\ e^T \qquad\qquad e^c_{ij} = \beta\ e^T_{ij} \tag{VI.29}$$

and the addition of the uniform strain field $\underset{\sim}{\varepsilon}^o$, it is easily found :

$$e_{ij} = (1 + \beta\ B)\ e^o_{ij} \qquad\qquad e = (1 + \alpha\ A)\ e^o \tag{VI.30}$$

where $\underset{\sim}{\varepsilon}$ is the uniform resulting strain within the sphere. The explicit solution may be given as :

$$\begin{aligned}\varepsilon_{ij} &= (1 + \beta\ B)\ \varepsilon^o_{ij} + \frac{\alpha A - \beta B}{3}\ \varepsilon^o_{kk}\ \delta_{ij} \\ &= \frac{5\mu(3k+4\mu)}{5\mu(3k+4\mu) + 6(k+2\mu)\ \delta\mu}\ \varepsilon^o_{ij} + \\ &+ \left[\frac{2\delta\mu(k+2\mu)}{5\mu(3k+4\mu) + 6(k+2\mu)\ \delta\mu} - \frac{\delta k}{3k + 4\mu + 3\delta k}\right] \varepsilon^o_{kk}\ \delta_{ij}\end{aligned} \tag{VI.31}$$

or, in the reversed form, as :

$$\varepsilon^o_{ij} = \left[1 + \frac{6(k+2\mu)\ \delta\mu}{5\mu(3k+4\mu)}\right] \varepsilon_{ij} + \frac{5\mu\ \delta k - 2(k+2\mu)\ \delta\mu}{5\mu(3k + 4\mu)}\ \varepsilon_{kk}\ \delta_{ij} \tag{VI.32}$$

Now, it is easy to verify that this result is nothing but the Lippmann-Schwinger equation (VI.12) for an isotropic elastic tensor $\delta \underset{\sim}{c}$, namely :

$$\delta c_{ijkl} = (\delta k - \frac{2\delta\mu}{3})\ \delta_{ij}\ \delta_{kl} + \delta\mu(\delta_{ik}\ \delta_{jl} + \delta_{il}\ \delta_{jk}) \qquad \text{(VI.33)}$$

if the Green operator $\underset{\sim}{\Gamma}$ refers to the isotropic infinite medium (k, μ) and is restricted to its local part $\underset{\sim}{\mathcal{E}}\ \delta(\vec{r}, \vec{r}')$, with $\underset{\sim}{\mathcal{E}}$ classically given by :

$$\mathcal{E}_{ijkl} = \frac{1}{15\mu(3k+4\mu)}\left[9(k+2\mu)\ I_{ijkl} - (3k+\mu)\ \delta_{ij}\ \delta_{kl}\right] \qquad \text{(VI.34)}$$

As a matter of fact, one can easily prove that :

$$(I_{ijkl} + \mathcal{E}_{ijmn}\ \delta c_{mnkl})\varepsilon_{kl} = \varepsilon_{ij} + \mathcal{E}_{ijmn}\left[(\delta k - \frac{2\delta\mu}{3})\ \varepsilon_{kk}\ \delta_{mn} + 2\delta\mu\varepsilon_{mn}\right]$$

$$= \varepsilon_{ij} + (\delta k - \frac{2\delta\mu}{3})\mathcal{E}_{ijmn}\ \varepsilon_{kk} + 2\delta\mu\ \mathcal{E}_{ijmn}\ \varepsilon_{mn} \qquad \text{(VI.35)}$$

$$= \left[1 + \frac{6(k+2\mu)\delta\mu}{5\mu(3k+4\mu)}\right]\varepsilon_{ij} + \frac{5\mu\ \ \delta k - \ 2(k+2\mu)\delta\mu}{5\mu(3k+4\mu)}\ \varepsilon_{kk}\ \delta_{ij}$$

If (VI.32) is considered, this is simply ε^{o}_{ij}. What means that, in this case, the long range $\underset{\sim}{F}$ contribution within the sphere vanishes with a uniform $\underset{\sim}{\varepsilon}$ and that the local $\underset{\sim}{\mathcal{E}}\ \delta(r, r')$ part only in acting on ε. This results is a partial demonstration (i.e. : in the restricted case of macrohomogeneity and macroisotropy) of the identification of the self-consistent scheme (with spherical elements) with the perfect disorder modelling, for linear elasticity. When the self-consistent procedure is performed from (VI.31) indeed, it obviously results in the equation :

$$<\underset{\sim}{\delta c}(\underset{\sim}{I} + \underset{\sim}{E}^{SC}\ \underset{\sim}{\delta c})^{-1}> = 0 \qquad \text{(VI.36)}$$

which coïncides with (VI.25), so that $\underset{\sim}{E}^{SC} = \underset{\sim}{E}^{PD}$. In (VI.36), the matrix is constituted with the effective medium, according to the self-consistent point of view, i.e. the medium with the moduli (k^{SC}, μ^{SC}) which enter the definition of $\underset{\sim}{E}^{SC}$.

For other cases (anisotropic elasticity, ellipsoïdal elements ...), the foregoing developments are no more valid in details, but the general procedure is the same. Let the basic inclusion/problem have a solution in the form :

$$\underset{\sim}{\varepsilon} = \underset{\sim}{S}\ \underset{\sim}{\varepsilon}^T \qquad \text{(VI.37)}$$

where $\underset{\sim}{\varepsilon}^T$ is a uniform stress-free strain and $\underset{\sim}{S}$ the generalized Eshelby tensor, which has been given an explicit form for isotropic elasticity only, but may be written in general as :

$$\underset{\sim}{S} = \underset{\sim}{C}.\int_V \underset{\sim}{\Gamma}\, dv \qquad \text{(VI.38)}$$

where V is the volume of the inclusion ($\underset{\sim}{S} = \underset{\sim}{C}\ \underset{\sim}{\xi}$ for a sphere). The uniform resultant strain within the ellipsoïdal inhomogeneity then reads :

$$\underset{\sim}{\varepsilon} = (\underset{\sim}{I} + \underset{\sim}{S}\ \underset{\sim}{C}^{-1}\ \underset{\sim}{\delta c})^{-1}\ \underset{\sim}{\varepsilon}^o \qquad \text{(VI.39)}$$

The resultant self-consistent equation is obtained thanks to the same arguments which have led from (VI.13) to (VI.20), where $\underset{\sim}{\Gamma}$ must be replaced by $\underset{\sim}{S}\underset{\sim}{C}^{-1}$, namely :

$$<\underset{\sim}{\delta c}(\underset{\sim}{I} + \underset{\sim}{S}\ \underset{\sim}{C}^{SC-1}\ \underset{\sim}{\delta c})^{-1}> = 0 \qquad \text{(VI.40)}$$

with $\underset{\sim}{\delta c} = \underset{\sim}{c} - \underset{\sim}{C}^{SC}$. Note that (VI.39) may be used in order to correlate internal stresses ($\underset{\sim}{\sigma} - <\underset{\sim}{\sigma}>$) and strain deviations ($\underset{\sim}{\varepsilon} - <\underset{\sim}{\varepsilon}>$), when $\underset{\sim}{C} = \underset{\sim}{C}^{SC}$

and $\underset{\sim}{\varepsilon}^o = <\underset{\sim}{\varepsilon}>$. A straightforward calculation leads to the "interaction law" :

$$\underset{\sim}{\sigma} = <\underset{\sim}{\sigma}> + \underset{\sim}{C}^{SC}(\underset{\sim}{S}^{-1} - \underset{\sim}{I})(<\underset{\sim}{\varepsilon}> - \underset{\sim}{\varepsilon}) \qquad \text{(VI.41)}$$

where $\underset{\sim}{C}^{SC}(\underset{\sim}{S}^{-1} - \underset{\sim}{I})$ plays the role of the "constraint tensor" defined by Hill.[28] This tensor clearly depends on the effective moduli, both through $\underset{\sim}{C}^{SC}$ and $\underset{\sim}{S}^{-1}$ (which depends on $\underset{\sim}{C}^{SC}$ itself). This point will now appear a crucial one when considering the case of plasticity.

4.3. <u>Extension of the self-consistent scheme to plasticity</u>. The first attempt at extending the self-consistent approach to plasticity, for an investigation of the elastic-plastic behaviour of polycrystals, was performed by Kröner, almost twenty five years ago.[29] Dealing with uniform isotropic elasticity, spherical inclusions and uniform plastic strain, without volumetric expansion, in the grains, this treatment was starting from (VI.37), with $\underset{\sim}{\varepsilon}^T = \underset{\sim}{\varepsilon}^P$ and $\underset{\sim}{S} = \beta\, \underset{\sim}{I}$. The resulting residual stress in the inclusion is then :

$$\underset{\sim}{\sigma} = 2\mu(\underset{\sim}{\varepsilon} - \underset{\sim}{\varepsilon}^P) = -\,2\mu(1-\beta)\,\underset{\sim}{\varepsilon}^P \qquad \text{(VI.42)}$$

Considering now the plastic strain deviation $(\underset{\sim}{\varepsilon}^P - <\underset{\sim}{\varepsilon}^P)>)$ instead of $\underset{\sim}{\varepsilon}^P$, according to an implicit picture of a uniformly plastified matrix, and adding a constant applied stress field $<\underset{\sim}{\sigma}>$, (VI.42) leads to the resulting interaction law :

$$\underset{\sim}{\sigma} = <\underset{\sim}{\sigma}> + 2\mu(1-\beta)(<\underset{\sim}{\varepsilon}^P> - \underset{\sim}{\varepsilon}^P) \qquad \text{(VI.43)}$$

This formula may be expressed in the form (VI.41) if the total strains are considered, namely :

$$\underset{\sim}{\sigma} = <\underset{\sim}{\sigma}> + \frac{2\mu(1-\beta)}{\beta}\,(<\underset{\sim}{\varepsilon}> - \underset{\sim}{\varepsilon}) \qquad \text{(VI.44)}$$

which leads to a constraint tensor $\underset{\sim}{L}^*_K = \frac{2\mu(1-\beta)}{\beta} \underset{\sim}{I}$, if $\underset{\sim}{L}^*_K$ is defined by :

$$\underset{\sim}{\sigma} = \langle\underset{\sim}{\sigma}\rangle + \underset{\sim}{L}^*_K(\langle\underset{\sim}{\varepsilon}\rangle - \underset{\sim}{\varepsilon}) \qquad \text{(VI.45)}$$

This means that, according to this model, $\underset{\sim}{L}^*_K$ is the same as in (VI.41) if $\underset{\sim}{C}^{SC}$ is replaced by the elastic tensor and $\underset{\sim}{S}$ is calculated from it, instead of considering the actual (unknown before performing the whole self-consistent calculation) elastic-plastic overall moduli. Consequently, this initial Kröner model is not really self-consistent, except within the elastic range, when residual stresses only are considered. Nevertheless, it has been applied by several authors, in order to investigate the plastic flow of polycrystals. Most of the corresponding results are close to the predictions which would be derived from the Taylor (or Lin-Taylor) model, except for a better account of transient stages at lower strain ; for example, Budiansky and Wu have simulated the tensile flow of isotropic FCC polycrystals : when non hardening single crystals are considered, the stress-strain curve exhibits a yield point at $\Sigma_o = 2\tau_o$ and then is progressively increasing towards the limit value $\Sigma = 3.06\tau_o$ while more and more slip systems, up to five system per grain, become active.[30] This means that the Taylor behaviour is joined as a limit one, what is not surprising when (VI.11) and (VI.43) are compared : according to (VI.28), common values of β are of the order of .5, so that Lin's and Kröner's interaction laws only differ by a factor 2 for the "constraint" coefficient (namely 2μ instead of μ). The corresponding internal stresses are still so high that they result in an almost uniform plastic strain distribution over the grains and the basic Taylor assumption is not really altered. If $\underset{\sim}{\ell}$ an $\underset{\sim}{L}^{eff}$ are the local and overall instantaneous elastic-plastic moduli respectively, it can be shown that the Kröner estimate of $\underset{\sim}{L}^{eff}$, $\underset{\sim}{L}_K$ say, is formally given by :

$$\underset{\sim}{L}_K = \langle \underset{\sim}{\ell} \rangle - \lambda(\langle \underset{\sim}{\ell}^2 \rangle - \langle \underset{\sim}{\ell} \rangle^2) + \underset{\sim}{\mathcal{O}}(\lambda^2) \tag{VI.46}$$

where $\lambda = \dfrac{\beta}{2\mu(1-\beta)}$ is of the order of $1/2\mu$, what means that $\underset{\sim}{L}_k$ differs only slightly from Lin's estimate $\langle \underset{\sim}{\ell} \rangle$.

A more rigorous, fully self-consistent formulation of the problem has been given by Hill who considered instantaneous local elastic plastic moduli $\underset{\sim}{\ell}$, defined by :

$$\dot{\underset{\sim}{\sigma}} = \underset{\sim}{\ell}\, \dot{\underset{\sim}{\varepsilon}} \tag{VI.47}$$

where $\underset{\sim}{\ell}$ is a complicated multi-branched tensor, depending on $\dot{\underset{\sim}{\varepsilon}}$ itself, and prejudged a similar form for the effective moduli, namely :

$$\langle \dot{\underset{\sim}{\sigma}} \rangle = \underset{\sim}{L}^{eff} \langle \dot{\underset{\sim}{\varepsilon}} \rangle \tag{VI.48}$$

The constraint tensor $\underset{\sim}{L}^*$ has to be derived from the inhomogeneity problem where the matrix behaviour is defined by $\underset{\sim}{L}^{eff} = \underset{\sim}{L}^{SC}$ and the grain one by $\underset{\sim}{\ell}$. So, $\underset{\sim}{L}^*$ clearly depends, as it must, on $\underset{\sim}{L}^{SC}$ and on the shape and orientation of the ellipsoïd.[28] From the corresponding interaction law :

$$\dot{\underset{\sim}{\sigma}} = \langle \dot{\underset{\sim}{\sigma}} \rangle + \underset{\sim}{L}^*(\langle \dot{\underset{\sim}{\varepsilon}} \rangle - \dot{\underset{\sim}{\varepsilon}}) \tag{VI.49}$$

and using (VI.47) and (VI.48), the self-consistent equation is readily obtained :

$$\underset{\sim}{L}^{SC} = \langle \underset{\sim}{\ell}(\underset{\sim}{\ell} + \underset{\sim}{L}^*)^{-1} (\underset{\sim}{L}^{SC} + \underset{\sim}{L}^*) \rangle \tag{VI.50}$$

Alternative equivalent equations may be used, such as the following ones :

$$
\begin{aligned}
&\langle(\underset{\sim}{\ell} + \underset{\sim}{L}^*)^{-1} (\underset{\sim}{L}^{SC} + \underset{\sim}{L}^*)\rangle = \underset{\sim}{I} \\
&\langle(\underset{\sim}{\ell} + \underset{\sim}{L}^*)^{-1} (\underset{\sim}{L}^{SC} - \underset{\sim}{\ell})\rangle = 0 \\
&\underset{\sim}{L}^{SC} = \langle(\underset{\sim}{\ell} + \underset{\sim}{L}^*)^{-1}\rangle^{-1} \langle(\underset{\sim}{\ell} + \underset{\sim}{L}^*)^{-1} \underset{\sim}{\ell}\rangle
\end{aligned}
\qquad (VI.51)
$$

All these forms derive from the general formula (VI.20), where $\underset{\sim}{\delta c}$ must be replaced by $\underset{\sim}{\delta\ell}$ and $\underset{\sim}{\Gamma}^{eff}$ by $(\underset{\sim}{L}^* + \underset{\sim}{L}^{SC})^{-1}$. None of them can disguise the fact that we have to deal with integral equations and that the dependence of $\underset{\sim}{L}^*$ on $\underset{\sim}{L}^{SC}$ cannot be put in a closed form in the general case, when $\underset{\sim}{L}^{SC}$ has a general anisotropy. A direct numerical treatment of this procedure has been performed by Hutchinson, in the case of a FCC polycrystal without texture, with non-hardening and isotropic hardening within the grains assumed spherical, in the case of a monotonically increasing uniaxial tensile stress.[31] The overall moduli $\underset{\sim}{L}^{SC}$ were considered as displaying transverse isotropy with respect to the tensile axis and a semi-numerical calculation of $\underset{\sim}{L}^*$ was performed, thanks to Kneer's ordinary integrals which can be used in this case (i.e. transverse isotropy and spherical voids).[32] The dependence of $\underset{\sim}{\ell}$ on $\underset{\sim}{\dot{\varepsilon}}$ makes necessary, at each incremental step of the tensile history, a double iterative procedure : a first one on $\underset{\sim}{L}^{SC}$, from which $\underset{\sim}{L}^*$ is derived, and, inside it, a second one on $\underset{\sim}{\ell}$, for each grain orientation, so as to yield a better estimate of $\underset{\sim}{L}^{SC}$, up to a satisfactory convergence (i.e. self-consistency). One of the most interesting results of this impressive study consists in the considerable effect which was so emphasized of the "softening" influence of an elastic-plastic account of $\underset{\sim}{L}^*$ on the resulting overall behaviour, with respect to the elastic estimate of (VI.44) : this stresses the fact that the "plastic accommodation" effect is the main physical phenomenon that the self-consistent modelling has to integrate and express in the field of plasticity with respect to the elastic case.

An other significant attempt to deal with a fully self-consistent treatment of polycrystal plasticity was performed more recently by Iwakuma and Nemat-Nasser.[33] They used a finite strain formulation which needed some further simplifications, namely : an averaged $\underset{\sim}{L}^*$ tensor instead of an inclusion shape dependent one, two-dimensional structures, initially isotropic polycrystals and spherical grains, uniaxial tension and crystals with two slip systems only. An interesting feature of this study lies in the correlation it shows between the inception of localization and the loss of convergence of the iterative scheme necessary for the evaluation of the overall instantaneous moduli thanks to the modified Green operator.

Nevertheless, such really self-consistent treatments look quite lengthy and, at the moment, they seem to be limited concerning either the field of possible applications or the physical content - or both. So, in order to deal with a larger variety of applications and to make a self-consistent type model almost as convenient as the Taylor one, it is necessary to introduce some further simplifications. It will be shown in the next section that this can be performed without a significant loss of self-consistency by using an isotropic approximation of the constraint tensor $\underset{\sim}{L}^*$, which makes the whole treatment much more tractable but saves, at the same time, the essential elastic-plastic nature of the accommodation process : this condition is quite necessary if what is wanted is to depart from the Taylor approach in order to deal with more general and disordered structures, and not, as it could be obtained elsewhere, a more complicated model than the Taylor one, but yielding similar predictions and conclusions ...

4.4. An isotropic approximation of the self-consistent plastic scheme. The main advantage of the initial Kröner model was the simplicity of its associated $\underset{\sim}{L}^*_K$ constraint tensor (VI.44), which was isotropic and constant, and so, a priori known ; its main drawback consisted in its elastic definition which led to excessive values of the corresponding parameter $2\mu(1-\beta)$ in (VI.43) and to a considerable loss of self-consistency. A reasonable, although not rigorous, compromise can be found by deriving an isotropic (constant or not) estimate of $\underset{\sim}{L}^*$, but saving its elastic-plastic character, what generally means a considerable lowering of its order of magnitude with respect to the elastic one.

This may be performed according to two distinct approaches. The first one, which is more defensible but restricted to specific loadings, refers to radial monotonic stress or strain paths for which an isotropic deformation rule may be derived according to the Hencky-Mises equations even if the incremental flow law is anisotropic, as it generally is.[34] This leads to derive a "secant" constraint tensor, $\underset{\sim}{L}^{*S}$ say, from "secant" elastic-plastic moduli μ^S and k^S, corresponding to isotropic overall secant moduli. For spherical inclusions, this leads to an "interaction law" relating finite quantities, instead of instantaneous ones, according to calculations quite similar to those which have led to (VI.31) but now evolving μ^S and k^S for the matrix. One finds easily :

$$L^{*S}_{ijkl} = \frac{\mu^S}{3(k^S+2\mu^S)} \left[\frac{9k^S+8\mu^S}{2} (\delta_{ik}\ \delta_{jl} + \delta_{il}\ \delta_{jk}) + \frac{3k^S+16\mu^S}{3}\ \delta_{ij}\ \delta_{kl} \right] \qquad \text{(VI.52)}$$

and then, with (μ, k) the uniform elastic moduli of the whole body :

$$\underset{\sim}{\sigma} = \langle\underset{\sim}{\sigma}\rangle + 2\mu\mu^S \frac{9k^S + 8\mu^S}{6\mu(k^S+2\mu^S) + \mu^S(9k^S+8\mu^S)} (\langle\underset{\sim}{\varepsilon}^P\rangle - \underset{\sim}{\varepsilon}^P) \qquad \text{(VI.53)}$$

Referring to (VI.43), this interaction law may take a form very similar to Kröner's one, namely :

$$\underset{\sim}{\sigma} = \langle\underset{\sim}{\sigma}\rangle + 2\alpha^S\mu(1-\beta)\ (\langle\underset{\sim}{\varepsilon}^P\rangle - \underset{\sim}{\varepsilon}^P) \qquad \text{(VI.54)}$$

where the "plastic accommodation secant factor" α^S is given by :

$$\alpha^S = \frac{9k^S+8\mu^S}{9k + 8\mu} \cdot \frac{5\mu^S(3k+4\mu)}{6\mu(k^S+2\mu^S) + \mu^S(9k^S+8\mu^S)} \qquad \text{(VI.55)}$$

Obviously, $\alpha^S = 1$ in the elastic range and at the very beginning of the plastic flow but it is steeply decreasing with increasing plastic strain ; this may be illustrated from an experimental tensile stress-strain curve yielding the secant Young modulus E^S = typical results on copper polycrystals show that α^S is about .15 as soon as $\langle\varepsilon^P\rangle$ is one percent. This comment allows a significant comparison between most of the models which have been reported up to now. They all obey the same form of interaction law, namely :

$$\underset{\sim}{\sigma} = \langle\underset{\sim}{\sigma}\rangle + K\mu(\langle\underset{\sim}{\varepsilon}^P\rangle - \underset{\sim}{\varepsilon}^P) \qquad \text{(VI.56)}$$

where : $K \to \infty$ for the Taylor model (since $\underset{\sim}{\varepsilon}^P = \langle\underset{\sim}{\varepsilon}^P\rangle$)

$K = 2$ for the Lin-Taylor model

$K = 2(1-\beta) \simeq 1$ for the Kröner model

$K = 2\alpha^S(1-\beta) \simeq \alpha^S$ for the present one (with $0 < \alpha^S \leq 1$)

$K = 0$ for the static model

Such a discrepancy for this K parameter according to the considered model stresses the fact that all of them use a very poor information on the material structure, i.e. the volume fractions of the constituents. These one-point correlation functions are associated with quite distant upper and lower bounds (for $K \to \infty$ and $K = 0$ respectively). From a practical point of view, considering common values of the elastic moduli, the yield stress and the hardening moduli of current metals, $K = 1$ is still very close to $K \to \infty$, whereas disordered plastic structures need much lower K values (say : of the order .1 or .01) as soon as plastic flow is really active.[35]

A severe limitation of this approximate self-consistent scheme derives from the monotonic loading restriction. When elastic unloading has to be considered, such as for cyclic loading, an incremental variant of this model may be preferred, according to which isotropic approximation is bearing upon the instantaneous constraint tensor $\underset{\sim}{L}^*$ instead of the secant one. Of course, an isotropic elastic-plastic instantaneous behaviour is hard to conceive, but this drawback may be less than the one which would derive from maintaining a secant elastic plastic approach during an elastic unloading. In this case, a "tangent" constraint tensor $\underset{\sim}{L}^{*T}$ should be considered, obeying (VI.52) except for changing k^S and μ^S into the tangent elastic plastic moduli k^T and μ^T. The resulting interaction law is :

$$\dot{\underset{\sim}{\sigma}} = \langle\dot{\underset{\sim}{\sigma}}\rangle + 2\alpha^T\mu(1-\beta)(\langle\dot{\underset{\sim}{\varepsilon}}^P\rangle - \dot{\underset{\sim}{\varepsilon}}^P) \qquad \text{(VI.57)}$$

where the "plastic accommodation tangent factor" α^T is defined by (VI.55) except for changing k^S and μ^S into k^T and μ^T too. Obviously, current values of α^T will now be still lower than α^S in the plastic range (typically of the order 10^{-2}) but will be allowed to reach unity during elastic unloading.

Some additional remarks may be made now, before giving illustrative examples of application of this model :

- except for the case of actually isotropic materials, the foregoing approximations clearly leads to weaken the self-consistency of the procedure. Nevertheless, they do not prevent from predicting an anisotropic overall behaviour, as it will be shown in the sequel : if an initial texture is considered, anisotropic $\underset{\sim}{L}^{SC}$ moduli will result from the calculations, but they are likely to be less anisotropic than they should be if a fully self-consistent treatment had been adopted ;

- spherical inclusions have been considered until now. But, as long as the isotropic approximation of the matrix behaviour is saved for the derivation of $\underset{\sim}{L}^*$, ellipsoïdal shapes may be dealt with without considerable heaviness of the calculations, thanks to Eshelby's results on the $\underset{\sim}{S}$ tensor (VI.37). Preliminary results have been obtained in this field which, in addition to their own interest concerning the investigation of the grain shape effects on the resulting overall behaviour, allow a closer comparison with the "relaxed constraint" Taylor models : whereas these models are generally worried about the way to decide how to gradually move from a "full constraint" to a "relaxed constraint" state, the self-consistent scheme may change the ellipticity of the inclusions in a progressive way, following the intensity of the plastic flow. Fig.

4 and 5 show examples of the variations of some of the $(\underset{\sim}{S}-\underset{\sim}{I})$ components during the plastic shape change of initially spherical grains undergoing a tensile and a rolling deformation respectively : thus, instead of changing the volume fraction of the flattened grains, e.g., during the flow or of parting each grain from the beginning into fully or partly constrained areas, the shape change here naturally follows the plastic deformation.[36] The relaxed constraint conditions are then only found as an asymptotic regime which, itself, is treated in a softer manner than according to a (partial) uniformity prescription ;

- the main conclusion of the preceding analysis lies in the very low value of the accommodation factor α^S or α^T with respect to Lin's or Kröner's estimates. Whether anisotropy is taken into account or not, and except during the very low plastic strain regime, the right order of magnitude of the constraint effect which is attached to the self-consistent procedure has been revealed. An inviting over-simplification then is coming out, namely : to deal with a constant accommodation factor over a wide range of plastic flow (as long as elastic unloading is excluded), in order to explore a larger application field with an approximate, but physically sound, modelling tool. This is the choice which has been fixed in studying the following examples, according to a formulation scheme which has now to be specified.

When the interaction law (VI.54) or (VI.57) is combined with the intragranular constitutive equations for a given grain family (as depicted in 2.3), and even if the α factor is given a known constant value, either the overall stress or strain (or their rates) is still unknown. In the case of isotropy, these quantities are proportionnal to

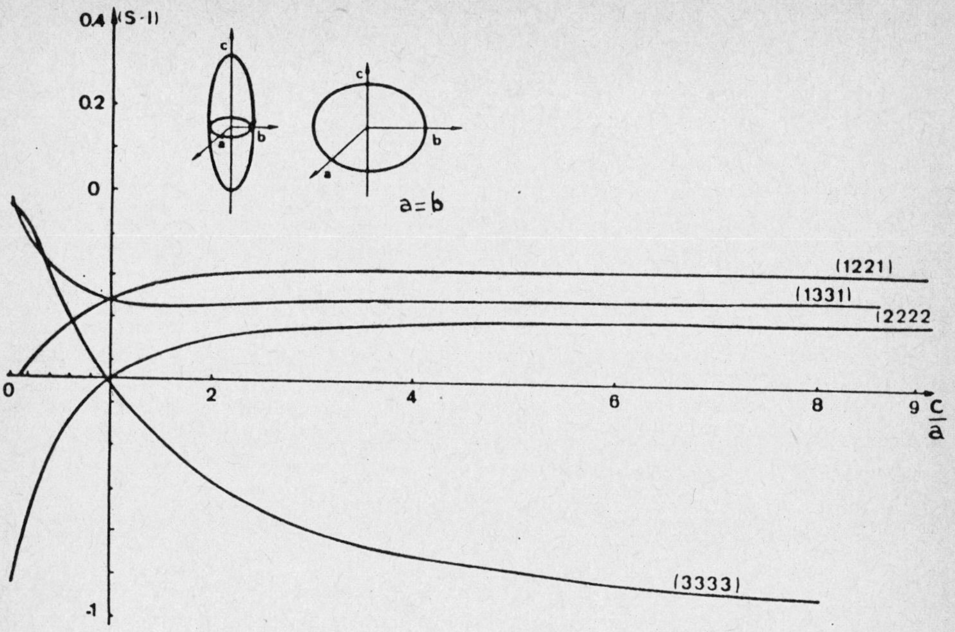

Fig. 4 : Some (S-I) components during a tensile deformation.

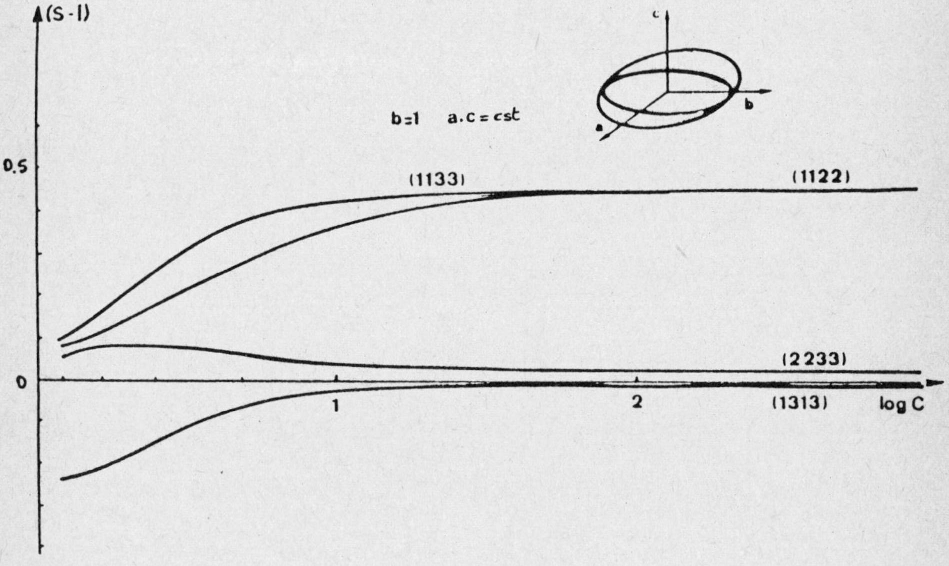

Fig. 5 : Some (S-I) components during a rolling deformation

each other so that they combine into one tensor which will be easy to finally part into its stress and strain components, according to Budiansky - Wu's formulation, e.g. :[30]

$$\dot{\underset{\sim}{Q}} = \langle\dot{\underset{\sim}{\sigma}}\rangle + 2\alpha^T\mu(1-\beta)\ \langle\dot{\underset{\sim}{\varepsilon}}^P\rangle \qquad \text{(VI.58)}$$

The $\underset{\sim}{Q}$ history may then be considered as given and, since each grain orientation can be dealt with separately, it will easily result in the $\langle\dot{\underset{\sim}{\sigma}}\rangle$ assigned history and the resultant $\langle\dot{\underset{\sim}{\varepsilon}}^P\rangle$ response (or the reverse). This is no more the case when anisotropy is present. Tedious iterations are then necessary in order to perform the right partition of $\underset{\sim}{Q}$. An alternative solution has been preferred in order to avoid the iterative procedure thanks to a simultaneous treatment of all the grain orientations.[37] Let f_I be the volume fraction of the grain family with the orientation Ω_I, after the texture function has been discretized and suppose the stress history is assigned. Then the interaction law (VI.57) may be written :

$$\dot{\underset{\sim}{\sigma}}_I + 2\alpha^T\mu(1-\beta)\ \dot{\underset{\sim}{\varepsilon}}^P_I - 2\alpha^T\mu(1-\beta)\sum_J f_J\ \dot{\underset{\sim}{\varepsilon}}^P_J = \langle\dot{\underset{\sim}{\sigma}}\rangle \qquad \text{(VI.59)}$$

so that, after the plastic strain rate tensors $\dot{\underset{\sim}{\varepsilon}}^P_J$ have been reduced to their constitutive resolved shear strain rates $\dot{\gamma}^g_J$ and the crystal flow equations between $\dot{\underset{\sim}{\sigma}}_I$ and $\dot{\underset{\sim}{\varepsilon}}^P_I$ have been introduced, we are left with a unique set of equations for all the $\dot{\gamma}^g_J$'s on the active slip systems of the whole polycrystal. A typical equation of this kind, when the rotation of the slip systems with respect to the applied stress axes is neglected and after the system has been symmetrized by adequate operations, looks like the following (with $\beta = \frac{1}{2}$) :

$$f_I \sum_h (H_I^{gh} + \alpha^T \mu \, R_I^g \, R_I^h) \, \dot{\gamma}_I^h - \alpha^T \mu \sum_J \sum_h f_I \, f_J \, R_I^g \, R_J^h \, \dot{\gamma}_J^h = f_I \, R_I^g \, \langle \dot{\sigma} \rangle \tag{VI.60}$$

where $\underset{\sim}{R}^g$ is defined in (VI.1). Of course, the number of equations and variables $\dot{\gamma}_I^g$ is ceaseless evolving, at each new activation or passivation event on any slip system of the polycrystal. Local and overall plastic strain rates are calculated at each step as well as local plastic spins $\underset{\sim}{\dot{\omega}}^P$. When texture development is considered, the lattice spin $\underset{\sim}{\dot{\omega}}^L$ must be calculated. Within this isotropic self-consistent approach, it is justified to use the simple relation :

$$\underset{\sim}{\dot{\omega}}^L = - \underset{\sim}{\dot{\omega}}^P + \langle \underset{\sim}{\dot{\omega}}^T \rangle \tag{VI.61}$$

where the macroscopic total spin $\langle \underset{\sim}{\dot{\omega}}^T \rangle$ is assigned as an exterior condition. More complex relations must be used as soon as anisotropy is present - even only when ellipsoïdal grains are considered.[38]

Fig. 6, 7, 8, 9 and 10 give illustrative examples of a comparison between calculations performed according to this treatment and the corresponding experimental data. Other results are reported in the literature concerning the determination of yield surfaces (initial as well as subsequent ones), the analysis of the Bauschinger effect, cyclic loading etc...[40,41,42] Even if a tight quantitative agreement is not always observed, most of the main features of polycrystal plasticity are recognized and described. Further advances are depending upon a better description of intragranular behaviour as well as the development of a finite strain formalism. Nevertheless, even the most simplified versions of the self-consistent scheme still look far from being able to be integrated into structural analysis programs.

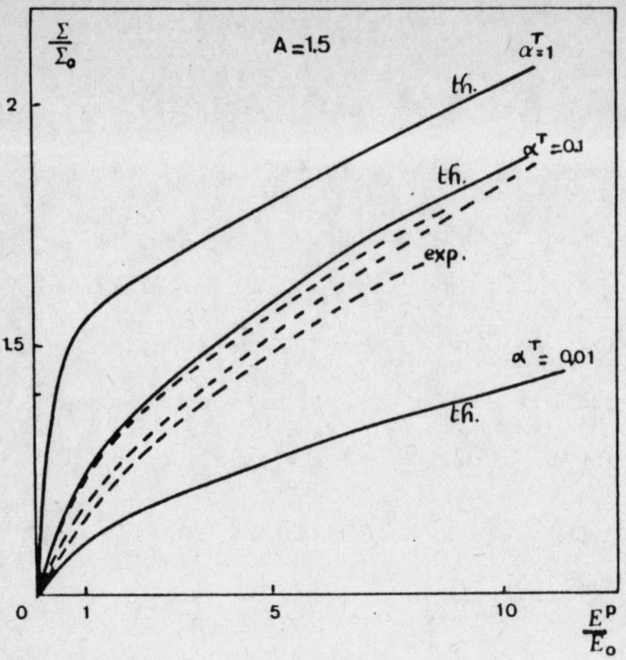

Fig. 6 : Simulation of a tensile test for various α values compared with experimental results on Cu polycrystals.

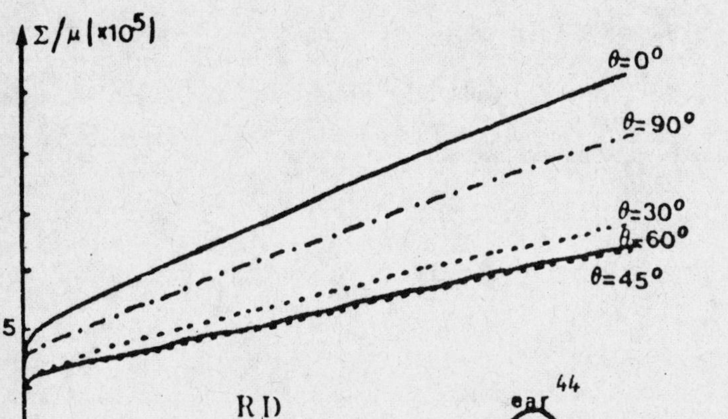

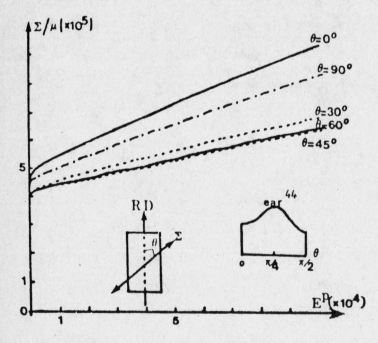

Fig. 7 : Simulation of tensile tests on samples cut from a cold rolled sheet, compared with the experimental earing behaviour.[39]

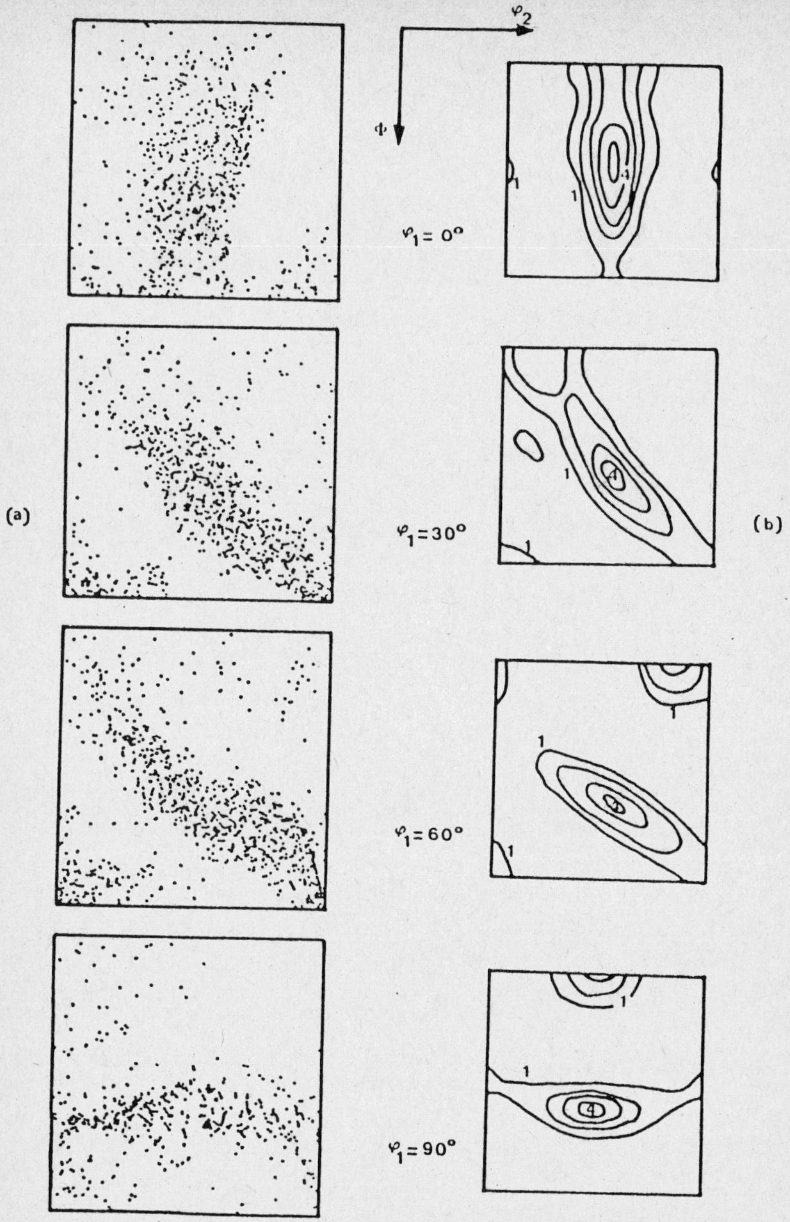

Fig. 8 : Simulation of the development of a rolling texture (a) compared with experimental data on BCC polycrystals (b).

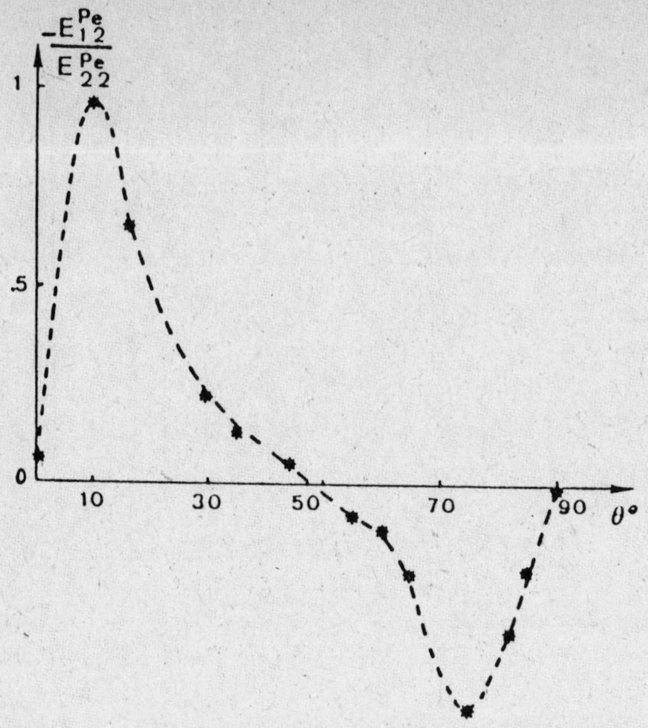

Fig. 9 : Simulation of the deviation of tensile stress and strain principal directions on samples cut from a cold-rolled sheet.

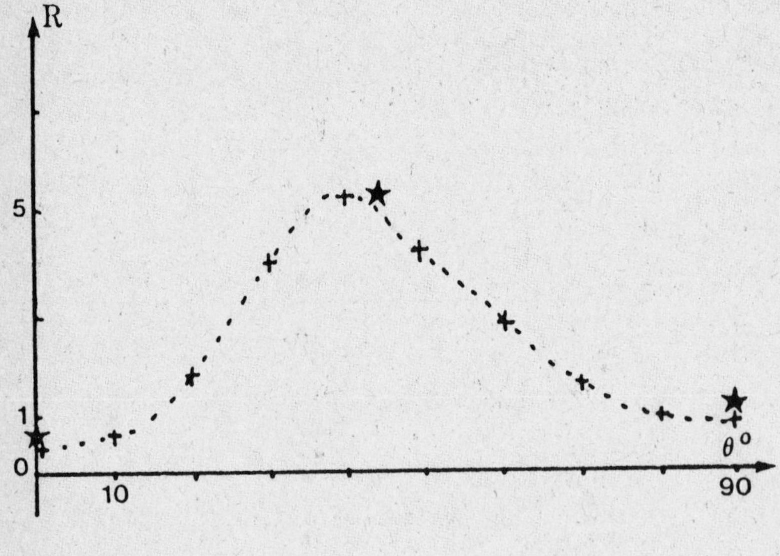

Fig. 10 : Simulation of the Lankford coefficient variation compared with experimental data (stars).[39]

The question of the relative merits of the self-consistent scheme with respect to the hereabove reported other models is, from a conceptual point of view, irrelevant.[43] Since all these models cannot deal with more than the phase volume fractions, nobody can decide which is the best or the worst one, as long as more statistical information on the material structure is not available. According as this structure is more ordered or more disordered, the best model would be this or that. This conclusion makes it all the more necessary to build new models which would be able to deal with richer statistical information. The next section aims at reporting some of the attempts which have been performed in order to give to the self-consistent approach such a capacity. In order to stress this point, the polycrystal will be replaced by a more simple, two-phase composite, without crystallographic definition. Moreover, we shall mostly restrict ourselves to the case of linear elasticity.

CHAPTER 5

EXTENDED SPACE DISTRIBUTION SENSITIVE SELF-CONSISTENT SCHEMES

Let us first briefly deal with the classical self-consistent scheme in the case of a two-phase isotropic material. The basic inclusion problem which is involved consists in a spherical inhomogeneity embedded within an infinite matrix with the effective behaviour : the sphere either consists in phase 1 or phase 2 with the respective probability c_1 and c_2 ($c_1 + c_2 = 1$). Let (μ, k) and (μ^I, k^I) be the shear and bulk moduli of the composite and of phase I (with I = 1,2) resp. According to the case, they could be elastic, secant elastic-plastic or tangent elastic plastic moduli, as discussed hereabove. Eqns (VI.30) and (VI.31) yield directly for strain or strain rate, e.g. for strain :

$$e^I_{ij} = \frac{5\mu(3k+4\mu)}{5\mu(3k+4\mu) + 6(\mu^I-\mu)(k+2\mu)} \langle e_{ij} \rangle \qquad \text{(VI.62)}$$

$$e^I = \frac{3k + 4\mu}{4\mu + 3k^I} \langle e \rangle$$

and the average equation is :

$$\langle \varepsilon \rangle = c_1\ \varepsilon^1 + c_2\ \varepsilon^2 \qquad \text{(VI.63)}$$

Combining (VI.62) and (VI.63) easily leads to the self-consistent equations :

$$\sum_I \frac{c_I}{\mu(9k+8\mu) + 6\mu^I(k+2\mu)} = \frac{1}{5\mu(3k+4\mu)}$$

$$\sum_I \frac{c_I}{3k^I+4\mu} = \frac{1}{3k+4\mu} \qquad \text{(VI.64)}$$

which are, obviously, valid for more than two phases. In case of incompressibility, they simply reduce to :

$$\sum_I \frac{c_I}{3\mu+2\mu^I} = \frac{1}{5\mu} \qquad \text{(VI.65)}$$

This method allows a very easy determination of the overall moduli, even in the case of plasticity : for example, when the tensile stress-strain curves of the pure phases are given, a simple graphic construction may be used to determine, at any stage of a tensile test on the composite, the stress and strain state in each phase and in the composite.[44]

But the limits of the self-consistent scheme may easily be pointed out if cavities or rigid inclusions with the volume fraction c are considered as one of the two phases. In the first case, (VI.65) yields :

$$\mu = \mu_2 \left(1 - \frac{5c}{3}\right) \qquad c \leq \frac{3}{5} \qquad \text{(VI.66)}$$

and μ vanishes for higher values of c, whereas in the second case, it gives :

$$\mu = \frac{\mu_2}{1 - \frac{5c}{2}} \qquad c \leq \frac{2}{5} \qquad \text{(VI.67)}$$

and the overall behaviour becomes a rigid one beyond. Such abrupt conclusions clearly conflict with many observed situations dealing with particuliar space arrangements of voids or rigid inclusions and exhibiting either a cohesion or a deformability for higher volume fractions than indicated in (VI.66) or (VI.67). This remark simply stresses the fact that the self-consistent scheme assumes specific space distributions - which are likely of the perfect disorder type - and may easily be contradicted by experimental results obtained on materials with quite different space distributions : for example, dispose a continuous layer of voids in the cross-section of any sample, and it will have no mechanical strength, however low the volume fraction may be... That is the reason why several attempts have been made in order to modify the self-consistent scheme so as to take into some account such and such geometrical characteristics.

Still dealing with the convenient case of a two-phase composite, one can mention the so-called "differential self-consistent scheme", according to which second phase inclusions are progressively introduced in the matrix so as to continuously raise their volume fraction up to the final value : this may, for instance, modify the hereabove mentioned critical values of voids or rigid inclusions, but do not really take geometrical characteristics of the phase arrangement into better account.[45,46] More relevant to our problematics, the "composite spheres model" is constituted with spheres of various sizes, down to infinitesimal ones, so as to completely fill the material volume : each sphere, with the radius b, consists in a smaller concentric one, with the radius a such that a/b is taken to be a constant, and a spherical shell around it. The smaller spheres represent one phase and the shells the second one ; the ratio of radii a/b determines the volume fractions of the phases. This model is well adapted to the case when one phase exists as inclusions embedded within a connex other phase. Hashin has given the solution for the effective bulk modulus and bounds for the effective shear modulus.[47]

The same kind of idea has been developed in a more self-consistent direction, with the so-called "three-phase model".[48] Here, a composite sphere is also considered, with $c = (a/b)^3$ is the volume fraction of inclusions, but it is now embedded within an infinite equivalent homogeneous medium, the behaviour of which is determined by requiring that this configuration is storing the same strain energy, under conditions of identical average strain, than the effective medium. Such a model, which yields definite effective values for the bulk, as well as for the shear modulus, thus integrates into the definition of the inclusion some specific geometrical informations on the mutual arrangement of the phases. It can be considered as a prototype of a whole class of extended self-consistent schemes dealing with more and more complex inclusions (always embedded within the sought equivalent homogeneous medium) which give a better and better description of the various constituent neighbourhoods of a given material at a definite scale, each typical configuration being weighted with its own frequency. Such procedures, which are dealing not only with basic elements, but also with their nearest (and then next nearest) neighbours configurations, have recently been considered too in similar situations of solid state physics.

Finally, we would like to suggest, as quite speculative preliminary attempts, another direction for extending the self-consistent classical approach towards a better sensitiveness to geometrical characteristics of the phase space distribution. At variance with the foregoing point of view, which is still dealing with a "one-site" approach, even if this site is going larger and larger and more and more detailed, we could adopt a "multi-site" analysis according to which several points (or inclusions) are considered at once in every possible situations and with every possible mechanical content for the considered multiphase material. Let us deal with a two-phase composite and a two-site description in order to make the argument clearer : a formal extension to more than two phases as well as to more than two sites would make no conceptual difference.

The kind of things we could do is suggested by (VI.59) where the proposed numerical formulation for the classical self-consistent scheme is pointing out the fact that, according to this scheme, the mechanical interactions between one phase (I) and another phase (J) do not depend on the relative position of these phases since everything occurs as if phase (I) was gathered into an inclusion and phase (J) diluted uniformly in the whole body up to infinity ; and the reverse would be true if the equation concerning (J) was considered.[43] We could then try to take pair interactions between (I) and (J) into account by replacing the scalar factor $2\alpha^T\mu(1-\beta)$ by an adequate fourth order tensor resulting from the solution of a "two inclusions/matrix" problem which has been already addressed in literature.[41] The difficulty lies in the way to perform this operation in a self-consistent manner. A possible answer may be given by using the Green operator formalism reported in section 4.1 of this chapter, especially the Lippmann-Schwinger equation (VI.12) as the starting point and the resulting equation (VI.18) for the effective moduli $\underset{\sim}{C}^{eff}$.[49] Let us now consider, after several authors, the "transition operator" $\underset{\sim}{T}$ defined in the following way :

$$\underset{\sim}{\sigma} = \underset{\sim}{C}^o \underset{\sim}{\varepsilon} + \underset{\sim}{T} \underset{\sim}{\varepsilon}^o \tag{VI.68}$$

so that :

$$\underset{\sim}{T} \underset{\sim}{\varepsilon}^o = (\underset{\sim}{c} - \underset{\sim}{C}^o) \underset{\sim}{\varepsilon} = \underset{\sim}{\delta c} \underset{\sim}{\varepsilon} = \underset{\sim}{\delta c}(\underset{\sim}{I} + \underset{\sim}{\Gamma} \underset{\sim}{\delta c})^{-1} \underset{\sim}{\varepsilon}^o \tag{VI.69}$$

An alternate definition of $\underset{\sim}{T}$ is then the following :

$$\underset{\sim}{T} = \underset{\sim}{\delta c}(\underset{\sim}{I} + \underset{\sim}{\Gamma} \underset{\sim}{\delta c})^{-1} \tag{VI.70}$$

and the equation (VI.18) may be written in the more condensed form :

$$\underset{\sim}{C}^{eff} = \underset{\sim}{C}^o + \langle \underset{\sim}{T} \rangle \langle \underset{\sim}{\delta c}^{-1} \underset{\sim}{T} \rangle^{-1} \tag{VI.71}$$

If we decide, now, to take the effective medium as the reference one, this equation simply writes :

$$\langle \underset{\sim}{T} \rangle_{eff} = 0 \tag{VI.72}$$

where the index "eff" specifies that $\underset{\sim}{\delta c}$, $\underset{\sim}{\Gamma}$ and then $\underset{\sim}{T}$ are respective to the effective medium. (Note that (VI.72) is simply a transcription of (VI.20)). Since we can get, from (VI.12) and (VI.69), the relation :

$$\underset{\sim}{T} = \underset{\sim}{\delta c}(I - \underset{\sim}{\Gamma}\underset{\sim}{T}) \qquad \text{(VI.73)}$$

eqn (VI.71) may also be put into the form :

$$\underset{\sim}{C}^{eff} = \underset{\sim}{C}^{o} + \langle\underset{\sim}{T}\rangle(\underset{\sim}{I} - \underset{\sim}{\Gamma}\langle\underset{\sim}{T}\rangle)^{-1} \qquad \text{(VI.74)}$$

which results in (VI.72) too.[50,51] Let us now consider, in view of the self-consistent statistical meaning (i.e. perfect disorder), a new "local" transition operator $\underset{\sim}{t}$, which has a definition analogous to $\underset{\sim}{T}$ (see VI.70) except for the fact that the local part $\underset{\sim}{E}$ of $\underset{\sim}{\Gamma}$ only is involved, namely :

$$\underset{\sim}{t} = \underset{\sim}{\delta c}(\underset{\sim}{I} + \underset{\sim}{E}\ \underset{\sim}{\delta c})^{-1} \qquad \text{(VI.75)}$$

From (VI.70) and VI.75), it is not difficult to conclude that :

$$\underset{\sim}{T} = \underset{\sim}{t} - \underset{\sim}{t}\underset{\sim}{F}\underset{\sim}{T} \qquad \text{(VI.76)}$$

where $\underset{\sim}{F}$ is the "long range" part of $\underset{\sim}{\Gamma}$.

If we consider now a "single site" situation, where the point-like inhomogeneity $\underset{\sim}{c}$ is embedded in a uniform matrix with the moduli $\underset{\sim}{C}^{SC1}$, where the superscript "SC1" indicates a "one site" self-consistent approach, $\underset{\sim}{T}$ vanishes everywhere except at this site where, according to (VI.76) and to the definition of $\underset{\sim}{F}$, it coïncides with $\underset{\sim}{t}$ at the same time. The effective equation (VI.72) now yields the classical self-consistent equation :

$$\langle\underset{\sim}{t}\rangle_{SC1} = \langle\underset{\sim}{\delta c}(\underset{\sim}{I} + \underset{\sim}{E}^{SC1}\underset{\sim}{\delta c})^{-1}\rangle = 0 \qquad \text{(VI.77)}$$

which looks like (VI.25). The implicit idea that self-consistency is identical with perfect disorder may be made explicit thanks to a formal series expansions of (VI.76), namely :

$$\underset{\sim}{T} = \underset{\sim}{t} - \underset{\sim}{t}\,\underset{\sim}{F}\,\underset{\sim}{t} + \underset{\sim}{t}\,\underset{\sim}{F}\,\underset{\sim}{t}\,\underset{\sim}{F}\,\underset{\sim}{t} - \ldots \qquad \text{(VI.78)}$$

and to the definition (VI.24) of perfect disorder, which, obviously, can be applied to the random operator $\underset{\sim}{t}$ too. Averaging of (VI.78) then yields :

$$\begin{aligned}
\langle\underset{\sim}{T}\rangle &= \langle\underset{\sim}{t}\rangle - \langle\underset{\sim}{t}\,\underset{\sim}{F}\,\underset{\sim}{t}\rangle + \langle\underset{\sim}{t}\,\underset{\sim}{F}\,\underset{\sim}{t}\,\underset{\sim}{F}\,\underset{\sim}{t}\rangle - \ldots \\
&= \langle\underset{\sim}{t}\rangle - \langle\underset{\sim}{t}\rangle\,\underset{\sim}{F}\,\langle\underset{\sim}{t}\rangle + \langle\underset{\sim}{t}\rangle\,\underset{\sim}{F}\,\langle\underset{\sim}{t}\rangle\,\underset{\sim}{F}\,\langle\underset{\sim}{t}\rangle - \ldots \\
&= \langle\underset{\sim}{t}\rangle(\underset{\sim}{I} - \underset{\sim}{F}\,\langle\underset{\sim}{t}\rangle + \underset{\sim}{F}\,\langle\underset{\sim}{t}\rangle\,\underset{\sim}{F}\,\langle\underset{\sim}{t}\rangle - \ldots.) \\
&= \langle\underset{\sim}{t}\rangle(\underset{\sim}{I} + \underset{\sim}{F}\langle\underset{\sim}{t}\rangle)^{-1}
\end{aligned} \qquad \text{(VI.79)}$$

so that (VI.72) leads to (VI.77) in the case of perfect disorder.

Let us now consider a "double-site" situation, which could be the basic pattern of a "double-site" self-consistent scheme : (1,2) with the moduli (c_1, c_2) embedded in a uniform medium with the moduli $\underset{\sim}{C}^{SC2}$. $\underset{\sim}{T}$ is now non zero both in (1) and (2) and it depends on the long-range interaction between (1) and (2), through $\underset{\sim}{F}_{12}$ and $\underset{\sim}{F}_{21}$, namely :

$$\begin{aligned}
\underset{\sim}{T}_1 &= \underset{\sim}{t}_1 - \underset{\sim}{t}_1\,\underset{\sim}{F}_{12}\,\underset{\sim}{T}_2 \\
&= \underset{\sim}{t}_1 - \underset{\sim}{t}_1\,\underset{\sim}{F}_{12}\,(\underset{\sim}{t}_2 - \underset{\sim}{t}_2\,\underset{\sim}{F}_{21}\,\underset{\sim}{T}_1)
\end{aligned} \qquad \text{(VI.80)}$$

so that $\underset{\sim}{T}_1$ may be expressed with $\underset{\sim}{t}_1$, $\underset{\sim}{t}_2$ and $\underset{\sim}{F}$ only :

$$\underset{\sim}{T}_1 = (\underset{\sim}{I} - \underset{\sim}{t}_1\,\underset{\sim}{F}_{12}\,\underset{\sim}{t}_2\,\underset{\sim}{F}_{21})^{-1}\,\underset{\sim}{t}_1(\underset{\sim}{I} - \underset{\sim}{F}_{12}\,\underset{\sim}{t}_2) \qquad \text{(VI.81)}$$

The two-site self-consistent equation results from (VI.72) and (VI.81) :

$$\langle\underset{\sim}{T}\rangle_{SC2} = \langle(\underset{\sim}{I} - \underset{\sim}{t}\,\underset{\sim}{F}\,\underset{\sim}{t}\,\underset{\sim}{F})^{-1}\,\underset{\sim}{t}(\underset{\sim}{I} - \underset{\sim}{F}\,\underset{\sim}{t})\rangle = 0 \qquad \text{(VI.82)}$$

where the important point lies in the fact that two-point interactions only have to be considered. As a matter of fact, if (VI.82) was formally written as a series expansion, it would look like :

$$\langle \underset{\sim}{t}_1 \rangle - \langle \underset{\sim}{t}_1 \underset{\sim}{F}_{12} \underset{\sim}{t}_2 \rangle + \langle \underset{\sim}{t}_1 \underset{\sim}{F}_{12} \underset{\sim}{t}_2 \underset{\sim}{F}_{21} \underset{\sim}{t}_1 \rangle - \\ - \langle \underset{\sim}{t}_1 \underset{\sim}{F}_{12} \underset{\sim}{t}_2 \underset{\sim}{F}_{21} \underset{\sim}{t}_1 \underset{\sim}{F}_{12} \underset{\sim}{t}_2 \rangle + \ldots \qquad = 0 \qquad \text{(VI.83)}$$

whereas the general equation resulting from (VI.72) and (VI.78) would be :

$$\langle \underset{\sim}{t}_1 \rangle - \langle \underset{\sim}{t}_1 \underset{\sim}{F}_{12} \underset{\sim}{t}_2 \rangle + \langle \underset{\sim}{t}_1 \underset{\sim}{F}_{12} \underset{\sim}{t}_2 \underset{\sim}{F}_{23} \underset{\sim}{t}_3 \rangle - \\ - \langle \underset{\sim}{t}_1 \underset{\sim}{F}_{12} \underset{\sim}{t}_2 \underset{\sim}{F}_{23} \underset{\sim}{t}_3 \underset{\sim}{F}_{34} \underset{\sim}{t}_4 \rangle + \ldots \qquad = 0 \qquad \text{(VI.84)}$$

which means that, according to (VI.83), pair interactions only are operating while interactions of a higher degree are neglected. The averaging procedure in (VI.82) or (VI.83) has now to deal with the probability to find one phase in (1), conditional upon finding (2) in the same phase or in the other phase etc... It is the delicate part of this procedure, and its application to an illustrative (non perfectly disordered) two-phase situation is still in progress ...

REFERENCES

1 TAYLOR, G.I., The mechanism of plastic deformation of crystals, *Proc. Roy. Soc.* London, A 145, 362, 1934.

2 FRANCIOSI, P., BERVEILLER, M. and ZAOUI, A., Latent hardening in copper and aluminium single crystals, *Acta Metall.* 28, 273, 1980.

3 MANDEL, J., Généralisation de la théorie de plasticité de W.T. Koiter, *Int. J. Solids Structures*, 1, 273, 1965.

4 HILL, R., Generalized constitutive relations for incremental deformation of metal crystals by multislip, *J. Mech. Phys. Solids*, 14, 95, 1966.

5 FRANCIOSI, P. and ZAOUI, A., Multislip tests on copper crystals : a junction hardening effect, *Acta Metall.*, 30, 2141, 1982.

6 ZAOUI, A., Aspects fondamentaux de la plasticité des polycristaux métalliques, *Dislocations et déformation plastique*, P. Groh, L.P. Kubin and J.L. Martin Ed., les Editions de physique, Paris, 1979.

7 MUSSOT, P., REY, C. and ZAOUI, A., Grain boundary sliding and strain compatibility, *Res Mech.*, 14, 69, 1985.

8 REY, C. and ZAOUI, A., Slip heterogeneities in deformed aluminium bicrystals, *Acta Metall.*, 28, 687, 1980.

9 REY, C. and ZAOUI, A., Grain boundary effects in deformed bicrystals, *Acta Metall.*, 30, 523, 1982.

10 REY, C., MUSSOT, P. and ZAOUI, A., Effects of interfaces on the plastic behaviour of metallic aggregates, *Proc. Int. Conf. on the structure and properties of internal interfaces*, Munich, in the press.

11 SACHS, G., Zur Ableitung einer Fliessbedingung, *Z. VDI*, 72, 734, 1928.

12 TAYLOR, G.I., Plastic strain in metals, *J. Inst. Metals*, 62, 307, 1938.

13 BISHOP, J.F.W. and HILL, R., A theory of the plastic distortion of a polycrystalline aggregate under combined stress, *Phil. Mag.*, 42, 414, 1951.

14 CHIN, G.Y. and MAMMEL, W.L., Generalization and equivalence of the minimum work (Taylor) and maximum work (Bishop-Hill) principles in crystal plasticity, *Trans. TMS-AIME*, 245, 1211, 1969.

15 AERNOUDT, P., Calculation of deformation textures according to the Taylor Model, *Textures of materials*, G. Gottstein and K. Lücke Ed, I, 45, 1978.

16 GRUMBACH, M., PARNIERE, P., ROESCH, L. and SAUZAY, C., Etude des relations quantitatives entre le coefficient d'anisotropie, les cornes d'emboutissage et la texture des tôles minces d'acier extra-doux, *Mem. Sci. Rev. Metallurg.*, 72, 241, 1975.

17 LIN, T.H., Analysis of elastic and plastic strains of a FCC crystal, *J. Mech. Phys. Solids*, 5, 143, 1957.

18 HONNEFF, M. and MECKING, H., A method for the determination of the active slip systems and orientation changes during single crystal deformation, *Textures of materials*, G. Gottstein and K. Lücke (ed), 265, Springer, 1978.

19 KOCKS, U.F. and CHANDRA, H., Slip geometry in partially constrained deformation, *Acta Metall.*, 30, 695, 1982.

20 VAN HOUTTE, P., On the equivalence of the relaxed Taylor theory and the Bishop-Hill theory for partially contrained plastic deformation of crystals, *Mat. Sci. Eng.*, 55, 69, 1982.

21 ASARO, R.J. and NEEDLEMAN, A., Texture development and strain hardening in rate dependent polycrystals, private communication, 1984.

22 BATDORF, S.B. and BUDIANSKY, B., A mathematical theory of plasticity based on the concept of slip, *NACA TN-1871*, 1949.

23. KRÖNER, E., Self-consistent scheme and graded disorder in polycrystal elasticity, *J. Phys. F*, 8, 2261, 1978.

24 HERSHEY, A.V., The elasticity of an isotropic aggregate of anisotropic cubic crystals, *J. Appl. Mech.*, 21, 236, 1954.

25 KRÖNER, E., Berechnung der elastischen Konstanten des Vielkristalls aus den Konstanten des Einkristalls, *Z. Phys.*, 151, 504, 1958.

26 ESHELBY, J.D., The determination of the elastic field of an ellipsoïdal inclusion and related problems, *Proc. Roy. Soc.*, London, A241, 376, 1957.

27 ESHELBY, J.D., Elastic inclusions and inhomogeneities, *Prog. in Sol. Mech.*, vol. II, I.N. Sneddon and R. Hill (ed), 87, North-Holland Pub. Co, Amsterdam, 1961.

28 HILL, R., Continuum micro-mechanics of elastoplastic polycrystals, *J. Mech. Phys. Solids*, 13, 89, 1965.

29 KRÖNER, E., Zur plastischen Verformung des Vielkristalls, *Acta Metall.*, 9, 155, 1961.

30 BUDIANSKY, B. and WU, T.T., Theoretical prediction of plastic strains of polycrystals, *Proc 4th U.S. Nat. Congr. Appl. Mech.*, 1175, 1962.

31 HUTCHINSON, J.W., Elastic-plastic behaviour of polycrystalline metals and composites, *Proc. Roy. Soc.* London, A 319, 247, 1970.

32 KNEER, G., "Uber die Berechnung der Elastizitätsmoduln vielkristalliner Aggregate mit Textur, *Phys. Stat. Sol.*, 9, 825, 1965.

33 IWAKUMA, T., and NEMAT-NASSER, S., Finite elastic-plastic deformation of polycrystalline metals, *Proc. Roy. Soc.*, London, A 394, 87, 1984.

34 BERVEILLER, M. and ZAOUI, A., An extension of the self-consistent scheme to plastically flowing polycrystals, *J. Mech. Phys. Solids*, 26, 325, 1979.

35 BRETHEAU, T. and CALDEMAISON, D., Test of mechanical interaction models between polycrystal grains by means of local strain measurements, *Deformation of polycrystals : Mechanisms and microstructures*, N. Hansen, A. Horsewell, T. Leffers and H. Lilholt (ed), 157, Risø National Laboratory, Roskilde (Denmark), 1981.

36 BERVEILLER, M. and ZAOUI, A., Extended self-consistent schemes, *Euromech 183*, Villetaneuse, France, 1984.

37 HIHI, A. BERVEILLER, M., and ZAOUI, A., Une nouvelle formulation de la modélisation autocohérente de la plasticité des polycristaux métalliques, *J. Mech. Theor. Appl.*, in the press.

38 BERVEILLER, M. and ZAOUI, A., Etude de l'anisotropie élastique, plastique et géométrique dans les polycristaux métalliques, *Mechanical behaviour of anisotropic solids*, J.P. Boehler (ed), 335, C.N.R.S. and Martinus Nyhoff Pub., The Hague, 1982.

39 HIRSCH, J., MUSICK, R. and LÜCKE, K., Comparison between earing, R-values and 3-dimensional orientation distribution functions of copper and brass sheets, *Textures of materials*, G. Gottstein and K. Lücke (ed.), II. 437, Springer, 1978.

40 WENG, G.J., A micromechanical theory of grain-size dependence in metal plasticity, *J. Mech. Phys. Solids*, 31, 193, 1983.

41 BERVEILLER, M., HIHI, A. and ZAOUI, A., Self-consistent schemes for polycrystalline and multiphase materials plasticity, *Deformation of polycrystals : Mechanisms and microstructures*, N. Hansen, A. Horsewell, T. Leffers and H. Lilholt (ed), 145, Risø National Laboratory, Roskilde (Denmark), 1981.

42 BERVEILLER, M. and ZAOUI, A., Modelling of the plasticity and the texture development of two-phase metals, *Deformation of multi-phase and particle containing materials*, J.B. Bilde-Sørensen, N. Hansen, A. Horsewell, T. Leffers and H. Lilholt (ed), 153, Risø National Laboratory, Roskilde (Denmark), 1983.

43 ZAOUI, A., Macroscopic plastic behaviour of microinhomogeneous materials, *Plasticity to-day*, A. Sawczuk and G. Bianchi (ed), 451, Elsevier Science Pub., London, 1984.

44 BERVEILLER, M. and ZAOUI, A., A simplified self-consistent scheme for the plasticity of two-phase metals, *Res. Mech. Letters*, 1, 119, 1981.

45 BOUCHER, S., Modules effectifs de matériaux composites quasihomogènes et quasiisotropes, constitués d'une matrice élastique et d'inclusions élastiques, *Revue M.*, 22, 1, 1976.

46 Mc LAUGHLIN, R., A study of the differential scheme for composite materials, *Int. J. Eng. Sci*, 15, 237, 1977.

47 HASHIN, Z., The elastic moduli of heterogeneous materials, *J. Appl. Mech.*, 29, 143, 1962.

48 CHRISTENSEN, R.M., and LO K.H., Solutions for effective shear properties in three phase sphere and cylinder models, *J. Mech. Phys. Solids*, 27, 4, 1979.

49 DEDERICHS, P.H. and ZELLER, R., Variational treatment of the elastic constants of disordered materials, *Z. Physik*, 259, 103, 1973.

50 BERVEILLER, M. and ZAOUI, A., Modeling of the plastic behavior of inhomogeneous media, *J. Eng. Mat. Tech.*, 106, 295, 1984.

51 EBERHARDT, O., The coherent potential approximation in random field theories, *Continuum models of discrete systems* (CMDS 3), E. Kröner and K.H. Anthony (ed), 425, University of Waterloo Press, 1980.